Introductory Calculus with Applications

Introductory Calculus with Applications

Second Edition

J. S. RATTI   M. N. MANOUGIAN   *University of South Florida*

Houghton Mifflin Company   **Boston**

Atlanta
Dallas
Geneva, Illinois

Hopewell, New Jersey
Palo Alto
London

Dedicated to our families

Printed in the U.S.A.
Library of Congress Catalog Card Number: 76-13096
ISBN: 0-395-24545-1

Contents

Preface

This edition, like its predecessor, is the by-product of our teaching an introductory course in calculus for nonmajors. Our objective is two-fold: first we bring out clearly some of the main ideas, language, and techniques of calculus for students with a basic high school mathematics program, and second we relate calculus to business, biology, and the social sciences.

We have used an intuitive approach to introduce many of the basic concepts of calculus. Explanations of the methods of calculus are supplemented with relevant theorems. Although the theoretical aspect of calculus is deemphasized, proofs of the simpler theorems are detailed and the more difficult theorems are stated clearly without proof. We feel that the student should learn the correct and precise definitions of the concepts and statements of the theorems. Many examples are given to illustrate the concepts discussed, and each section contains numerous exercises ranging from direct substitutions and elementary applications to more challenging problems.

Each page is divided into two parts, the main text and a wide margin. The margin is used for notes, comments, and remarks to aid the student in his or her learning process. Historical developments of some of the basic concepts are presented and biographical sketches of some of the leaders of mathematics are included.

We are grateful to the thousands of students and to the instructors who used the first edition. Many of the constructive comments and suggestions made by the users of the first edition were incorporated during the preparation of this text. We have preserved the style and the spirit of the first edition, and we have made several changes that we hope are improvements. In this second edition the chapters on basic algebraic concepts (Chapter 1) and on limits (Chapter 3) have been rewritten. In Chapter 1 we have added manipulative-type problems involving rational exponents, radicals, and rational expressions. Linear equations are also discussed. In Chapter 3 we have taken a more intuitive approach to the limit concept.

We have added a chapter on differential equations with applications and sections on improper integrals and double integrals.

We have added several routine problems to the exercises. Each chapter is now concluded by a set of review questions and problems for homework assignments and student self-evaluations. Starred exercises are those of greater difficulty.

At the end of each chapter we have added problems that can be handled easily by using pocket calculators. If pocket calculators are not accessible, these problems could easily be omitted without affecting the continuity of the book.

The entire book can be covered in a year's course, three quarters, or two semesters. Portions of the text, such as the review of algebra (Chapters 1 and 2), the trigonometric functions (Chapter 10), and the sections involving pocket calculators may be omitted. The text may thus be used in a one-semester or two-quarter course.

We are grateful to Professor Joseph J. Liang for providing the sections on pocket calculators and for reading and checking the final manuscript. We wish to thank Bert K. Waits, John Schiller, and Rolland G. Crouch for reading parts of the manuscript and offering helpful suggestions. We also thank Witold A.J. Kosmala for preparing and checking the answers to the exercises.

J.S.R.
M.N.M.

Introductory Calculus with Applications

Basic Algebraic Concepts (optional)

CHAPTER 1

1.1 Sets

The originator of the theory of *sets* is the famous mathematician Cantor. The concept of a set plays a central role in the foundations of mathematics. The idea of a set (or collection) of objects is very common in our everyday life. For example,

1. The set of car styles produced by Ford Motor Company
2. The set of the current governor of Florida
3. The set of member countries in the United Nations in 1976
4. The set of counting numbers
5. The set of all 200-year-olds in your mathematics class

We shall describe the basic property a set must have: *A set is any well-defined collection of objects. Any object in the set is called an element (or member) of the set.*

Set

The chief characteristic of a set is that it is well defined; that is, given any particular object, it must be clear whether the object is an element of the set or not. For example, the collection of all good students at your university is considered ill defined, because it is unlikely that we all can agree on its members. On the other hand, the collection of all those students who maintained a grade point average of 3.6 or higher during 1972–1973 at your university is well defined.

Notation We shall follow the usual custom of denoting sets by uppercase letters such as A, B, C, X, Y, Z. The elements of a set will be denoted by lowercase letters such as a, b, c, x, y, z. The elements that make up sets are listed within braces and are separated by commas. For example,

$$A = \{1, 3, 5, 7\}$$
$$B = \{\text{Maria, Sue, Vito}\}$$
$$X = \{\text{a, e, i, o, u}\}$$

The elements of the set A are 1, 3, 5, and 7; the elements of the set B are Maria, Sue, and Vito; and the elements of the set C are the vowels of the English alphabet.

We indicate in symbols that A is a set and x is an element of A by writing

$$x \in A$$

(read "x is an element of A" or "x belongs to A"). If A is a set and x is not an element of A, then we write

$$x \notin A$$

(read "x is not an element of A" or "x does not belong to A").

A set may have only one element, such as item 2 in the list of sets that began this section; a finite number of elements, such as items 1 and 3; or no elements, such as item 5. The sets in items 1, 2, and 3 are called *finite* sets. The set described in item 5 is called the *empty* (or *null* or *void*) set. The null set is denoted by the symbol $\emptyset$. A set that is not a finite set or the empty set is called an *infinite* set. Thus the set in item 4 is an infinite set.

A given set may be expressed in tabular form or in set-building notation. In tabular form, a set is specified by listing its elements. Some examples are

---

Georg F. L. P. Cantor

Georg F. L. P. Cantor (1845–1918) was born on March 3, 1845 in St. Petersburg, Russia. His father, Georg W. Cantor, was a Danish merchant and his mother, Maria Bohm, was a talented artist. The family, which was of Jewish descent, converted to Christianity and moved to Frankfurt, Germany, in 1856. Cantor received his university education at Zurich and at the University of Berlin. At Berlin his instructors were the famous mathematicians Kummer, Weierstrass, and Kronecker. Cantor received his Ph.D. in 1867. He had an active professional career, but it was spent at a mediocre university, and he never realized his ambition for a professorship at the University of Berlin.

At the age of 29, Cantor published his first revolutionary paper on the theory of infinite sets, a work that provided a common language for most of mathematics. Kronecker's attack of Cantor's work as nonsense may have contributed to the latter's first nervous breakdown in 1884. Cantor was subject to recurring breakdowns throughout the rest of his life, and died in a mental institution in January 1918.

Today Cantor's work is recognized as a fundamental contribution to all of mathematics, particularly in the foundations of analysis.

$$A = \{-5, 8, 17\}$$
$$B = \{1, 4, 7, 10, 13, \ldots, 100\}$$
$$C = \{2, 4, 6, 8, \ldots\}$$

Here the set A consists of only three elements, so we list all three. The set B consists of a relatively large number of elements, so we employ the three-dots (ellipsis) notation. This notation indicates that some elements have not been listed; to calculate these elements, use the rule illustrated by the first few elements. For the set B, it means that we calculate the sixth and succeeding elements by adding 3 to the preceding element until we reach the last element, 100. Similarly, the set C indicates all even natural numbers, since no last element is listed.

In set-building notation, a set is defined by stating a condition or a property which is satisfied by the elements of the set and by no other objects. For example, the set C defined previously could be written as

$$C = \{x \mid x \text{ is an even natural number}\}$$

Here x is a letter which merely represents an element of the set. The vertical bar $|$ can be translated "such that," and after the bar we write the condition on x that must be satisfied by x if it is to be included as an element of the set C. For example, $50 \in C$ and $51 \notin C$.

Caution. *It is assumed that the reader can recognize the pattern (as conceived by the writer) from the first few numbers.*

A natural number is even if it is divisible by 2.

Definition 1.1 *Two sets A and B are said to be* equal *written*

$$A = B$$

◀ *Equality of sets*

if and only if they have the same (identical) elements. In other words, every element that belongs to the set A is also an element of the set B, and every element that belongs to the set B is also an element of the set A. If A is not equal to B, we write

$$A \neq B$$

We illustrate these ideas in the following examples.

Example 1.1 Let $A = \{1, 3, 5, 7, 9\}$ and $B = \{3, 9, 1, 7, 5\}$. Are these sets equal?

SOLUTION Since every element which belongs to A also belongs to B and conversly, the two sets A and B are equal; that is, $A = B$ by Definition 1.1.

Example 1.2 Let $A = \{2, 4, 6\}$ and $B = \{6, 4\}$. Are these sets equal?

SOLUTION Since $2 \in A$ and $2 \notin B$, then $A \neq B$ by Definition 1.1.

Suppose that U is the set of all students in your university and C is the set of all students in your calculus class. Clearly, if $x \in C$, then

$x \in U$. In other words, all the elements of C are also elements of U. We say that C is a *subset* of U.

Subset ▶ **Definition 1.2**   *If every element of a set A is also an element of the set B, then A is called a* subset *of B, and we write*

$$A \subseteq B$$

(*read "A is a subset of B" or "A is contained in B"*).

If C is not a subset of D, we write $C \nsubseteq D$. The empty set $\emptyset$ is assumed to be a subset of every set.

Example 1.3   Let $A = \{1, 3, 5\}$ and $B = \{1, 2, 3, 4, 5\}$. Is

Note. *It should be clear from Definition 1.1 that A ⊆ A.*

(a) $A \subseteq B$;   (b) $B \subseteq A$?

SOLUTION
(a)   Since every element of A is also an element of B, then by Definition 1.2, $A \subseteq B$.
(b)   Since $2 \in B$ and $2 \notin A$, therefore $B \nsubseteq A$.

Proper subset

In this example we find that $A \subseteq B$ and that B has at least one element that is not an element of A. In such cases we say that A is a *proper subset* of B and we write $A \subset B$ (without the horizontal bar).

It is noteworthy that in general $B \nsubseteq A$ does not imply that $A \subseteq B$ as can be seen by considering the sets

$$A = \{1, 3, 5\} \quad \text{and} \quad B = \{3, 5, 7\}$$

Example 1.4   Let $A = \{a, b, c\}$. Write all the proper subsets of A.

SOLUTION   The subsets of A are

$$\{a\}, \quad \{b\}, \quad \{c\}, \quad \{a, b\}, \quad \{a, c\}, \quad \{b, c\}, \quad \emptyset$$

Geometrically, a set S may be represented by a closed curve in the plane and we represent the elements of S by points inside the curve. (See Figure 1.1) Thus, $a \in S$ while $b \notin S$. Such figures were first used by the Swiss mathematician Euler to illustrate principles of logic. Later, an English logician, John Venn (1834–1923) also used diagrams to illustrate mathematical ideas. Today diagrams such as Figure 1.1 are referred to as *Venn diagrams*.

Venn diagrams

Figure 1.1

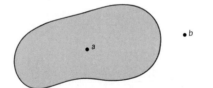

Let us consider the set of all students in your university; call it U.

Let C and Z represent the sets of students taking calculus and zoology, respectively. We illustrate this in a Venn diagram (Figure 1.2). The

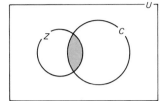

Figure 1.2

rectangle represents the set U. This is the overall set to which the discussion will be confined, and is called the *universal set*. The two sets C and Z are represented by the two circles as shown. In this case we are assuming that there are students taking both calculus and zoology. These students are represented by points in the shaded region. Thus we have obtained a new set from the sets C and Z. The set composed of students who are in both C and Z is called the *intersection* of C and Z. Another set that is obtained from C and Z is the set of students who are taking at least one of the two courses: calculus or zoology. This set is called the *union* of C and Z. In general, we have

Note. *In general the* universal *set, denoted by U, is a set that contains all the elements that are being considered in a given discussion.*

Definition 1.3   The intersection *of a set A and a set B, written A ∩ B, is the set of all elements that are both in A and in B. (See Figure 1.3.)*

◄ *Intersection*

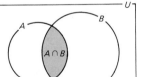

Figure 1.3

Few men have left a greater mark in the development of mathematics than the Swiss mathematician Leonhard Euler (1707–1783). He was born in Basel, Switzerland, and died in St. Petersburg (now Leningrad), Russia. Euler, one of the master analysts of the eighteenth century, has been called the most prolific mathematician in history. He was introduced to mathematics by his father, a Calvinist pastor who had studied mathematics with the first Jacob Bernoulli. At the age of 17, Euler received his master's degree, and at 19 he wrote his first paper for the Academy of Sciences in Paris.

From 1732 to 1741 Euler held a Chair of Mathematics in the Academy of Sciences in St. Petersburg. He then accepted an offer of the King of Prussia to be director of the department of mathematics at the Academy of Berlin, where he stayed 25 years. At the invitation of Catherina II, he returned to St. Petersburg. Although the last years of his life were spent in total blindness, his productivity continued undiminished. He was the author of more than 700 papers and 32 books in various branches of mathematics.

Leonhard Euler

Union ▶   *Definition 1.4* *The* union *of a set A and a set B, written A ∪ B, is the set of all elements that belong either to A or to B or to both A and B. (See Figure 1.4.)*

Figure 1.4

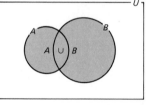

Example 1.5 Let $A = \{1, 2, 4, 5\}$, $B = \{2, 4, 6\}$, and $C = \{-1, 0\}$. Find

(a) $A \cap B$ (b) $A \cup B$ (c) $A \cap C$

SOLUTION

(a) By Definition 1.3, we have

$$A \cap B = \{1, 2, 4, 5\} \cap \{2, 4, 6\}$$
$$= \{2, 4\}$$

(b) By Definition 1.4, we have

$$A \cup B = \{1, 2, 4, 5\} \cup \{2, 4, 6\}$$
$$= \{1, 2, 4, 5, 6\}$$

(c) By Definition 1.3, we find that

$$A \cap C = \{1, 2, 4, 5\} \cap \{-1, 0\}$$
$$= \emptyset$$

In part (c) we have no element common to both A and C. Thus the intersection of A and C is the empty set. In such a case we say that A and C are *disjoint* sets.

Sets and Venn diagrams are helpful in analyzing a multitude of problems. We illustrate one such problem.

Example 1.6 In a survey of 60 freshman students the following information was obtained.

30 took English	17 took English and mathematics
28 took mathematics	10 took mathematics and chemistry
21 took chemistry	8 took English and chemistry

3 took all three subjects

Find the number of students that took

(a) None of the three subjects
(b) Mathematics, but not English or chemistry
(c) English, but not mathematics or chemistry
(d) Chemistry, but not English or mathematics

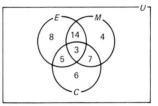

Figure 1.5

SOLUTION   In Figure 1.5, let U represent the set of 60 students that were interviewed; E, M, and C represent the sets of students taking English, mathematics, and chemistry, respectively. First we put in the number of students who took all 3 subjects. We place a 3 in the region representing $E \cap M \cap C$. We have 8 students taking English and chemistry, so the set $E \cap C$ must have 8 elements. Since we already have 3 elements in $E \cap M \cap C$, we place a 5 in the remainder of the region representing $E \cap C$. In a similar manner we find the numbers in the remaining regions. For instance, the region $M \cap C$ must have 10 elements and we have already accounted for 3. Thus the remainder of the region of $M \cap C$ must have 7 elements. Completing the diagram in this fashion, we find that

(a)   There are 13 students who are not taking any of the three subjects listed.

(b)   There are 4 students who are taking mathematics, but not English or chemistry.

(c)   There are 8 students who are taking English, but not mathematics or chemistry.

(d)   There are 6 students who are taking chemistry, but not mathematics or English.

1.   Write in words the following set notations.

  (a) $P \subset Q$ (b) $y \in A$ (c) $S \not\subset T$
  (d) $x \notin Q$ (e) $\emptyset$ (f) $\{0\}$

2.   Which of the following are true?

  (a) $2 \in \{1, 2, 3\}$ (b) $\{2\} \in \{1, 2, 3\}$
  (c) $\{2, 3\} = \{3, 2\}$ (d) $\{2, 3\} = \{2, 2, 3, 2, 3\}$
  (e) $\{a\} \in \{\{a\}, \{\{a\}\}, a\}$ (f) $\{\{2\}\} = \{2, \{2\}\}$

3.   Give, if possible, a property which describes each of the following sets.

  (a) $\{2, 4, 6, 8\}$ (b) $\{1, 4, 9, 16, 25, 36\}$
  (c) $\{0, 1, 2, 3, 4, 5, 6, 7, 8\}$ (d) $\{2, 3, 5, 7, 11, 13, 17, 19\}$
  (e) $\{1, 3, 5, 7, \ldots, 91\}$ (f) $\{1, 3, 5, 7, \ldots\}$
  (g) $\{25, 36, 49, 64, 81\}$ (h) $\{-1, 1\}$

4.   Exhibit the elements of each of the following sets.

  (a) $\{x \mid x$ is an even number less than $13\}$
  (b) $\{y \mid y$ is a continent$\}$
  (c) $\{z \mid z$ is an ocean$\}$
  (d) $\{x \mid x$ is a month$\}$
  (e) $\{n \mid n$ is a natural number greater than 5 and less than $10\}$
  (f) $\{h \mid h$ is an island state of the United States$\}$

5. Find all possible solutions of x and y in each of the following cases.
   (a) $\{2x, y\} = \{4, 8\}$ (b) $\{2x\} = \{10\}$
   (c) $\{2x, 3\} = \{20, 3\}$ (d) $\{x, x^2\} = \{4, 2\}$

6. Are the following sets equal?
   $A = \{x|x$ is a letter of the word "follow"$\}$
   $B = \{x|x$ is a letter of the word "wolf"$\}$
   $C = \{x|x$ is a letter of the word "flow"$\}$

7. List all the proper subsets of $A = \{a, b, c, d\}$.

8. Give all the subsets of the following sets.
   (a) $\{1, 3, 5\}$ (b) $\{0\}$
   (c) $\{1, 2\}$ (d) $\{0, 1, 2, 3, 4\}$

9. If a set A has n elements, how many subsets of the set A do we have?

In Problems 10 through 23, let $A = \{1, 2, 3, 4, 5, 6, 7\}$, $B = \{1, 2, 3, 5\}$, and $C = \{4, 6, 8\}$. Determine the given set.

10. $A \cap B$ 11. $A \cap C$ 12. $B \cap C$
13. $A \cup B$ 14. $(A \cup C) \cap B$ 15. $(A \cup B) \cap C$
16. $(A \cap C) \cup B$ 17. $(A \cap B) \cup C$ 18. $(A \cup B) \cup C$
19. $A \cap (B \cup C)$ 20. $A \cup (B \cap C)$ 21. $A \cup (B \cup C)$
22. $(A \cup B) \cap (B \cup C)$ 23. $(A \cap B) \cap C$

24. Given that $A \cap B = \{2, 4\}$, $A \cup B = \{2, 3, 4, 5\}$, $A \cap C = \{2, 3\}$, and $A \cup C = \{1, 2, 3, 4\}$, find A, B, and C.

Complement of a set ▶ **Definition 1.5** *Let U be the universal set and let A be a subset of U. Then the* complement *of A, denoted by A', is the set of all elements in U that are not in A.*

In Problems 25 through 35, let $U = \{1, 2, 3, 4, 5, 6, 7, 8, 9, 10\}$, $A = \{2, 5, 7, 10\}$, $B = \{1, 3, 5, 7, 9\}$, and $C = \{2, 4, 6, 8\}$. Describe the given set.

25. A' 26. B' 27. C'
28. $A' \cup B'$ 29. $(B \cup A)'$ 30. $B' \cap A'$
31. $(A \cap B)'$ 32. $B \cup C'$ 33. $B \cap C'$
34. U' 35. $(A')'$

Cardinal number ▶ **Definition 1.6** *The* cardinal number *of a set A, denoted by $n(A)$, is the number of elements in the set A.*

36. Find $n(A)$ if
    (a) $A = \{0, 1, 2, 3\}$
    (b) $A = \{x|x$ is a letter of the word "zigzag"$\}$

37. Give an example of a set A such that
    (a) $n(A) = 5$ (b) $n(A) = 0$ (c) $n(A) = 1$

38. In a survey of 100 students the following information was obtained.
    29 read *Newsweek*   11 read *Newsweek* and *Playboy*

25 read *Business Week*   10 read *Newsweek* and *Business Week*
47 read *Playboy*   6 read *Playboy* and *Business Week*
                    4 read all three

How many students read
(a)   none of the above magazines?
(b)   only *Playboy*?
(c)   only *Business Week*?

39.   In a survey of some freshman students the results were as follows.
      28 like classical music   13 like classical music and rock
      39 like rock   7 like rock and jazz
      24 like jazz   5 like all three
      12 like classical music and jazz 3 like none
      (a)   How many students were surveyed?
      (b)   How many like classical only?
      (c)   How many like exactly two of the three?
      (d)   How many do not like rock?

1.2   The Real Numbers

One of the most important sets in mathematics is the set R of real numbers. There are many ways of describing this set. Let us first consider the intuitive approach. Early in our education we first become familiar with the set N of *natural numbers*. Since these numbers are used for counting, they are also called *counting numbers*. We know these numbers by their names: one, two, three, . . . , and denote them by the symbols 1, 2, 3, ... Thus

$$N = \{1, 2, 3, \ldots\}$$

Natural numbers

The set N is a subset of the set W of *whole numbers*, denoted by

$$W = \{0, 1, 2, 3, \ldots\}$$

Whole numbers

We then encounter the set I of *integers*, which contains W as a subset,

$$I = \{\ldots, -3, -2, -1, 0, 1, 2, 3, \ldots\}$$

Integers

The system of numbers encompassing the integers is the set Q of *rational numbers*. A *rational number* is defined as any number which can be expressed in the form p/q, where p and q are integers and $q \neq 0$. Thus we may write

$$Q = \left\{ \frac{p}{q} \middle| \ p, q \in I \text{ and } q \neq 0 \right\}$$

Rational numbers

Since every integer n can be written as $n/1$, it follows that n is a rational number. Hence the set of integers is a subset of the set of rational numbers. Other examples of rational numbers are

$$\frac{3}{5}, \quad \frac{-5}{9}, \quad \frac{3}{-16}, \quad -\frac{7}{22}, \quad 0.25, \quad 1.378, \quad \text{etc.}$$

Note. $0.25 = \frac{25}{100}$ *and* $1.378 = \frac{1378}{1000}$

One reason for extending a number system is our need to solve certain equations. Let us consider the equation

$$a + x = b \qquad (1.1)$$

where a and b are natural numbers.

In Equation (1.1), if $a = 3$ and $b = 5$, it is easy to see that x must be equal to 2, which is a natural number. However, if $a = 5$ and $b = 3$, then there is no solution of $5 + x = 3$ in the set of natural numbers. In order that Equation (1.1) always have a solution, we extend the set N to the set I.

Similarly, it is necessary to introduce the rational numbers if we wish to solve equations of the form

$$ax = b \qquad (1.2)$$

where $a, b \in I$ and $a \neq 0$.

For example, if $a = 2$ and $b = 1$, then the equation

$$2x = 1 \qquad (1.3)$$

has no solution in the set of integers. However $x = \frac{1}{2}$ (a rational number) is a solution of (1.3).

Now let us consider the solution of the equation

$$x^2 = 2 \qquad (1.4)$$

It can be proved (see Problem 27 in the exercises) that Equation (1.4) has no solution in the set of rational numbers. We then extend the set of rational numbers to the set R of *real numbers*, so that Equation (1.4) has a solution in R.

We present the set R of real numbers by associating this set with points on a line in such a way that to each real number x there corresponds one and only one point on the line and conversely. Such an association is called a *one-to-one correspondence*. To do this, we draw a line (imagine the line to be extended indefinitely in both directions) and call this line the *number line* or the *real line*. We associate with each point on this line a number, and the collection of all these numbers is called the *set of real numbers*.

One-to-one correspondence

Number line

To begin, we choose any two points on this line and label them 0 and 1. By agreement we choose 1 to be to the right of 0 (see Figure 1.6).

Figure 1.6

The direction traveled when going from 0 to 1 is called the *positive direction*, and the opposite direction is called the *negative direction* of the line.

The fixed straight segment from 0 to 1 is called the unit segment, and its length is chosen as the unit length. By reproducing this unit of length successively on both sides of 0 and 1, we get the familiar graphical representation of the set of integers $I = \{. . . , -3, -2, -1, 0, 1, 2, 3, . . .\}$ on the number line (see Figure 1.7).

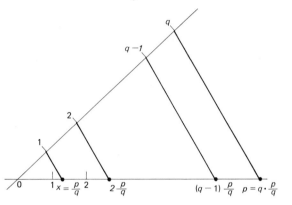

Figure 1.7

We now locate the points corresponding to the rational numbers. Let x be a positive rational number. Then $x = p/q$, where p and q are integers and $q \neq 0$. We divide the line segment from 0 to p into q equal parts (this can be done with a compass and straightedge; see Figure 1.8); then the endpoint of the first subdivision is the number $x = p/q$. The negative rational numbers can be similarly located.

Figure 1.8

Let us now imagine that all the rational numbers have been identified, or placed on the line. It turns out that the rational numbers do not fill up the line: we exhibit a point on the line which is not a rational number. In Figure 1.9 let $x = OQ =$ length of the diagonal of the square of side length 1. By using a compass, we can draw an arc of the circle having radius x and center at 0. The point P where the arc intersects the number line through 0 and 1 is the point associated with the number x. It is easy to see from the Pythagorean theorem that $x^2 = 1^2 + 1^2 = 2$ and (see Problem 27) x is not rational.

The Pythagorean theorem: If a right triangle has legs of length b and c and hypotenuse of length a, then $a^2 = b^2 + c^2$.

Pythagoras (569–500 B.C.?), a half-mythical figure, was a mathematician, philosopher, and metaphysician. During extensive travels he learned much from the priests of Egypt and Babylon, and he founded a secret brotherhood of mathematicians in southern Italy. Among his contributions to mathematics was the introduction of axiom (or postulates) and proof into the mathematical language. He also discovered that it is impossible to find two natural numbers such that the square of one is equal to twice the square of the other (i.e., he showed that $\sqrt{2}$ is not rational).

According to one legend, his school was burned down in a protest against his teachings, and Pythagoras died in the flames.

Pythagoras

Figure 1.9

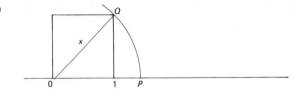

Those numbers on the real line which are not rational numbers are called *irrational* numbers. Consequently, the set of real numbers is the union of two disjoint sets: the set of rational numbers and the set of irrational numbers.

The relationships among the sets of numbers previously considered is as follows.

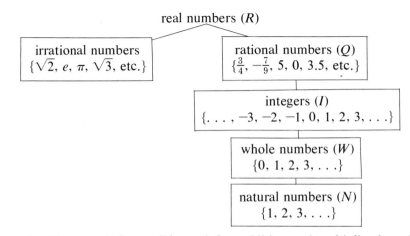

At this stage it is possible to define addition and multiplication of real numbers and prove the familiar properties of real numbers. We shall not follow this route. Instead, we shall assume that in the set R of real numbers, the operations of addition (denoted by the plus sign) and multiplication (denoted by the center dot or juxtaposition) have been defined. We shall briefly review some of the familiar properties of R.

Closure property ▶ *P1 Closure Property The set R is closed with respect to addition and multiplication. This means that if x and y are real numbers, then so are the unique numbers x + y and x · y (also denoted xy). The numbers x + y and xy are called the sum and product, respectively, of x and y.*

Commutative laws ▶ *P2 Commutative Laws for Addition and Multiplication For every pair x, y in R,*

$$x + y = y + x \quad \text{and} \quad x \cdot y = y \cdot x$$

Examples $\qquad\qquad 2 + 3 = 3 + 2, \qquad 2 \cdot 3 = 3 \cdot 2$

P3 Associative Laws for Addition and Multiplication *For every x, y, z in R,*   ◀ *Associative laws*

$$(x + y) + z = x + (y + z) \quad \text{and} \quad (x \cdot y) \cdot z = x \cdot (y \cdot z)$$

Examples $(2 + 3) + 7 = 2 + (3 + 7),$ $(2 \cdot 3) \cdot 7 = 2 \cdot (3 \cdot 7)$

P4 Distributive Laws *For every x, y, z in R,*   ◀ *Distributive laws*

$$(x + y) \cdot z = x \cdot z + y \cdot z \quad \text{and} \quad x \cdot (y + z) = x \cdot y + x \cdot z$$

Examples
$$(2 + 3) \cdot 5 = 2 \cdot 5 + 3 \cdot 5, \qquad 2 \cdot (3 + 5) = 2 \cdot 3 + 2 \cdot 5$$

or

$$5 \cdot 5 = 10 + 15, \qquad\qquad 2 \cdot 8 = 6 + 10$$

P5 Identity Elements *The real numbers zero and one are called the*   ◀ *Identity*
additive identity and multiplicative identity, respectively. In other
words, for every x in R,

$$x + 0 = 0 + x = x \quad \text{and} \quad x \cdot 1 = 1 \cdot x = x$$

P6 Inverse Elements *For each x in R, there is a unique element*   ◀ *Inverses*
called the **negative** *of x in R, denoted by* $-x$ *such that*

$$x + (-x) = (-x) + x = 0$$

For each real number $x \neq 0$, there is a unique element called the
reciprocal of x in R, denoted by $1/x$ or x^{-1} such that

$$x \cdot \frac{1}{x} = \frac{1}{x} \cdot x = 1$$

Many other properties of real numbers can be proved by using Properties P1 through P6. (The properties P1 through P6 may be assumed by the reader as having been proved or may be taken as axioms.) We shall list some of the important properties. Let x, y, and z be real numbers.

Cancellation law for addition

(i)   If $x + z = y + z$, then $x = y$.
     Example If $2 + 3 = 2 + a$, then $a = 3$.

Cancellation law for multiplication

(ii)   If $x \cdot z = y \cdot z$ and $z \neq 0$, then $x = y$.
     Example If $7 \cdot 2 = a \cdot 2$, then $a = 7$.

Law of zero products

(iii) If $x \cdot y = 0$, then either $x = 0$ or $y = 0$.
     Example If $2 \cdot x = 0$, then $x = 0$.

(iv) $x \cdot 0 = 0$

(v) $(-1)x = -x$

(vi) $-(-x) = x$

(vii) $-(x + y) = -x + (-y)$

(viii) $(-x)y = -(xy) = x(-y)$

(ix) $(-x)(-y) = xy$

The other two operations of arithmetic, subtraction and division, may be defined in terms of addition and multiplication, respectively.

Subtraction and division ▶   **Definition 1.7** *The operation of subtraction (denoted by the minus sign) is defined by*

$$x - y = x + (-y)$$

and the operation of division (denoted by the division sign) is defined by

$$x \div y = x \cdot \frac{1}{y}, \qquad y \neq 0$$

Note. $a \div 0$ or $\dfrac{a}{0}$ is not defined.   *Often $x \div y$ is expressed as $\dfrac{x}{y}$ or x/y, and we refer to it as the* quotient *of x by y, or the* fraction *x over y. The number x is called the* numerator *and y is called the* denominator *of the fraction.*

The following rules of operation for fractions can be established. We assume that all denominators are nonzero real numbers.

(x) $\dfrac{-x}{y} = \dfrac{x}{-y} = -\dfrac{x}{y}$

(xi) $\dfrac{-x}{-y} = \dfrac{x}{y}$

(xii) $y\left(\dfrac{x}{y}\right) = x$

(xiii) $\dfrac{1}{x} \cdot \dfrac{1}{y} = \dfrac{1}{xy}$

(xiv) $\dfrac{1}{1/x} = x$

(xv) $\dfrac{x}{y} = \dfrac{z}{u}$ if and only if $xu = yz$

(xvi) $\dfrac{xy}{zy} = \dfrac{x}{z}$

(xvii) $\dfrac{x}{y} \cdot \dfrac{z}{u} = \dfrac{xz}{yu}$

(xviii) $\dfrac{x}{y} + \dfrac{z}{y} = \dfrac{x+z}{y}$

(xix) $\dfrac{x}{y} + \dfrac{z}{u} = \dfrac{xu+yz}{yu}$

(xx) $\dfrac{x/y}{z/u} = \dfrac{xu}{yz}$

In Problems 1 through 12, justify each statement by using one and only one of the properties (P1 through P6) of real numbers.

Exercises 1.2

1. $4 + 7$ is a real number

2. $3 \cdot 6$ is a real number

3. $5 + 9 = 9 + 5$

4. $(-4) + 0 = -4$

5. $1 \cdot 7 = 7$

6. $4(5 + 7) = 4 \cdot 5 + 4 \cdot 7$

7. $3 \cdot 2 = 2 \cdot 3$

8. $3 + (-3) = 0$

9. $5 \cdot \left(\frac{1}{5}\right) = 1$

10. $(2 + 5) + 11 = 2 + (5 + 11)$

11. $0 \cdot 1 = 0$

12. $(5 \cdot 6) \cdot 7 = 5 \cdot (6 \cdot 7)$

*13. Prove the cancellation law (i) for addition.

*14. Prove the cancellation law (ii) for multiplication.

15. Prove the law of zero products (iii). [*Hint:* Suppose that $x \neq 0$, and multiply both sides by $1/x$.]

In Problems 16 through 22, perform the indicated operations.

16. $8 - (-4)$

17. $-5 - (-6 + 2)$

18. $3 - [-2 - (-4)]$

19. $\frac{2}{3} + \frac{4}{3}$

20. $\frac{2}{5} - \frac{1}{5}$

21. $\frac{1}{2} - \frac{1}{3}$

22. $\dfrac{\frac{3}{5} - \frac{1}{2}}{\frac{4}{5} - \frac{2}{3}}$

In Problems 23 through 26, perform a geometrical construction similar to that of Figure 1.8 to obtain the rational numbers.

23. $\frac{2}{3}$

24. $\frac{8}{5}$

25. $-\frac{3}{7}$

26. $-\frac{10}{3}$

*27. Prove that there is no rational number x such that $x^2 = 2$. [*Hint:* Suppose that $x = p/q$, where p and q are integers and they have no common factor, and obtain a contradiction.]

*28. Suppose that x is an irrational number and that a and b are rational numbers. Prove the following statements.
(a) ax is irrational if $a \neq 0$
(b) $a + bx$ is irrational if $b \neq 0$

29. Is the set of rational numbers closed under addition and multiplication?

30. Is the set of irrational numbers closed under addition and multiplication?

1.3 Inequalities

Our representation of the real numbers on the real line shows that the set of real numbers is the union of three disjoint sets.

1. The positive real numbers: those real numbers to the right of zero on the number line.
2. The negative real numbers: those real numbers to the left of zero on the number line
3. Zero

The set of positive real numbers is closed under the operations of addition and multiplication.

We assume that the reader knows some of the elementary properties of positive real numbers. For example, the sum of two positive numbers is a positive real number, the product of two positive real numbers is a positive real number, etc. We are ready now for the following definition.

$a > b; a < b$ ▶ *Definition 1.8 For any two real numbers a and b, we say that a is greater than b and write $a > b$ if and only if $a - b$ is positive. Further, we say that a is less than b and write $a < b$ if and only if $a - b$ is negative.*

The notation $a \geq b$ indicates that $a > b$ or $a = b$. Similarly, $a \leq b$ means $a < b$ or $a = b$. To say that $a > b$ is equivalent to saying that a is to the right of b on the number line. Similarly, $a < b$ if and only if a is to the left of b on the number line. It is easy to see from the Definition 1.8 that a real number a is positive if and only if $a > 0$.

As illustrations of the previous definition, we have $5 > 2$, since $5 - 2 = 3$, which is positive. Similarly, $-7 < -4$ (or $-4 > -7$), since $-4 - (-7) = -4 + 7$, which is positive. Similarly, $1 > 0, 2 > 1, 3 > 2$, and so on. Thus we have the following ordering of the integers.

$$\cdots < -3 < -2 < -1 < 0 < 1 < 2 < 3 \cdots$$

An expression involving $<$, $\leq$, $>$, or $\geq$ is called an inequality. We list here some of the important properties of inequalities, which the reader should learn thoroughly.

Property 1 is called the law of trichotomy. ▶ *Property 1 For each pair of real numbers a and b, exactly one of the following relations hold.*

$$a < b, \qquad a > b, \qquad a = b$$

Property 2 is called the law of transitivity. ▶ *Property 2 If $a > b$ and $b > c$, then $a > c$.*
Illustration $7 > 5$ and $5 > 2$, then $7 > 2$.

Property 3 *For any real number c, if a > b, then a + c > b + c.* ◄
Illustration $6 > -2$, then $6 + 4 > -2 + 4$.

Property 4 *For any real number c > 0, if a > b, then ac > bc.* ◄
Illustration $5 > 2$, then $5 \cdot 3 > 2 \cdot 3$.

Remember that the sense of
the inequality is unchanged if
we multiply both sides of the
inequality by a positive
number. However, if we
multiply both sides by a
negative number, then the
sense of the inequality
is reversed.

Property 5 *For any real number c < 0, if a > b, then ac < bc.* ◄
Illustration $5 > 2$, then $5(-3) < 2(-3)$.

We shall prove Properties 1 and 4. The reader will be asked to prove
the remaining properties in the exercises. We note that the converse of
Properties 3, 4, and 5 is also true.

PROOF OF PROPERTY 1 If a and b are real numbers, then $a - b$ is a
real number. As mentioned before, the set of real numbers is the disjoint union of (i) the set of positive real numbers, (ii) the set of negative
real numbers, and (iii) the set consisting of zero. Consequently, $a - b$
is either positive, negative, or zero. Hence either $a > b$, $a < b$, or $a = b$.

PROOF OF PROPERTY 4 Let a, b, and c be real numbers such that
$a > b$ and $c > 0$. Since $a > b$, we have $a - b > 0$. Now since $a - b$ and
c are positive real numbers and the product of positive real numbers
is positive, we have

$$(a - b)c > 0$$

or

$$ac - bc > 0$$

Hence

$$ac > bc$$

See Problem 25 in Exercises 1.4.

Conversely, suppose that $ac > bc$. Since $c > 0$, we have $1/c > 0$
(Why?). Thus using the first part of this property, we have

$$(ac)\,\frac{1}{c} > (bc)\,\frac{1}{c}$$

or

$$a > b$$

Similar results hold for the symbol $<$. For example, we can prove
the following.

1. If $a < b$ and $b < c$, then $a < c$.
2. If $a < b$, then $a + c < b + c$.
3. If $a < b$ and $c > 0$, then $ac < bc$.
4. If $a < b$ and $c < 0$, then $ac > bc$.

Now we shall introduce some standard shorthand notations to describe subsets of the number line. The compound inequality $a < c < b$ means that $a < c$ and $c < b$ simultaneously. For example, $-3 < 4 < 8$ is correct. However, $3 < x < 2$ is incorrect use of this notation, since $3 < x < 2$ suggests that $3 < x$ and $x < 2$ simultaneously. Thus we would conclude that $3 < 2$, which is false. In other words, $a < c < b$ means that c is between a and b on the number line and $a < b$.

Intervals   The sets described in Figure 1.10 are called *intervals*. For $a < b$, we write

1. $(a, b) = \{x \mid a < x < b\}$
2. $[a, b] = \{x \mid a \le x \le b\}$
3. $[a, b) = \{x \mid a \le x < b\}$
4. $(a, b] = \{x \mid a < x \le b\}$
5. $[a, \infty) = \{x \mid a \le x\}$
6. $(a, \infty) = \{x \mid a < x\}$
7. $(-\infty, a] = \{x \mid x \le a\}$
8. $(-\infty, a) = \{x \mid x < a\}$
9. $(-\infty, \infty) = \{x \mid x \text{ is a real number}\}$

Figure 1.10

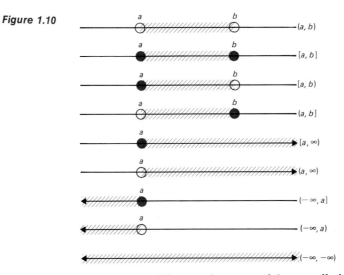

Endpoints   **Remark 1**   The numbers a and b are called the *endpoints* of the interval. A square bracket indicates that the endpoint is included in the interval, whereas a parenthesis means that the endpoint does not lie in the interval.

It must be carefully noted that the symbol ∞ is not a number. It is simply part of a short form of notation for the given intervals.

The intervals

$$(a, b), \quad (a, \infty), \quad (-\infty, a), \quad \text{and} \quad (-\infty, \infty)$$

Open intervals   are called *open intervals*.

Remark 2   We note that if l is an open interval, then given any x in l, there are points to the right and left of x in l. Such a point is called an *interior point*.

The intervals

$$[a, b], \quad [a, \infty), \quad (-\infty, a], \quad \text{and} \quad (-\infty, \infty)$$

are called *closed intervals*.

Remark 3 Note that $(-\infty, \infty)$ is both an open and a closed interval. The intervals

$$(a, b] \quad \text{and} \quad [a, b)$$

are neither open nor closed intervals. Sometimes they are called half-open or half-closed intervals.

In many practical situations we encounter inequalities of the form

$$2x + 5 < x + 10 \qquad (1.5)$$

or

$$5x^2 + 8x - 2 < 0 \qquad (1.6)$$

We are asked to solve (1.5) or (1.6); i.e., we are asked to find those values of x for which these inequalities hold. In other words, we want to rewrite the definitions of the sets

$$\{x \mid 2x + 5 < x + 10\} \quad \text{or} \quad \{x \mid 5x^2 + 8x - 2 < 0\}$$

in a simpler form which shows at once what the elements are. Such sets are subsets of the real numbers and generally consist of intervals or unions of intervals. They will be called solution sets.

Example 1.7 Solve $2x + 5 < x + 10$.

SOLUTION   By Property 1.3,

$$2x + 5 - 5 < x + 10 - 5$$

or

$$2x < x + 5$$

Again, by Property 1.3,

$$2x - x < x + 5 - x$$

or

$$x < 5$$

Thus the solution set of the given inequality is $(-\infty, 5)$. The graph of this solution set is given in Figure 1.11.

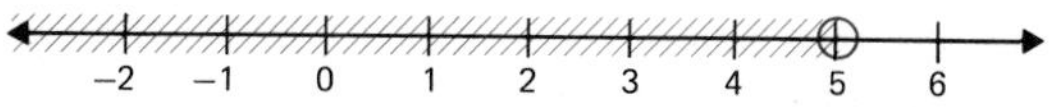

Figure 1.11

Example 1.8 Find all real numbers x which satisfy the compound inequality

$$2 \le 5x - 3 < 7 \qquad (1.7)$$

SOLUTION   Adding 3 to each member of the inequality, we have

$$2+3 \leq 5x-3+3 < 7+3$$

or

$$5 \leq 5x < 10 \tag{1.8}$$

Saying that inequalities (1.7) and (1.9) are equivalent means that (1.7) holds if and only if (1.9) holds; i.e., (1.7) and (1.9) have the same solution set.

Now, multiplying each member of the inequality (1.8) by $\frac{1}{5}$, we have

$$1 \leq x < 2 \tag{1.9}$$

Since these steps are reversible, inequality (1.9) is equivalent to (1.7). Thus the solution set of (1.7) is the interval $[1, 2)$, as illustrated in Figure 1.12.

Figure 1.12

Example 1.9   Solve the inequality

$$x^2 + x - 2 > 0 \tag{1.10}$$

SOLUTION   Factoring this expression, we obtain an equivalent inequality,

$$(x + 2)(x - 1) > 0 \tag{1.11}$$

Consequently, x is a solution of the given inequality if and only if the product $(x + 2)(x - 1)$ is positive. In order for this to happen, the factors $x + 2$ and $x - 1$ must have the same sign. Therefore we examine the sign of each factor. Since $x + 2 > 0$ if and only if $x > -2$, the factor $x + 2$ is positive whenever $x \in (-2, \infty)$ and is negative whenever $x \in (-\infty, -2)$. Similarly, the factor $x - 1$ is positive whenever $x \in (1, \infty)$ and negative whenever $x \in (-\infty, 1)$. Let us display in Figure 1.13 the intervals in which the factors $x + 2$ and $x - 1$ are positive or negative.

Figure 1.13

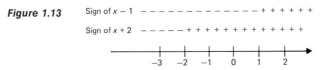

Sign of $x - 1$

Sign of $x + 2$

From Figure 1.13, we observe that both factors are positive when $x \in (1, \infty)$ and both factors are negative when $x \in (-\infty, -2)$. Hence the solution set is $(-\infty, -2) \cup (1, \infty)$. The solution set is sketched in Figure 1.14.

Figure 1.14

Example 1.10   Solve the following inequality.

$$\frac{2x}{x-4} \le 1 \qquad\qquad (1.12)$$

SOLUTION   The following inequalities are equivalent.

$$\frac{2x}{x-4} \le 1$$

$$\frac{2x}{x-4} - 1 \le 0$$

$$\frac{2x - x + 4}{x-4} \le 0 \qquad\qquad (1.13)$$

$$\frac{x+4}{x-4} \le 0$$

Now $\dfrac{x+4}{x-4} = 0$ if and only if $x+4 = 0$ or $x = -4$.

Hence $x = -4$ is in the solution set.

The quotient $(x+4)/(x-4)$ is negative if and only if $x+4$ and $x-4$ have opposite signs. In Figure 1.15, we note that $x+4$ and $x-4$ have

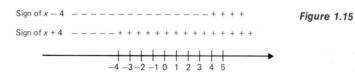

Figure 1.15

opposite signs in the interval $(-4, 4)$. Since -4 is also a solution, it follows that the solution set of the given inequality is $[-4, 4)$. The solution set is sketched in Figure 1.16.

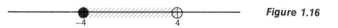

Figure 1.16

In Problems 1 through 8, rewrite the given statement using the symbols $>$, $\ge$, $<$, and $\le$.

Exercises 1.3

1. 5 is less than 8

2. -3 is not greater than 2

3. 6 is not less than 3

4. 8 is a positive number

5. -2 is a negative number   6. x is nonnegative

7. x is nonpositive

8. $2x - 5$ is greater than or equal to 5 and less than 9

9. Prove Property 2.   10. Prove Property 3.

11. Prove Property 5.

In Problems 12 through 17, express each of the given intervals in set notation.

12. $(2, 5)$ 13. $(3, 11]$ 14. $[4, 9]$

15. $[-3, 6)$ 16. $[-5, \infty)$ 17. $(-\infty, -3]$

In Problems 18 through 23, express each of the given sets in interval notation.

18. $\{x | 2 < x < 5\}$ 19. $\{x | -3 \le x < 5\}$ 20. $\{x | -2 < x\}$

21. $\{x | x > 3\}$ 22. $\{x | x \le 4\}$ 23. $\{x | x \ge -3\}$

In Problems 24 through 28, let a and b be real numbers. Prove the given inequality by showing your work step by step.

24. If $a \ne 0$, then $a^2 > 0$ 25. $a > 0$ if and only if $1/a > 0$

26. If $0 < a < b$, then $1/a > 1/b$ 27. $a > 0$ if and only if $-a < 0$

28. $a > b$ if and only if $-a < -b$

In Problems 29 through 56, solve the given inequality and graph the solution set on the number line.

29. $2x - 3 < 5$ 30. $-2x + 4 < 7$

31. $2x - (x + 2) \ge 1$ 32. $-3x - 2(3x - 4) \le -4$

33. $7x - 4 < 4x + 11$ 34. $3(x - 2) + 5 \le -3x + 4$

35. $3x - 2 \ge -2x + 10$ 36. $2(x + 4) - 3(2x - 3) \ge -8(x - 2)$

37. $1 < 7x + 4 < 5$ 38. $2 < 3x - 4 \le 8$

39. $-2 \le 2x - 3 \le 1$ 40. $(x - 3)(x + 5) < 0$

41. $(x - 2)(x + 3) > 0$ 42. $(2x - 1)(x - 3) \le 0$

43. $(3x - 2)(2x + 5) \ge 0$ 44. $x^2 \le 5x$

45. $x^2 \ge 2x$ 46. $x^2 - 6x + 8 \ge 0$

47. $x^2 + x - 12 \le 0$ 48. $x^2 - x < 6$

49. $\dfrac{1}{x} \ge 2$ 50. $7 - \dfrac{3}{x} < 12$

51. $\dfrac{x + 1}{x + 2} \ge 0$ 52. $\dfrac{x}{x - 2} > 3$

53. $1 - \dfrac{3x}{2x - 3} \le 5$ 54. $(x - 1)(x - 2)(x - 3) > 0$

55. $(2x - 1)(x - 2)(x + 1) \le 0$ 56. $\dfrac{1}{2x - 7} \le \dfrac{3}{5 - 3x}$

1.4 Absolute Value

The absolute value of x (also called the numerical value of x), denoted by $|x|$, is an important concept often useful in the study of inequalities.

Definition 1.9   *For each real number x, the* absolute value *of x is defined by*

$$|x| = \begin{cases} x & \text{if } x \geq 0 \\ -x & \text{if } x < 0 \end{cases}$$

◀ *Absolute value*

For example,

$$|3| = 3, \qquad |-3| = -(-3) = 3$$

$$|0| = 0, \qquad |7 - 13| = |-6| = -(-6) = 6$$

Now $|3| = |-3|$. Thus if we look on the number line, we observe that 3 and -3 are equidistant from 0 (see Figure 1.17). Geometrically,

Figure 1.17

then, the absolute value of a number x is its distance from 0, without regard to direction. In general, $|a - b|$ denotes the distance between a and b without regard to direction (see Figure 1.18).

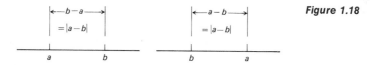

Figure 1.18

Definition 1.9 gives the following properties. For any real numbers x and y,

$$|x| \geq 0 \quad \text{and} \quad -|x| \leq x \leq |x| \tag{1.13}$$

$$|x| = 0 \quad \text{if and only if } x = 0 \tag{1.14}$$

$$|xy| = |x|\,|y| \tag{1.15}$$

$$|x - y| = |y - x| \tag{1.16}$$

$$\left|\frac{x}{y}\right| = \frac{|x|}{|y|}, \quad y \neq 0 \tag{1.17}$$

We now prove an important theorem.

▶   *Theorem 1.1   For $a > 0$, $|x| < a$ if and only if $-a < x < a$.*

PROOF   Since the theorem is an "if and only if" statement, we have to prove it in two parts.

I.   Assume that $|x| < a$. We must show that $-a < x < a$.

Case 1   For the case in which $x \geq 0$, $|x| = x$, and therefore $x = |x| < a$ implies that $x < a$. Also, a is given to be positive. Thus $-a < 0 \leq x$. Combining, we get

$$-a < 0 \leq x = |x| < a$$

or

$$-a < x < a$$

Case 2   For the case in which $x < 0$, $|x| = -x$. Since $|x| < a$, we conclude that $-x < a$ or $x > -a$ or $-a < x$. Also, since $x < 0$, we have $0 < -x$. Therefore, $-a < 0 < -x$ or $-a < -x$ or $x < a$. Combining, we get

$$-a < x < a$$

In both cases

$$|x| < a \quad \text{implies that} \quad -a < x < a$$

II.   Suppose that $-a < x < a$. We must show that $|x| < a$.

Case 1   Let $x \geq 0$. Then $|x| = x$, so $x < a$ implies that $|x| < a$.

Case 2   Let $x < 0$. Then $|x| = -x$. Since $-a < x$, we have $-x < a$ (Why?). Consequently,

$$|x| = -x < a \quad \text{or} \quad |x| < a$$

▶   *Theorem 1.2   For $a > 0$, $|x| \leq a$ if and only if $-a \leq x \leq a$.*

▶   *Theorem 1.3   For $a > 0$, $|x| > a$ if and only if $x > a$ or $x < -a$.*

▶   *Theorem 1.4   For $a > 0$, $|x| \geq a$ if and only if $x \geq a$ or $x \leq -a$.*

The proofs of Theorems 1.2, 1.3, and 1.4 are similar to that of Theorem 1.1 and are left as exercises for the reader. The preceding theorems also hold true if x is replaced by an algebraic expression involving x.

We shall now find the solution of inequalities involving absolute values.

Example 1.11 Solve for x:

$$|3x - 5| \leq 4 \tag{1.18}$$

SOLUTION By Theorem 1.2, the inequality (1.18) is equivalent to

$$-4 \leq 3x - 5 \leq 4 \tag{1.19}$$

The inequality (1.19) can easily be seen to be equivalent to

$$\tfrac{1}{3} \leq x \leq 3 \tag{1.20}$$

Thus the solution set for (1.18) is the closed interval $[\tfrac{1}{3}, 3]$ (see Figure 1.19).

Figure 1.19

Example 1.12 Solve for x:

$$\left|\frac{2 - 3x}{x + 1}\right| \leq 5, \qquad x \neq -1 \tag{1.21}$$

SOLUTION By Theorem 1.2, this inequality is equivalent to

$$-5 \leq \frac{2 - 3x}{x + 1} \leq 5 \tag{1.22}$$

To solve (1.22), we wish to multiply each of its members by $x + 1$. Consequently, we consider the two cases.

Case 1 $x + 1 > 0$, which means that $x > -1$. Multiplying both sides of (1.22) by $x + 1$, we get

$$-5x - 5 \leq 2 - 3x \leq 5x + 5 \tag{1.23}$$

To solve (1.23), we first consider

$$-5x - 5 \leq 2 - 3x \tag{1.24}$$

The solution of (1.24) is

$$x \geq -\tfrac{7}{2} \tag{1.25}$$

Now we can solve the second inequality in (1.23),

$$2 - 3x \leq 5x + 5 \tag{1.26}$$

The solution of (1.26) is (see Figure 1.20)

$$x \geq -\tfrac{3}{8} \tag{1.27}$$

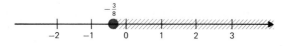

Figure 1.20

Therefore, if $x > -1$, the original inequality (1.21) holds if and only if $x \geq -\frac{7}{2}$ and $x \geq -\frac{3}{8}$. Since all three inequalities, $x > -1$, $x \geq -\frac{7}{2}$, and $x \geq -\frac{3}{8}$, must hold for every x in the solution set, we must have $x \geq -\frac{3}{8}$, or the interval $[-\frac{3}{8}, \infty)$.

Case 2 $x + 1 < 0$ or $x < -1$. Multiplying each member of (1.22) by $x + 1$ and reversing the direction of the inequalities (since $x + 1$ is negative), we get

$$-5x - 5 \geq 2 - 3x \geq 5x + 5 \tag{1.28}$$

The solution of the first inequality in (1.28)

$$-5x - 5 \geq 2 - 3x \tag{1.29}$$

is

$$x \leq -\frac{7}{2} \tag{1.30}$$

The solution of the second inequality in (1.28)

$$2 - 3x \geq 5x + 5 \tag{1.31}$$

is

$$x \leq -\frac{3}{8} \tag{1.32}$$

Since all three inequalities, $x < -1$, $x \leq -\frac{7}{2}$, and $x \leq -\frac{3}{8}$, must be satisfied, the solution of (1.21) in this case is $x \leq -\frac{7}{2}$, or the closed interval $(-\infty, -\frac{7}{2}]$. (See Figure 1.21.)

Figure 1.21

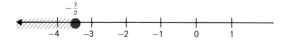

Combining the solutions of Case 1 and Case 2, we get the solution of (1.21): $(-\infty, -\frac{7}{2}] \cup [-\frac{3}{8}, \infty)$. This solution is illustrated in Figure 1.22.

Figure 1.22

Example 1.13 Solve for x:

$$|3x + 7| > 5 \tag{1.33}$$

SOLUTION By Theorem 1.3 this inequality is equivalent to

$$3x + 7 > 5 \quad \text{or} \quad 3x + 7 < -5 \tag{1.34}$$

Considering the first inequality in (1.34), we have

$$3x + 7 > 5 \quad \text{or} \quad 3x > -2 \quad \text{or} \quad x > -\frac{2}{3}$$

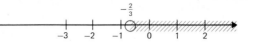

Figure 1.23

Thus the interval $(-\frac{2}{3}, \infty)$ is a solution set. (See Figure 1.23.)
From the second inequality, we have

$$3x + 7 < -5 \quad \text{or} \quad 3x < -12 \quad \text{or} \quad x < -4$$

Therefore the interval $(-\infty, -4)$ is a solution set. (See Figure 1.24.)

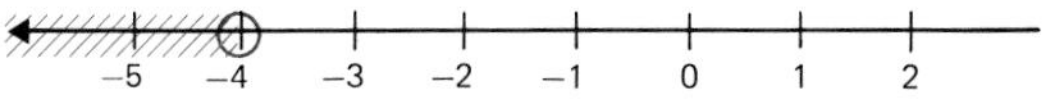

Figure 1.24

The given inequality (1.33) will be satisfied if either of the inequalities (1.34) are satisfied. Consequently, the solution set of (1.33) is $(-\infty, -4) \cup (-\frac{2}{3}, \infty)$. (See Figure 1.25.)

Figure 1.25

Theorem 1.5 *If x and y are any pair of real numbers, then*   ◀ *The triangle inequality*

$$|x + y| \leq |x| + |y| \tag{1.35}$$

PROOF From the definition of absolute value, we can see that for any real number x,

$$-|x| \leq x \leq |x| \tag{1.36}$$

Similarly, for any real number y,

$$-|y| \leq y \leq |y| \tag{1.37}$$

Adding inequalities (1.36) and (1.37), we get

$$-(|x| + |y|) \leq x + y \leq |x| + |y| \tag{1.38}$$

and (1.38) is equivalent to

$$|x + y| \leq |\,|x| + |y|\,| \quad \textbf{(Why?)} \tag{1.39}$$

But

$$|\,|x| + |y|\,| = |x| + |y|$$

so

$$|x + y| \leq |x| + |y|$$

Exercises 1.4   In Problems 1 through 24, solve for x.

1. $|x - 2| < 3$

2. $|x + 4| \leq 5$

3. $|x + \frac{1}{2}| > 2$

4. $|x - \frac{3}{4}| \geq \frac{1}{4}$

5. $|2x - 3| \leq 4$

6. $|3x + 5| < 8$

7. $|3x - 7| > 5$

8. $|4x + 5| \geq 9$

9. $2 < |7x - 2|$

10. $5 \geq 2|x - 3|$

11. $|5x - \frac{7}{2}| < \frac{5}{2}$

12. $|7x - \frac{8}{3}| \leq \frac{2}{3}$

13. $\left|\dfrac{x + 1}{x - 1}\right| < 3$

14. $\left|\dfrac{x - 3}{x + 4}\right| < 5$

15. $\left|\dfrac{x - 1}{2x - 3}\right| \leq 2$

16. $\left|\dfrac{3x + 1}{x - 4}\right| \leq 4$

17. $\left|\dfrac{x - 1}{x + 1}\right| > 1$

18. $\left|\dfrac{2x + 1}{x - 4}\right| \geq 3$

19. $\left|\dfrac{3 - 4x}{6x + 1}\right| \leq 1$

20. $3 \geq \left|\dfrac{x + 5}{3x - 2}\right|$

21. $|7 - 3x| \geq |5x|$

22. $|x - 4| \leq |2x + 3|$

23. $\left|\dfrac{1}{x - 2}\right| \leq \left|\dfrac{1}{x + 1}\right|$

24. $\left|\dfrac{1}{2x - 5}\right| \geq \left|\dfrac{3}{x - 1}\right|$

25. Prove Theorem 1.3.

26. Prove that $|x - y| \leq |x| + |y|$. [*Hint:* Replace y by $-y$ in Theorem 1.5.]

27. Prove that $|x| - |y| \leq |x - y|$. [*Hint:* Write $x = y + (x - y)$ and use Theorem 1.5.]

28. Prove that $|x - y| \leq |x - z| + |z - y|$.

1.5   Rational Exponents and Radicals

Expressions such as $2 \times 2 \times 2 \times 2 \times 2$ arise frequently in mathematical work. We use the following notation:

$$2 \times 2 \times 2 \times 2 \times 2 = 2^5$$

This notation was introduced by Descartes in his book *La Géométrie* (1637). In general, we have

▶   *Definition 1.10*   *If b is a real number and n is a natural number, then*

$$b^n = \underbrace{b \cdot b \cdot b \cdots b}_{n \text{ times}}$$

The number b is called the base, *and the number n is called the* exponent. *For instance,*

$$3^4 = 3 \cdot 3 \cdot 3 \cdot 3 = 81$$
$$(-2)^3 = (-2)(-2)(-2) = -8$$
$$\left(\tfrac{1}{2}\right)^5 = \left(\tfrac{1}{2}\right)\left(\tfrac{1}{2}\right)\left(\tfrac{1}{2}\right)\left(\tfrac{1}{2}\right)\left(\tfrac{1}{2}\right) = \tfrac{1}{32}$$

If $n = 1$, then, by definition, $b^1 = b$.

The following theorem can be easily established using Definition 1.10.

Theorem 1.6 *Let $a, b \in R$ and let m and n be natural numbers. Then* ◄

(a) $b^m b^n = b^{m+n}$ (b) $(b^m)^n = b^{mn}$ (c) $(ab)^n = a^n b^n$

and if $b \neq 0$,

(d) $\left(\dfrac{a}{b}\right)^n = \dfrac{a^n}{b^n}$ (e) $\dfrac{b^m}{b^n} = \begin{cases} b^{m-n} & \text{if } m > n \\ 1 & \text{if } m = n \\ \dfrac{1}{b^{n-m}} & \text{if } m < n \end{cases}$

For example,

(i) $2^3 \cdot 2^2 = 2^5 = 32$ (ii) $(2^3)^2 = 2^6 = 64$

(iii) $(2 \cdot 3)^3 = 2^3 \cdot 3^3 = 216$ (iv) $\left(-\dfrac{2}{3}\right)^4 = \dfrac{(-2)^4}{3^4} = \dfrac{16}{81}$

(v) $\dfrac{2^{17}}{2^{14}} = 2^{17-14} = 2^3 = 8,$ $\dfrac{3^{12}}{3^{14}} = \dfrac{1}{3^{14-12}} = \dfrac{1}{3^2} = \dfrac{1}{9}$

From Theorem 1.6, part (e), we find that $b^m/b^m = 1$. Thus

$$\frac{b^m}{b^m} = b^{m-m} = b^0$$

René Descartes (1586–1650) was born in La Haye, France. His delicate health as a child kept him away from physical activities and forced him into intellectual pursuits. After becoming disenchanted with his friends and his studies as a young man, he involved himself in the many wars plaguing Europe. He went to Holland and learned the trade of war under Prince Maurice of Orange; he enlisted under the Elector of Bavaria and fought against Bohemia, fought in the battle of Prague, and was at the siege of La Rochelle with the king of France.

Between wars, Descartes occupied himself with mathematics, philosophy, and almost all aspects of the sciences and theology. At the age of 32, he went to Holland, and from there corresponded extensively with the leading scientists and philosophers of Europe. In 1637 he published the masterpiece known in short as the *Method*, which introduced the world to analytic geometry. In 1646 Queen Christina of Sweden brought Descartes to Stockholm as her tutor. There he died of an inflammation of the lungs in February 1650.

René Descartes

---

and we define $b^0 = 1$. For example,

$$3^0 = 1, \qquad (-101)^0 = 1, \quad \text{and} \quad \left(\tfrac{1}{213}\right)^0 = 1$$

Care should be exercised when $b = 0$. The expression 0^0 is not defined. Now, from the expression

$$\frac{b^m}{b^n} = b^{m-n}$$

we get a negative exponent in b^{m-n} if $m < n$. Thus

$$\frac{2^7}{2^{10}} = 2^{7-10} = 2^{-3}$$

But we have seen that $2^7/2^{10} = 1/2^3$. Therefore we define $2^{-3} = 1/2^3$. In general, we have

▶ *Definition 1.11 Let $b \in R$, $b \neq 0$, and n be a natural number. Then*

$$b^{-n} = \frac{1}{b^n}$$

With Definition 1.11, we may restate Theorem 1.6 as follows.

▶ *Theorem 1.7 Let $a, b \in R$ and let m and n be integers. Then properties (a) through (e) in Theorem 1.6 hold.*

Example 1.14 Simplify each of the following expressions and eliminate negative exponents. Assume that x and y are not zero.

(a) $\dfrac{7x^2 y^{-2}}{28x^{-3}y}$

(b) $(x^5 y^{-4})^{-4}$

(c) $\left[\dfrac{x^{-4}(-271)^{16}y^3}{(301y^2)^{10}}\right]^0$

(d) $\dfrac{x^{-1} - y^{-1}}{(xy)^{-1}}$

SOLUTION

(a) $\dfrac{7x^2 y^{-2}}{28x^{-3}y} = \dfrac{7}{28} \cdot x^2 \cdot x^3 \cdot y^{-2} \cdot y^{-1} = \dfrac{1}{4} x^5 y^{-3} = \dfrac{x^5}{4y^3}$

(b) $(x^5 y^{-4})^{-4} = x^{5(-4)}y^{(-4)(-4)} = x^{-20}y^{16} = \dfrac{y^{16}}{x^{20}}$

By definition, $b^0 = 1$ (c) $\left[\dfrac{x^{-4}(-271)^{16}y^3}{(301y^2)^{10}}\right]^0 = 1$

(d) $\dfrac{x^{-1} - y^{-1}}{(xy)^{-1}} = \dfrac{1/x - 1/y}{1/xy} = \dfrac{(y - x)/xy}{1/xy} = \dfrac{y - x}{xy} \cdot \dfrac{xy}{1} = y - x$

Now we extend the definition of exponents to the set of rational numbers. We have seen that $4^3 = 64$; that is, if 4 is multiplied by itself three times, we get 64. What if we reverse this process and we ask the question, what number multiplied by itself 3 times is 64? We denote this process by

$$64^{1/3} = x$$

In other words, x is a number such that $x^3 = 64$. Clearly, one such number is $x = 4$. We call 4 the *cube root* of 64. In general, we have

Definition 1.12 *If $b \in R$ and n is a positive integer, then the* principal   ◄ *Principal nth root*
nth root of b, denoted by $b^{1/n}$, is defined by

$$b^{1/n} = x, \qquad \text{where } x \in R, \, n \text{ is odd, and } x^n = b$$

$$b^{1/n} = x, \qquad \text{where } x \geq 0, \, n \text{ is even, and } x^n = b$$

If n is even and $b < 0$, there is no real nth root of b. If $b > 0$ and n is even, then there are two nth roots of b that are real numbers.

For example,

$$(81)^{1/4} = 3 \quad \text{since} \quad 3^4 = 81$$

Also, we find that $(-3)^4 = 81$. Thus the fourth roots of 81 are 3 and -3. However, the positive number 3 is the principal fourth root of 81.

Example 1.15 Compute the values of the following.

(a) $(-64)^{1/3}$ (b) $\left(\frac{1}{32}\right)^{1/5}$ (c) $(-4)^{1/2}$

SOLUTION

(a) $(-64)^{1/3} = -4$,   since $(-4)^3 = -64$

(b) $\left(\frac{1}{32}\right)^{1/5} = \frac{1}{2}$,   since $\left(\frac{1}{2}\right)^5 = \frac{1}{32}$

(c) $(-4)^{1/2}$ does not exist in the real number system

Another notation used to denote the principal nth root of a number employs the *radical* sign $\sqrt{}$. We define the radical as

$$b^{1/n} = \sqrt[n]{b}$$

where b is the *radicand*, n is the *index* of the radical and the expression $\sqrt[n]{b}$ is called a *radical expression of order n*. The principal square root of b $(b \geq 0)$ is denoted by $\sqrt{b}$ rather than $\sqrt[2]{b}$.

Definition 1.13 *Let m and n be integers, n be positive, and m and n*   ◄
have no common factors. If $b \in R$ for which $b^{1/n}$ exists, then

$$b^{m/n} = (b^{1/n})^m = \left(\sqrt[n]{b}\right)^m$$

For example,

$$(-8)^{4/3} = [(-8)^{1/3}]^4 = (-2)^4 = 16$$

However, $(-9)^{3/2}$ is not defined since $(-9)^{1/2}$ is not defined.

Now we may extend Theorem 1.7 as follows.

▶ **Theorem 1.8**   *Let a, b ∈ R and r and s be rational numbers. Then properties (a) through (e) in Theorem 1.6 hold.*

Naturally, the properties (a) through (e) hold provided that each individual factor exists.

As a consequence of the definition of the radical, we have the following properties that are valid if each of the roots exists.

(i) $\sqrt[n]{a}\,\sqrt[n]{b} = \sqrt[n]{ab}$

(ii) $\dfrac{\sqrt[n]{a}}{\sqrt[n]{b}} = \sqrt[n]{\dfrac{a}{b}}, \qquad b \neq 0$

(iii) $\sqrt[n]{b^m} = (\sqrt[n]{b})^m$

(iv) $\sqrt[m]{\sqrt[n]{b}} = \sqrt[n]{\sqrt[m]{b}} = \sqrt[mn]{b}$

Example 1.16   Assume that x, y, and b are positive and simplify each of the following expressions.

(a) $(-8)^{2/3}\,9^{-3/2}$

(b) $\left(\dfrac{4x^8}{y^4}\right)^{3/2}$

(c) $\dfrac{x^{1/2} + x^{3/2}}{\sqrt{x}}$

(d) $\sqrt[3]{64x^7}$

(e) $\dfrac{\sqrt{27b^3}}{\sqrt{3b}}$

(f) $\sqrt{\sqrt[3]{81}}$

SOLUTION   Using Theorem 1.8 and the properties of radicals, we find that

(a) $(-8)^{2/3}\,9^{-3/2} = \dfrac{[(-8)^{1/3}]^2}{(9^{1/2})^3} = \dfrac{(-2)^2}{3^3} = \dfrac{4}{27}$

(b) $\left(\dfrac{4x^8}{y^4}\right)^{3/2} = \dfrac{4^{3/2}(x^8)^{3/2}}{(y^4)^{3/2}} = \dfrac{2^3 x^{12}}{y^6} = \dfrac{8x^{12}}{y^6}$

(c) $\dfrac{x^{1/2} + x^{3/2}}{\sqrt{x}} = x^{-1/2}[x^{1/2} + x^{3/2}] = x^0 + x^1 = 1 + x$

(d) $\sqrt[3]{64x^7} = 64^{1/3}x^{7/3} = 4x^2 x^{1/3} = 4x^2\sqrt[3]{x}$

(e) $\dfrac{\sqrt{27b^3}}{\sqrt{3b}} = \sqrt{\dfrac{27b^3}{3b}} = \sqrt{9b^2} = 3b$

(f) $\sqrt{\sqrt[3]{81}} = \sqrt[3]{\sqrt{81}} = \sqrt[3]{9}$

Finally, we note that if $x \in R$ we can use radicals to express the absolute value of x. We have

$$|x| = \sqrt{x^2}$$

Thus

$$\sqrt{(-3)^2} = |-3| = 3$$
$$\sqrt{2^2} = |2| = 2$$

In Problems 1 through 44, simplify each of the given expressions and eliminate negative and zero exponents. Assume that all the variables are positive.

Exercises 1.5

1. $b^3 b b^6$

2. $\left(-\frac{1}{2}\right)^3$

3. $\left(\frac{2}{5}\right)^2$

4. $x^{2n} x^3$

5. $x^{4n} x^n$

6. $\dfrac{b^{2n}}{b^n}$

7. $\left(\dfrac{a^4}{y}\right)^2$

8. $\dfrac{2^{16}}{4^8}$

9. $(3^{-1} 2^{-2})^{-3}$

10. $(3^2)^{-2}(2^{-3})^{-1}$

11. $\dfrac{4^{-2} - 1}{2^{-2} - 1}$

12. $\dfrac{13^0 - 14^0}{13^0 + 14^0}$

13. $\left(\dfrac{13^2 - 3^6}{16^5}\right)^0$

14. $((2^3)^2)^3$

15. $\dfrac{\left(\frac{x^3 y^2}{x^2 y^3}\right)^4}{\left(\frac{x^5 y^2}{xy^5}\right)^2}$

16. $\dfrac{2x^{-5} y^0 z^3}{-8x^5 yz^{-3}}$

17. $\left(\dfrac{-x^{-2} y^{-3} z^2}{z^3 x^2 y^{-3}}\right)^{-2}$

18. $\left(\dfrac{2^2 x^3 y^{-5} z^{-6}}{5x^3 y^5 z^6}\right)^2$

19. $x^2(x^{-2} + x^{-1})$

20. $(b^{-1} + b^{-2})b^{-3}$

21. $\dfrac{x^{-1} y^{-1}}{x^{-1} + y^{-1}}$

22. $\dfrac{x^{-1} + y^{-1}}{(x + y)^{-1}}$

23. $x^k x^{2k+1}(x^{3k+2})^{-1}$

24. $64^{1/2}$

25. $(-27)^{2/3}$

26. $\left(\frac{25}{9}\right)^{-3/2}$

27. $\left(-\frac{1}{8}\right)^{2/3}$

28. $\left(\frac{4}{81}\right)^{-1/2}$

29. $x^{2/3} x^{7/3}$

30. $b^{1/2} b^{5/3}$

31. $\dfrac{x^{1/2}}{x^{-4/3}}$

32. $x^{-1/2} x^{3/2} x^{5/7}$

33. $\left(\dfrac{a^4}{b^2}\right)^{1/2}$

34. $\left(\dfrac{x^2 y^4}{16}\right)^{-1/2}$

35. $\dfrac{x^{2n}}{x^{n/2}}$

36. $(a^{2n})^{2/n}(b^2)^{n/2}$

37. $\sqrt{\frac{4}{9} a^8 b^{10}}$

38. $\sqrt[3]{\dfrac{125}{-8x^3}}$

39. $\sqrt[4]{27}\,\sqrt[4]{9}$

40. $\sqrt{72}$

41. $\sqrt{\sqrt{8}}$

42. $\sqrt{9\sqrt{81}}$

43. $\sqrt{b}(b^{3/2} - b^4)$

44. $\dfrac{[\sqrt{2x} + (2x)^{-1/2}]^{-1}}{\sqrt{2x}}$

*45.   Prove Theorem 1.6.

*46.   Show that if $x^r = x^s$ and $x \neq 0, 1$, then $r = s$ (r and s are rational numbers).

1.6 Rational Expressions

In later chapters we shall need to simplify and combine algebraic expressions of the form

$$\frac{2x^2 + x}{(2x + 1)(x - 3)}, \quad \frac{x^2}{x + 2x^2}, \quad \frac{y + \dfrac{1}{y}}{y^2 + 1}, \quad \frac{\dfrac{3}{x} - \dfrac{1}{x}}{x + \dfrac{1}{x}}, \quad \text{etc.}$$

Rational expressions   Such expressions are called *rational expressions*. The rules for simplifying and combining rational expressions are the same as those for numerical fractions in arithmetic. Since division by zero is undefined, we shall always *assume that the denominator of each rational expression is nonzero*. Let us first consider the multiplication and division of two fractions.

Multiplication   To multiply two fractions (rational expressions) a/b and c/d, we multiply their numerators and also their denominators. Thus

$$\frac{a}{b} \cdot \frac{c}{d} = \frac{ac}{bd} \tag{1.40}$$

Example 1.17   Find $\dfrac{x - 1}{2x + 1} \cdot \dfrac{x - 2}{x + 3}$

SOLUTION $\dfrac{x - 1}{2x + 1} \cdot \dfrac{x - 2}{x + 3} = \dfrac{(x - 1)(x - 2)}{(2x + 1)(x + 3)}$

$$= \frac{x^2 - 3x + 2}{2x^2 + 7x + 3}$$

Division   To divide the fraction a/b by c/d, we multiply the first fraction a/b by the reciprocal of the second fraction c/d. The *reciprocal* of c/d is d/c, which is obtained by inverting c/d. Thus

$$\frac{a}{b} \div \frac{c}{d} = \frac{a}{b} \cdot \frac{d}{c} = \frac{ad}{bc} \tag{1.41}$$

Example 1.18   Divide $\dfrac{x - 1}{x + 1}$ by $\dfrac{2x - 3}{x + 2}$

SOLUTION   By (1.41), we have

$$\frac{x-1}{x+1} \div \frac{2x-3}{x+2} = \frac{x-1}{x+1} \cdot \frac{x+2}{2x-3}$$

$$= \frac{(x-1)(x+2)}{(x+1)(2x-3)}$$

$$= \frac{x^2+x-2}{2x^2-x-3}$$

Suppose that $c \neq 0$. Then we find that the fraction a/b is equal to the fraction ac/bc. Since   *Simplification*

$$\frac{ac}{bc} = \frac{a}{b} \cdot \frac{c}{c} = \frac{a}{b} \cdot 1 = \frac{a}{b}.$$

Thus when we wish to simplify rational expressions, we can use this procedure to cancel the common factor c. The fractions a/b and ac/bc are called *equivalent fractions*.   *Equivalent fractions*

Example 1.19   Simplify $\dfrac{x^2+x-6}{x^2-x-2}$

SOLUTION $\dfrac{x^2+x-6}{x^2-x-2} = \dfrac{(x+3)(x-2)}{(x+1)(x-2)}$

$$= \frac{x+3}{x+1}$$

As pointed out earlier, we are assuming that $x-2 \neq 0$, that is, $x \neq 2$.

Example 1.20   Simplify $\dfrac{x^2-3x+2}{x^2-2x} \div \dfrac{x^2-1}{x^3+x^2}$

SOLUTION   We first factor the numerators and denominators of each of the fractions and then change division to multiplication.

$$\frac{x^2-3x+2}{x^2-2x} \div \frac{x^2-1}{x^3+x^2} = \frac{(x-1)(x-2)}{x(x-2)} \div \frac{(x+1)(x-1)}{x^2(x+1)}$$

$$= \frac{(x-1)(x-2)}{x(x-2)} \cdot \frac{x^2(x+1)}{(x+1)(x-1)}$$

$$= \frac{(x-1)(x-2)x^2(x+1)}{x(x-2)(x+1)(x-1)} = x$$

To add (or subtract) two fractions having the same denominator,   *Addition and subtraction* we add (or subtract) the numerators and retain the common denominator.

$$\frac{a}{b} + \frac{c}{b} = \frac{a+c}{b} \tag{1.42}$$

$$\frac{a}{b} - \frac{c}{b} = \frac{a-c}{b} \tag{1.43}$$

Example 1.21 Combine the two fractions in each of the following cases to form one fraction.

(a) $\dfrac{x+1}{2x^2+1} + \dfrac{x^2-1}{2x^2+1}$

(b) $\dfrac{x^2-3}{x+3} - \dfrac{x-2}{x+3}$

(c) $\dfrac{x^2+2x+1}{x-3} + \dfrac{x^2+2}{3-x}$

SOLUTION   (a) $\dfrac{x+1}{2x^2+1} + \dfrac{x^2-1}{2x^2+1} = \dfrac{(x+1)+(x^2-1)}{2x^2+1} = \dfrac{x+x^2}{2x^2+1}$

(b) $\dfrac{x^2-3}{x+3} - \dfrac{x-2}{x+3} = \dfrac{(x^2-3)-(x-2)}{x+3}$

$$= \frac{x^2-3-x+2}{x+3} = \frac{x^2-x-1}{x+3}$$

(c) Here we observe that the two fractions do not have the same denominator. However, we can obtain the same denominator by multiplying the numerator and the denominator of the second fraction by -1. Consequently,

$$\frac{x^2+2}{3-x} = \frac{-1(x^2+2)}{-1(3-x)} = \frac{-x^2-2}{x-3}$$

Therefore

$$\frac{x^2+2x+1}{x-3} + \frac{x^2+2}{3-x} = \frac{x^2+2x+1}{x-3} + \frac{-x^2-2}{x-3}$$

$$= \frac{(x^2+2x+1)+(-x^2-2)}{x-3}$$

$$= \frac{x^2+2x+1-x^2-2}{x-3}$$

$$= \frac{2x-1}{x-3}$$

Now to add or subtract two fractions which do not have the same denominator, we change them to equivalent fractions having the same denominator by multiplying the numerator and denominator of each fraction by a suitable expression. We then follow the rules just given.

In general:

$$\frac{a}{b} + \frac{c}{d} = \frac{ad}{bd} + \frac{bc}{bd} = \frac{ad + bc}{bd} \qquad (1.44)$$

$$\frac{a}{b} - \frac{c}{d} = \frac{ad}{bd} - \frac{bc}{bd} = \frac{ad - bc}{bd} \qquad (1.45)$$

Example 1.22 Perform the indicated operations.

(a) $\dfrac{x+1}{x-1} + \dfrac{x}{x-2}$ 　　(b) $x - \dfrac{1}{x-1} + \dfrac{x+1}{x+2}$

SOLUTION   (a) $\dfrac{x+1}{x-1} + \dfrac{x}{x-2} = \dfrac{(x+1)(x-2)}{(x-1)(x-2)} + \dfrac{x(x-1)}{(x-1)(x-2)}$

$$= \frac{(x+1)(x-2) + x(x-1)}{(x-1)(x-2)}$$

$$= \frac{x^2 - x - 2 + x^2 - x}{x^2 - 3x + 2}$$

$$= \frac{2x^2 - 2x - 2}{x^2 - 3x + 2}$$

(b) $x - \dfrac{1}{x-1} + \dfrac{x+1}{x+2}$

$$= \frac{x}{1} - \frac{1}{x-1} + \frac{x+1}{x+2}$$

$$= \frac{x(x-1)(x+2)}{1(x-1)(x+2)} - \frac{1(x+2)}{(x-1)(x+2)} + \frac{(x-1)(x+1)}{(x-1)(x+2)}$$

$$= \frac{x(x-1)(x+2) - (x+2) + (x-1)(x+1)}{(x-1)(x+2)}$$

$$= \frac{x(x^2 + x - 2) - x - 2 + (x^2 - 1)}{x^2 + x - 2}$$

$$= \frac{x^3 + x^2 - 2x - x - 2 + x^2 - 1}{x^2 + x - 2} = \frac{x^3 + 2x^2 - 3x - 3}{x^2 + x - 2}$$

Sometimes a fraction may have numerator or denominator, or both, that already are, or can be combined to form, rational expressions. Such fractions are called *complex fractions*. To simplify a complex fraction, we combine the numerator or denominator, or both, as necessary so that the numerator and denominator both become single fractions. Then we divide the numerator by the denominator.

Complex fractions

Example 1.23 Simplify $\dfrac{\dfrac{1}{x} - \dfrac{2x+1}{x+2}}{(x+1) + \dfrac{1}{x-1}}$

SOLUTION

$$\frac{\dfrac{1}{x} - \dfrac{2x+1}{x+2}}{(x+1) + \dfrac{1}{x-1}} = \frac{\dfrac{1(x+2)}{x(x+2)} - \dfrac{x(2x+1)}{x(x+2)}}{\dfrac{(x+1)(x-1)}{x-1} + \dfrac{1}{x-1}}$$

$$= \frac{\dfrac{(x+2) - x(2x+1)}{x(x+2)}}{\dfrac{(x+1)(x-1) + 1}{x-1}}$$

$$= \frac{\dfrac{-2x^2 + 2}{x(x+2)}}{\dfrac{x^2 - 1 + 1}{x-1}} = \frac{-2x^2 + 2}{x(x+2)} \div \frac{x^2}{x-1}$$

$$= \frac{(-2x^2 + 2)}{x^2 + 2x} \cdot \frac{x-1}{x^2} = \frac{(-2x^2 + 2)(x-1)}{(x^2 + 2x)x^2}$$

$$= \frac{-2x^3 + 2x^2 + 2x - 2}{x^4 + 2x^3}$$

Exercises 1.6

In Problems 1 through 28, perform the indicated operations and simplify.

1. $\dfrac{x}{x+1} \cdot \dfrac{2x+2}{x^2}$

2. $\dfrac{x-y}{x+y} \cdot \dfrac{(x+y)^2}{x^2 - y^2}$

3. $\dfrac{3x-3}{(x+1)^2} \cdot \dfrac{x+1}{x-1}$

4. $\dfrac{x^2 + 3x + 2}{2x-1} \cdot \dfrac{4x-2}{x^2 + 4x + 3}$

5. $\dfrac{2a-4}{a^2-4} \div \dfrac{a}{a+2}$

6. $\dfrac{y^2 + y - 6}{y^2 + 2y - 3} \div \dfrac{y^2 - 4}{y^2 + y - 2}$

7. $\dfrac{a^2 - 2a + 1}{a^2 + a - 2} \div \dfrac{a^2 + 2a - 3}{a+2}$

8. $\dfrac{9y^2 - 16}{3y + 4} \div \dfrac{3y^2 - 7y + 4}{y}$

9. $\dfrac{2x+1}{x} + \dfrac{1-x}{x}$

10. $\dfrac{x}{x+y} + \dfrac{y}{x+y}$

11. $1 - \dfrac{1}{x}$

12. $\dfrac{1}{y} - \dfrac{1}{y+1}$

13. $\dfrac{1}{x} - \dfrac{1}{x^2}$

14. $\dfrac{1}{2x} - \dfrac{1}{3x} + \dfrac{1}{4x}$

15. $\dfrac{1}{1-x} + \dfrac{1}{1-x^2}$

16. $\dfrac{1}{x-y} + \dfrac{x+y}{x^2 - y^2}$

17. $x - 2 + \dfrac{1}{x+2}$

18. $\dfrac{1}{2x-y} + \dfrac{3}{y-2x}$

19. $\left(\dfrac{3}{y+1} - \dfrac{1}{y}\right)\left(\dfrac{y+1}{y}\right)$

20. $\dfrac{x-1}{2x+1} + \dfrac{1}{x-1} + x$

21. $\dfrac{1/x}{1 + 1/x}$

22. $\dfrac{1 - 1/x}{3 + 1/x}$

23. $\dfrac{(x+h)^2 - x^2}{h}$

24. $\dfrac{(x+h)^3 - x^3}{h}$

25. $\dfrac{\dfrac{1}{x+h} - \dfrac{1}{x}}{h}$

26. $\dfrac{\dfrac{1}{(x+h)^2} - \dfrac{1}{x^2}}{h}$

27. $\dfrac{\dfrac{1}{x}}{1 + \dfrac{1}{1 + 1/x}}$

28. $\dfrac{\dfrac{y-2}{y-1} - \dfrac{y-1}{y-2}}{\dfrac{2}{y-1} - \dfrac{2}{y-2}}$

1.7   Linear Equations in One Variable

In the study of elementary algebra, one of the more important tasks is that of solving equations. In this section we shall review some procedures involved in solving linear equations. Many real-world problems can be described as mathematical models involving linear equations.

The following statements are algebraic equations in the variable x:

$$3x = 6 \qquad\qquad (1.46)$$

$$x^2 - 9 = 0 \qquad\qquad (1.47)$$

$$\frac{2}{x} - \frac{1}{x-1} = \frac{x-2}{x(x-1)} \qquad\qquad (1.48)$$

If we replace x by 2 in Equation (1.46), the resulting statement $3(2)=6$ is true. However, if we replace x by any other real number, the statement is not true. We call 2 a *solution* (or *root*) of Equation (1.46). If we set $x = 3$ in Equation (1.47), we get $(3)^2 - 9 = 0$, which is a true statement. Also, we find that if we set $x = -3$, we get $(-3)^2 - 9 = 0$, which is again true. Thus both 3 and -3 are solutions of Equation (1.47). It can be shown that every real number except 0 and 1 is a solution of Equation (1.48). Thus an equation may have a finite number of solutions or an infinite number of solutions. There are equations such as $x + 1 = x + 5$ that have no solution. The set of all solutions of an equation is called the *solution set* of the equation. Thus the solution set of Equation (1.46) is $\{2\}$ and that of Equation (1.47) is $\{3, -3\}$. If an equation has no solution, we designate the solution set by the empty set $\emptyset$. An equation that is satisfied by all values of the variable for which the members of the equation are defined is called an *identity*. Thus Equation (1.48) is an identity. An equation that is not an identity is called a *conditional equation*.

Note.   *An equation that has no solution is called* inconsistent.

Solution set

Identity

Conditional equation

By *solving* an equation, we mean determining the solution set of the equation. Certain equations can be solved by inspection. If an equation cannot be solved easily, we perform algebraic operations and reduce the given equation to an equivalent equation which can be readily solved.

Equivalent equations ▶ *Definition 1.14   Two equations are said to be* equivalent *if they have the same solution set.*

For example, the equations

$$2x - 5 = 7 \quad \text{and} \quad 2x = 12 \quad \text{and} \quad x = 6$$

are equivalent equations because they have the same solution set, namely, {6}. On the other hand, the equations

$$x = 2 \quad \text{and} \quad x^2 - 2x = 0$$

are not equivalent because the solution set of the first is {2} while that of the second is {0, 2}.

We can often solve an equation by converting the given equation through a succession of equivalent equations until we obtain an equation for which the solution set is apparent. The following theorem is frequently used in generating equivalent equations over the set of real numbers.

▶ *Theorem 1.9   Let E(x), F(x) and G(x) be algebraic expressions in the variable x. Then for all values of x for which E(x), F(x) and G(x) are real numbers,*

$$E(x) = F(x)$$

is equivalent to each of the following:

(a) $E(x) + G(x) = F(x) + G(x)$

(b) $E(x) \cdot G(x) = F(x) \cdot G(x),$ provided that $G(x) \neq 0$

We shall use this theorem to solve linear equations.

Linear equation ▶ *Definition 1.15   An equation of the form*

$$ax + b = 0$$

where a, b ∈ R and a ≠ 0, is called a linear (*or* first degree) *equation in one variable. Also, any equation equivalent to this equation is called a linear equation.*

We can show that such an equation has one and only one solution in *R*. We obtain a succession of equivalent equations as follows:

$$ax + b = 0 \tag{1.49}$$

By Theorem 1.9(a).
$$ax + b - b = 0 - b \tag{1.50}$$

$$ax = -b \tag{1.51}$$

By Theorem 1.9(b).
$$\frac{1}{a}(ax) = \frac{1}{a}(-b) \tag{1.52}$$

$$x = -\frac{b}{a} \tag{1.53}$$

The last equation has the unique solution $-b/a$. Since Equations (1.49) through (1.53) are equivalent, (1.49) has the unique solution $-b/a$.

Example 1.24 Solve each of the following linear equations.

(a) $2[x - (2x - 1)] = 8$ (b) $\dfrac{3x}{4} = 2 - \dfrac{x}{3}$

SOLUTION (a) We simplify the algebraic expression and obtain a succession of equivalent equations.

$$2[x - (2x - 1)] = 8 \tag{1.54}$$
$$2[x - 2x + 1] = 8$$
$$2[-x + 1] = 8$$
$$-2x + 2 = 8$$
$$-2x = 6$$
$$-\tfrac{1}{2}(-2x) = -\tfrac{1}{2}(6)$$
$$x = -3 \tag{1.55}$$

The reader should check this answer.

Since Equation (1.55) is equivalent to (1.54), then the unique solution of (1.54) is -3.

(b) Noting that the least common denominator (LCD) is 12, we obtain

$$\frac{3x}{4} = 2 - \frac{x}{3} \tag{1.56}$$
$$12\left(\frac{3x}{4}\right) = 12\left(2 - \frac{x}{3}\right)$$
$$9x = 24 - 4x$$
$$9x + 4x = 24 - 4x + 4x$$
$$13x = 24$$
$$\tfrac{1}{13}(13x) = \tfrac{1}{13}(24)$$
$$x = \tfrac{24}{13} \tag{1.57}$$

The unique solution of (1.57) is $\frac{24}{13}$ and (1.57) is equivalent to Equation (1.56). Therefore the unique solution of (1.56) is $\frac{24}{13}$.

Check If $x = \frac{24}{13}$, we find

$$\frac{3}{4}\left(\frac{24}{13}\right) \stackrel{?}{=} 2 - \frac{1}{3}\left(\frac{24}{13}\right)$$
$$\frac{18}{13} \stackrel{?}{=} 2 - \frac{8}{13}$$

$$\frac{18}{13} \overset{?}{=} \frac{26-8}{13}$$

$$\frac{18}{13} = \frac{18}{13}$$

Thus the solution set is $\left\{ \frac{24}{13} \right\}$.

Note. *Extraneous solutions are those numbers that do not satisfy the original equation.*

If both sides of an equation are multiplied by an expression involving a variable, care should be exercised. There is a chance that extraneous solutions may be introduced. To avoid extraneous solutions, we must be certain that the candidate for a solution does not give us a zero in the denominator of the original equation. We illustrate this in the following example.

Example 1.25 Solve

$$\frac{1}{x-1} + \frac{2}{x+1} = \frac{2}{x^2-1} \tag{1.58}$$

SOLUTION   We see that the LCD is $x^2 - 1 = (x-1)(x+1)$. Multiplying both sides of Equation (1.58) by $x^2 - 1$, we get

$$(x^2-1)\left(\frac{1}{x-1} + \frac{2}{x+1}\right) = (x^2-1)\frac{2}{x^2-1}$$

$$x + 1 + 2(x-1) = 2$$

$$x + 1 + 2x - 2 = 2$$

$$3x - 1 = 2$$

$$3x = 3$$

$$x = 1 \tag{1.59}$$

If we replace x by 1, in Equation (1.58), we obtain

$$\frac{1}{0} + \frac{2}{2} = \frac{2}{0}$$

Note. *In Theorem 1.9(b) we required that $G(x) \neq 0$.*

Since division by zero is not defined, we reject $x = 1$ as a solution of (1.58). Note that Equation (1.58) was multiplied by $x^2 - 1$ which is zero when $x = 1$. Thus Equation (1.59) is not equivalent to (1.58). In this case Equation (1.58) has no solution and so its solution set is $\emptyset$.

Many real-world problems are stated in words. Such problems are called *word problems*. In many cases we can transform the given information into an equation involving known and unknown numbers to be determined. The following examples illustrate the procedure involved in solving word problems.

Example 1.26   If the length of a rectangle is 1 less than twice the width and its perimeter is 22, what are the dimensions of the rectangle?

SOLUTION   Let x represent the width of the rectangle. Then $2x - 1$ is the length of the rectangle (see Figure 1.26). The total distance

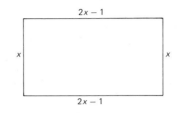

Figure 1.26

around the rectangle is the perimeter. Therefore we have the equation

$$x + (2x - 1) + x + (2x - 1) = 22 \tag{1.60}$$

Solving Equation (1.60), we get

$$6x - 2 = 22$$

$$6x = 24$$

$$x = 4$$

Hence the width is 4 and the length is $2(4) - 1 = 7$.

Note. *In solving word problems use the following steps.*

1. *Read the problem carefully.*
2. *Draw a picture (if possible).*
3. *Introduce variables for the unknowns and write each unknown in terms of one variable only.*
4. *Write an equation relating the unknown and known numbers.*
5. *Solve the equation.*
6. *Check your solution.*

   Check   The perimeter is

$$4 + 7 + 4 + 7 = 22$$

Example 1.27   Juanita invests part of $5000 at the rate of 6% annually and the remaining part at 7% annually. If her annual income from the total investment is $328.00, how much has she invested at each rate?

SOLUTION   We use the formula

$$I \quad = \quad Pr$$

$$(\text{income}) = (\text{principal})(\text{rate})$$

Let x represent the amount invested at 6%. Then $5000 - x$ is the amount invested at 7%. Now the income from the 6% investments is $\left(\frac{6}{100}\right)x$, and the income from the 7% investment is $\left(\frac{7}{100}\right)(5000 - x)$. Thus we have the equation

Note. $6\% = \frac{6}{100}$

$$\frac{6}{100}x + \frac{7}{100}(5000 - x) = 328 \tag{1.61}$$

Solving equation (1.61), we get

$$6x + 35,000 - 7x = 32,800$$

$$-x = -2,200$$

$$x = 2,200$$

Therefore Juanita has invested $2200 at 6% and $5000 - $2200 = $2800 at 7%.

Check $2200\left(\frac{6}{100}\right) + 2800\left(\frac{7}{100}\right) = (22)6 + (28)7$

$$= 132 + 196 = 328.$$

Example 1.28   How many quarts of 55% antifreeze solution should be added to how many quarts of 20% antifreeze solution to give 35 quarts of a 40% antifreeze solution?

SOLUTION   Let x denote the number of quarts of 55% antifreeze solution. Then $(35 - x)$ is the number of quarts of 20% antifreeze solution. Now in x qt of a 55% solution, we have $\left(\frac{55}{100}\right)x$ qt of antifreeze, and in $(35 - x)$ qt of a 20% solution, we have $\left(\frac{20}{100}\right)(35 - x)$ qt of antifreeze. On the other hand, in 35 qt of a 40% solution, we have $\left(\frac{40}{100}\right)(35)$ qt of antifreeze. Thus we have the equation

$$\frac{55}{100}x + \frac{20}{100}(35 - x) = \frac{40}{100}(35) \qquad (1.62)$$

Solving Equation (1.62), we get

$$55x + 700 - 20x = 1400$$

$$35x = 700$$

$$x = 20$$

Therefore 20 qt of 55% solution must be mixed with $35 - 20 = 15$ qt of 20% solution to give 35 qt of 40% solution.

Check   35 qt of 40% solution will give 14 qt of pure antifreeze. Also, 20 qt of 55% solution will give $20 \cdot \left(\frac{55}{100}\right) = 11$ qt of pure antifreeze and 15 qt of 20% solution will give $\left(\frac{20}{100}\right)(15) = 3$ qt of pure antifreeze. Thus we have $11 + 3 = 14$.

Exercises 1.7   Which pairs of equations are equivalent in problems 1 through 4?

1. $4x + 5 = 8$
    $x = \frac{3}{4}$

2. $r = 1$
    $r^2 = r$

3. $2x - 1 = 6$
    $7x + \frac{1}{3} = 13$

4. $\frac{5}{x} = \frac{2}{x}$
    $5x = 2x$

Solve the given equation in problems 5 through 26.

5. $5x + 6 = 12$

6. $2x - 1 = 6$

7. $3x + 4 = -5$

8. $4 = 12x - 17$

9. $2x + 4 = 5 - 3x$

10. $x + 5 = 2 + 3(x - 2)$

11. $5x - 6(x + 4) = 2x - 9$

12. $2(x - 5) + 4x - 6 = 2x$

13. $\dfrac{x}{2} + \dfrac{x}{3} = 7$

14. $\dfrac{1}{2} - 2x = -x + \dfrac{2}{3}$

15. $\dfrac{7}{50} - x = \dfrac{3}{10} - \dfrac{4x}{25}$

16. $\dfrac{2x}{3} - \dfrac{5x}{4} = \dfrac{1}{2}$

17. $\dfrac{2x - 1}{7} = \dfrac{x + 1}{5}$

18. $\dfrac{1 - 3x}{5} = \dfrac{1}{15} - \dfrac{x - 3}{3}$

19. $\dfrac{3}{x} - 5 = \dfrac{6}{x}$

20. $\dfrac{1}{2x} + \dfrac{1}{3x} = \dfrac{1}{6} - \dfrac{2}{x}$

21. $\dfrac{x}{x - 2} - 7 = \dfrac{2}{x - 2}$

22. $\dfrac{2}{x - 3} - 3 = \dfrac{x + 2}{x - 3}$

23. $\dfrac{3}{x - 1} = \dfrac{2}{x + 1}$

24. $\dfrac{x}{x - 2} - 2 = \dfrac{2}{x - 2}$

25. $\dfrac{1}{x + 4} - \dfrac{5}{x - 4} = \dfrac{7}{x^2 - 16}$

26. $\dfrac{5}{2x + 5} - \dfrac{5}{2(2x + 5)} = -\dfrac{3}{5}$

27. Find three consecutive integers whose sum is 54.

28. The length of a rectangle is 1 ft more than twice its width. If the perimeter of the rectangle is 26 ft, find its dimensions.

29. The sum of the present ages of Mario and Anna is 44. In five years Mario will be twice as old as Anna. Find their present ages.

30. Bob is 40 years old and Paul, his son, is 8. In how many years will Bob be twice as old as his son?

31. Fred invests $15,000, part at the rate of 5% annually and the remaining at 6% annually. If Fred's annual income from this investment is $855, how much does he have invested at each rate?

32. Carlos has 27 coins in nickels, dimes, and quarters. The total value of the coins is $3.95. If he has two more dimes than quarters, how many of each coin does he have?

33. The total value of 21 coins in dimes and quarters is $3.30. Find the number of each type of coin.

34. How many gallons of a 40% alcohol solution should be added to how many gallons of a 20% alcohol solution to obtain 50 gal of a 25% alcohol solution?

35. If we have 12 oz of a 45% alcohol solution, how many ounces of pure alcohol should we add to obtain a 60% alcohol solution?

36. A merchant has two grades of sugar, one worth $1.65 and the other $1.80 per pound. Find the weight of each grade used by the merchant to obtain a mixture of 50 lb of sugar worth $1.70 per pound.

Note.   *Use the formula*
distance = (average speed)(time)
d = st

37. Two cars start together and travel in the same direction. If one car travels at half the speed of the other and at the end of three hours they are 96 mi apart, find how fast each car was traveling.

38. Irving drove his car from city A to city B at an average speed of 50 mph. On the return trip, he drove at an average speed of 60 mph. Find the distance between the two cities if the return trip took one hour less.

39. Theresa can do a certain job in 6 hr. Alan can do the same job in 8 hr. If Alan and Theresa work together on the same job, in how many hours will the job be finished?

Chapter 1 Review   Part 1   (*Oral*)

Define or discuss each of the following.

1. Equality of sets
2. Subset and proper subset
3. Union and intersection of sets
4. Null set
5. Universal set
6. Counting numbers
7. Integers
8. Rational numbers
9. Properties assumed for the real numbers
10. One-to-one correspondence
11. Positive and negative numbers
12. Properties concerning inequalities of real numbers
13. Closed interval
14. Open interval
15. Absolute value
16. Triangle inequality
17. Principal nth root of a real number
18. Identity
19. Conditional equation
20. Equivalent equations

Part 2   (*Written*)

1. Give the proper subsets of $\{-1, 0, 1\}$.
2. What is the union of $\{a, b, c\}$ and $\{b, d\}$?
3. What is the intersection of $\{1, 2, 3\}$ and $\{2, 4, 6\}$?

4. Let the set A have three elements and the set B have four elements. What is the maximum number of elements in

   (a) $A \cup B$,   (b) $A \cap B$?

5. What is the union of the set of rational and irrational numbers? What is their intersection?

Perform the indicated operations in Problems 6 through 10.

6. $[3 - (-5)] + (-4)$ 7. $-3 - [-1 - (-2)]$

8. $\frac{1}{5} - \frac{2}{5} + \frac{6}{5}$ 9. $\frac{3}{2} - \frac{4}{3}$

10. $\dfrac{\frac{2}{3} - \frac{1}{2}}{\frac{4}{5} \quad \frac{5}{3}}$

11. What is the point halfway between $2\frac{2}{3}$ and 4?

Solve the given inequality in Problems 12 through 19.

12. $3x - 5 < 4x + 2$ 13. $\frac{1}{2}x - \frac{2}{3} < \frac{3}{2}x + 2$

14. $-3 < 1 - 2x < 5$ 15. $\dfrac{3x - 5}{2} \geq 5x - 1$

16. $(2x - 1)(x + 5) \geq 0$ 17. $(x + 1)(2x - 3) < 0$

18. $\dfrac{5}{x} - 3 < 4$ 19. $\dfrac{x + 1}{2x - 3} \geq 0$

Solve for x in Problems 20 through 23.

20. $|3x - 1| = 15$ 21. $|3 - 4x| \leq 9$

22. $|x - 1| > 4$ 23. $\left|\dfrac{2 - x}{x}\right| \geq 1$

Simplify the given expressions in Problems 24 through 29.

24. $\dfrac{(-8)^{1/3} + 4^{3/2}}{\sqrt{9}}$ 25. $\dfrac{2^{-2} - 1}{4^{-2} - 1}$

26. $x^{2/3} x^{3/2} x^{7/4}$ 27. $\sqrt[3]{\dfrac{-8y^3}{27}}$

28. $\dfrac{2x - 4}{(x - 1)^2} \cdot \dfrac{(x - 1)}{x - 2}$

29. $\dfrac{(x - 2)(x + 3)}{x(2 - x)} \div \dfrac{(3 + x)(x - 1)}{x^2}$

30. Add $\dfrac{x - 1}{x + 2} + \dfrac{x}{x - 1} - \dfrac{2x - 1}{x}$

31. How much pure copper should be mixed with 20 lb of an alloy containing 60% copper to obtain an alloy containing 85% copper?

32. A piece of wire 54 cm long is to be bent into the shape of a rectangle whose length is twice its width. Find the dimensions of the rectangle.

33. Twelve years from now Millie will be twice as old as she was 6 years ago. How old is Millie now?

34. Ms. Gomez has $100,000 invested in 8% mortgages, 10% stocks, and 6% bonds. If the amount invested in stocks is twice the amount invested in bonds and if the total income from mortgages, stocks, and bonds is $8420, how much money is invested in each?

The Function Concept

CHAPTER 2

2.1 The Rectangular Coordinate System

In this chapter we shall introduce an important mathematical concept —that of a *function*, a most fundamental notion for the study of calculus. The word function was introduced into the language of mathematics in the seventeenth century by Leibniz, one of the originators of calculus. Functions have to do with pairs of numbers. A pair of numbers in which the order is specified is called an *ordered pair* of numbers.

Ordered pair

Gottfried Wilhelm von Leibniz (1646–1716), known mainly as a philosopher and metaphysician, was one of the greatest thinkers of modern times. As a mathematician, he contributed greatly to the development of differential and integral calculus and of symbolic logic. Leibniz was also a lawyer, theologian, diplomat, geologist, and historian; he wrote with ease in Greek, Latin, French, English, and German. He was the inventor of a calculating machine, and founded the Academy of Sciences of Berlin.

Leibniz, an only child, was born into a scholarly family. He was only 6 when his father (a professor of philosophy) died. At the age of 17, he received his bachelor's degree in philosophy and at 20, his doctorate in law. His last years were saddened by ill health, and by the controversy and neglect that accompanied the bitter dispute over discovery of calculus. Newton and Leibniz each discovered calculus by different means. Newton made his discovery a few years before Leibniz, although Newton's results were published after those of Leibniz.

Gottfried Wilhelm von Leibniz

We denote an ordered pair by the symbol (a, b), not to be confused with the interval (a, b). The number a is called the *first component* and the number b is called the *second component* of the ordered pair (a, b).

We shall introduce functions in terms of the notion of sets. The reader is familiar with certain operations involving sets. Here we shall define another operation, the *Cartesian product* of two sets A and B.

Cartesian product: $A \times B$ ▶
$= \{(a, b) | a \in A, b \in B\}$

Definition 2.1 *If A and B are two sets, then the Cartesian product of A and B, denoted by $A \times B$, is the set of all ordered pairs (a, b) such that $a \in A$ and $b \in B$.* Thus

$$A \times B = \{(a, b) | a \in A, b \in B\}$$

Example 2.1 Suppose that $A = \{x, y, z\}$ and $B = \{o, p\}$. Find $A \times B$ and $B \times A$.

Note If $a \neq b$, then (a, b) $\neq (b, a)$. Hence $A \times B \neq B \times A$. In general, $(a, b) = (c, d)$ if and only if $a = c$ and $b = d$.

SOLUTION $A \times B = \{(x, o), (x, p), (y, o), (y, p), (z, o), (z, p)\}$

and $\qquad B \times A = \{(o, x), (o, y), (o, z), (p, x), (p, y), (p, z)\}$

Ordered pairs and Cartesian products have many everyday applications. Imagine a travel agent working out a schedule for a client who is leaving from Miami and wishes to go to either Rome or Athens. The travel agent presents the client with choices involving travel to either Athens or Rome with a stop at either New York, London, or Paris. The different possibilities may be described using set notation.

Let $S = \{N, L, P\}$ denote the set of cities New York, London, and Paris, and let $T = \{A, R\}$ denote the set of cities Athens and Rome. The possible routes may be described by the Cartesian product

$$S \times T = \{(N, A), (N, R), (L, A), (L, R), (P, A), (P, R)\}$$

Thus the client is presented with six possibilities. One of the possibilities, for instance, is the ordered pair (L, A). This ordered pair denotes travel from Miami to London to Athens. We can show the set $S \times T$ in tabular form.

	New York N	*London L*	*Paris P*
Athens A	(N, A)	(L, A)	(P, A)
Rome R	(N, R)	(L, R)	(P, R)

Now consider the set R consisting of all real numbers. The Cartesian product $R \times R$, which may be denoted also by R^2, is the set of all possible ordered pairs of real numbers

$$R \times R = \{(x, y) | x, y \in R\}$$

Geometrically, R^2 may be interpreted as the set of all points in the Cartesian (or geometric) plane. The *Cartesian* (or *rectangular*) *coordinate system* is ordinarily used to set up a correspondence between ordered pairs of real numbers and points in the plane. The formulation of this correspondence is attributed to the French mathematician-philosopher Descartes.

We begin by constructing a line in the plane from left to right which we call the *x-axis*, as shown in Figure 2.1. Then a vertical line is drawn,

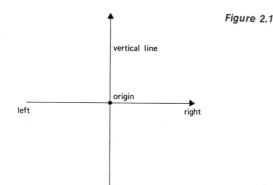

Figure 2.1

called the *y-axis*. The two lines are referred to as the *coordinate axes* and their point of intersection is called the *origin*. A convenient unit of length is chosen and a number scale is marked off on each of the axes. Usually the number zero is placed at the origin; on the *x*-axis the positive numbers are to the right (as in Chapter 1), and on the *y*-axis the positive numbers extend upward. Figure 2.2 shows a Cartesian

Coordinate axes

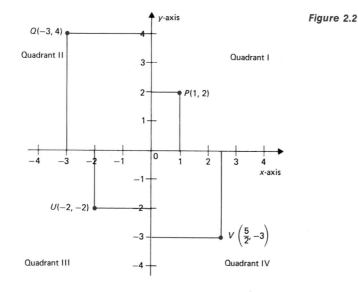

Figure 2.2

coordinate system including some points. The axes divide the plane into four parts called *quadrants*. The first quadrant is the set

$$\{(x, y) | x > 0 \text{ and } y > 0\}$$

the second quadrant is the set

$$\{(x, y) | x < 0 \text{ and } y > 0\}$$

the third quadrant is the set

$$\{(x, y) | x < 0 \text{ and } y < 0\}$$

and the fourth quadrant is the set

$$\{(x, y) | x > 0 \text{ and } y < 0\}$$

The points P, Q, U, and V are in the first, second, third, and fourth quadrants, respectively. The association of these points in the plane with the corresponding ordered pairs is illustrated in Figure 2.2. For example, the *coordinates* of the point P are the ordered pair of numbers $(1, 2)$; the abscissa is 1 and the ordinate is 2.

Coordinates

To each element in R^2 there corresponds exactly one point in the plane and conversely, to each point in the plane there corresponds exactly one element of R^2.

One of the oldest and most famous results in mathematics is the *Pythagorean theorem*. The ancient Greeks proved that the area of the square constructed on the hypotenuse of a right triangle is equal to the sum of the areas of the squares constructed on the other two sides (see Figure 2.3). If a, b, and c represent the lengths of the sides as

Figure 2.3

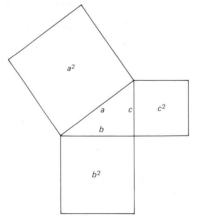

indicated, then

$$a^2 = b^2 + c^2$$

We shall use this result to find the distance between two points in R^2.

In Chapter 1 we defined the absolute value of a number and defined $|a - b|$ as the distance between the two points a and b on the number line. Here we shall extend the notion of distance to R^2. The distance

between two points in R^2 is the length of the line segment joining them. We shall establish a formula for this distance in terms of the coordinates of the two endpoints.

Consider two points on the x-axis, $A(x_1, 0)$ and $B(x_2, 0)$. We have defined the distance $d(A, B)$ between A and B to be

$$d(A, B) = |x_2 - x_1|$$

Consider the points $A'(x_1, b)$ and $B'(x_2, b)$, where $b \neq 0$ (see Figure 2.4). Since the second components of A' and B' are equal, the line

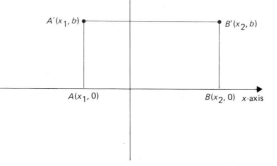

Figure 2.4

segment $A'B'$ is parallel to the x-axis. The reader will note that the four points A, B, B', and A' are the vertices of a rectangle. Hence the line segment $A'B'$ may be assigned a length

$$d(A', B') = d(A, B) = |x_2 - x_1|$$

Similarly (see Figure 2.5), we have

$$d(C', D') = d(C, D) = |y_2 - y_1|$$

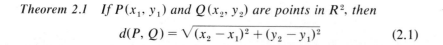

Figure 2.5

Theorem 2.1 *If $P(x_1, y_1)$ and $Q(x_2, y_2)$ are points in R^2, then*   ◀ *The distance formula*

$$d(P, Q) = \sqrt{(x_2 - x_1)^2 + (y_2 - y_1)^2} \qquad (2.1)$$

PROOF   Let S be the point (x_2, y_1). Then the points P, Q, and S are the vertices of a right triangle as shown in Figure 2.6. Since the line

Figure 2.6

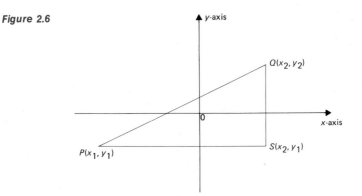

segment PS is parallel to the x-axis and the line segment QS is parallel to the y-axis, we have

$$d(P, S) = |x_2 - x_1| \tag{2.2}$$

and

$$d(Q, S) = |y_2 - y_1| \tag{2.3}$$

Hence by the Pythagorean theorem,

$$\begin{aligned}
[d(P, Q)]^2 &= [d(P, S)]^2 + [d(Q, S)]^2 \\
&= |x_2 - x_1|^2 + |y_2 - y_1|^2 \\
&= (x_2 - x_1)^2 + (y_2 - y_1)^2
\end{aligned}$$

and

$$d(P, Q) = \sqrt{(x_2 - x_1)^2 + (y_2 - y_1)^2}$$

This proves Theorem 2.1.

It makes no difference whether the coordinates of P or those of Q are chosen as (x_1, y_1). The reader should show that
$d(P, Q) = d(Q, P).$

Example 2.2   Find the distance between $P(2, -1)$ and $Q(5, 2)$.

SOLUTION   Using Equation (2.1), we have (see Figure 2.7)

$$\begin{aligned}
d(P, Q) &= \sqrt{(x_2 - x_1)^2 + (y_2 - y_1)^2} \\
&= \sqrt{(5 - 2)^2 + [2 - (-1)]^2} \\
&= \sqrt{9 + 9} = 3\sqrt{2}
\end{aligned}$$

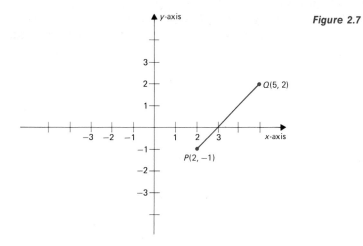

Figure 2.7

Example 2.3 Show that the triangle with vertices $P(3, -2)$, $Q(4, 3)$, and $S(-6, 5)$ is a right triangle.

SOLUTION   Using the distance formula, we have

$$d(P, Q) = \sqrt{(4 - 3)^2 + (3 + 2)^2} = \sqrt{26}$$

$$d(Q, S) = \sqrt{(-6 - 4)^2 + (5 - 3)^2} = \sqrt{104}$$

and

$$d(P, S) = \sqrt{(-6 - 3)^2 + (5 + 2)^2} = \sqrt{130}$$

We observe that

$$(\sqrt{130})^2 = (\sqrt{104})^2 + (\sqrt{26})^2$$

or

$$[d(P, S)]^2 = [d(Q, S)]^2 + [d(P, Q)]^2$$

Hence, by the converse of the Pythagorean theorem, the triangle with vertices P, Q, and S is a right triangle with the line segment PS as its hypotenuse (see Figure 2.8).

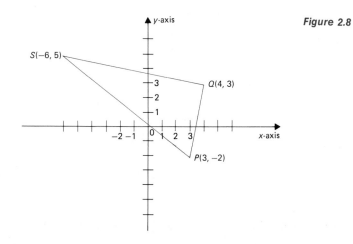

Figure 2.8

Now we shall show how to find the coordinates of the midpoint of a line segment. Consider the line segment PQ in Figure 2.9. Let $M(x, y)$

Figure 2.9

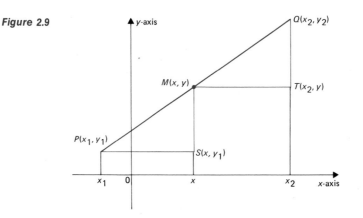

be the midpoint of the line segment PQ. We wish to express x, y in terms of x_1, y_1, x_2, y_2. With the help of results in plane geometry, the reader will see that the triangles PMS and MQT are congruent. Therefore

Congruent triangles: Two triangles are said to be congruent if and only if the corresponding sides and angles are equal.

$$d(P, S) = d(M, T) \quad \text{or} \quad |x - x_1| = |x_2 - x|$$

Further, since x is to the right of x_1 and x_2 is to the right of x, we have

$$x - x_1 = x_2 - x$$

Solving for x, we obtain

$$x = \frac{x_1 + x_2}{2} \tag{2.4}$$

Also,

$$d(S, M) = d(T, Q)$$

or

$$y - y_1 = y_2 - y$$

$$y = \frac{y_1 + y_2}{2} \tag{2.5}$$

Midpoint formula   Hence the coordinates of the midpoint of the line segment PQ are given by (2.4) and (2.5).

Example 2.4 Find the midpoint of the line segment with endpoints $A(-4, 2)$ and $B(3, 4)$.

SOLUTION   Let $M(x, y)$ be the midpoint of the line segment AB. Then

$$x = \frac{-4 + 3}{2} = -\frac{1}{2} \quad \text{and} \quad y = \frac{2 + 4}{2} = 3$$

Therefore the midpoint is $M\left(-\frac{1}{2}, 3\right)$.

1. On a Cartesian coordinate system indicate the points with the following coordinates.

   (a) $(3, 2)$ (b) $\left(-\frac{3}{2}, 2\right)$
   (c) $(0, 5)$ (d) $(5, 0)$
   (e) $(-3, -3)$ (f) $(-2, -5)$
   (g) $(1, -3)$

In Problems 2 through 5, find the Cartesian products $A \times B$ and $B \times A$ and represent the resulting set in a rectangular coordinate system.

2. $A = \{1, 2\}, \quad B = \{-1, -2\}$ 3. $A = \{0, 1, 2\}, \quad B = \{-1, 0\}$

4. $A = \{0, 1, 2\}, \quad B = \{0, 1, 2\}$ 5. $A = \{-1, 1\}, \quad B = \{2, 3, 4\}$

6. Let $A = \{a, b, c, d\}$ and $B = \{r, s, t\}$. Find
   (a) $n(A \times B)$ (b) $n(B \times A)$

Note $n(A)$ *is the number of elements in the set A.*

In Problems 7 through 14, find the distance between the points with the given coordinates. Also find the coordinates of the midpoint.

7. $A(1, 3), \quad B(4, 8)$ 8. $A(2, 5), \quad B(-1, 4)$

9. $A(-3, -3), \quad B(-1, -6)$ 10. $A(-2, 5), \quad B(-1, -2)$

11. $A(-2, 0), \quad B(-3, 1)$ 12. $A\left(2, \frac{1}{2}\right), \quad B\left(\frac{1}{2}, -3\right)$

13. $A(-1, 3), \quad B(-1, 7)$ 14. $A(5, 2), \quad B(-3, 2)$

Show that ABC is a right triangle in Problems 15 through 20.

15. $A(-3, 2), \quad B(5, 5), \quad C(0, -6)$

16. $A(2, 0), \quad B(6, 3), \quad C(5, -4)$

17. $A(0, 0), \quad B(3, 2), \quad C(-4, 6)$

18. $A(4, 3), \quad B(1, 4), \quad C(-2, -5)$

19. $A(1, 2), \quad B(3, 4), \quad C(1, 6)$

20. $A(0, -3), \quad B(-1, -1), \quad C(-3, -2)$

In Problems 21 through 24, identify the triangle ABC as isosceles, equilateral, or right.

21. $A(-6, 6), \quad B(-2, 5), \quad C(-5, 2)$

22. $A(2, 1), \quad B(5, 5), \quad C(-2, 4)$

23. $A(1, -1), \quad B(-1, 1), \quad C(-\sqrt{3}, -\sqrt{3})$

24. $A(-4, 6) \quad B(0, 5), \quad C(-3, 2)$

Hint: *If P, Q, and S are collinear, then*

$$d(P, Q) + d(Q, S) = d(P, S)$$

25. Which of the following sets of points are collinear (lie on the same line)?
    (a) $(0, 5)$, $(4, 1)$, $(12, -7)$
    (b) $(-3, 2)$, $(2, 0)$, $(1, -2)$
    (c) $(3, -5)$, $\left(\frac{1}{2}, 0\right)$, $(0, 1)$

26. The distance from the point P to the point $(2, 0)$ is $\sqrt{45}$. If the abscissa of P is -4, find the ordinate of P.

27. Show that the points $A(3, -16)$, $B(-5, -1)$, $C(10, 7)$, and $D(18, -8)$ are the vertices of a square. Find the length of the diagonals.

28. If the vertices of a triangle are $A(1, 2)$, $B(-4, 5)$, and $C(3, 7)$, find the length of each of the medians (a *median* is a line segment drawn from a vertex to the midpoint of the opposite side).

29. A pilot is flying from Dullsville to Middale to Pleasantville. With reference to an origin, Dullsville is located at $(2, 4)$, Middale at $(8, 12)$, and Pleasantville at $(20, 3)$, all numbers being in 100-mile units.
    (a) Locate the positions of the three cities on a coordinate system.
    (b) Compute the distance traveled by the pilot.
    (c) Compute the direct distance from Dullsville to Pleasantville.

2.2 Relations and Functions

In many real-life situations, expressing the relationship among things is very common and important. Relationships may be expressed by simple statements such as "Willie is a brother of Rose," "3 is less than 5," and "the price of a stock increases as its earnings increase," or by very complex mathematical formulas. One way to describe the relationships is by setting up ordered pairs.

Relation ▶ *Definition 2.2 A relation from a set A to a set B is defined as a subset of $A \times B$. The set of all first components in a relation is called its* domain *and the set of all second components in a relation is called its* range.

In many situations, the set B is the same as the set A. If $B = A$, we usually say "relation on A" instead of "relation from A to A." We shall be primarily concerned with the relations on the set R of real numbers. Thus a relation S on R is any set of ordered pairs in $R \times R$, that is, $S \subseteq R \times R$.

Since a relation is a subset of ordered pairs in $R \times R$, we can use the rectangular coordinate system to represent a relation as a set of points in the plane. This leads us to our next definition:

Graph ▶ *Definition 2.3 If S is a relation on R, then the* graph *of S is the set of all points in a coordinate plane that correspond to the ordered pairs in S.*

Example 2.5 Let S be the relation given by

$$S = \{(1, 2), (-1, 0), (-2, -1), (0, -1), (1, -2)\}$$

Find the domain and range of S; sketch its graph.

SOLUTION   domain of $S = \{1, -1, -2, 0\}$

range of $S = \{2, 0, -1, -2\}$

The graph of S consisting of five points is sketched in Figure 2.10.

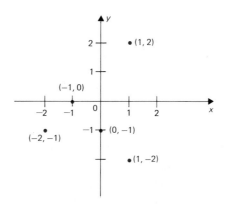

Figure 2.10

An important kind of relation is the one that is called a *function*. Consider the following examples.

Example 2.6 In an experiment the effect of a bactericidal agent on the number N of viable bacteria with respect to time t was measured. The following table indicates the result of the experiment.

t (*minutes*)	0	5	10	15	20	25	30
N = *number of viable bacteria per milliliter (approx)*	10^{12}	10^{10}	10^{8}	10^{6}	10^{4}	10^{2}	1

We may express this result as a relation consisting of the set of ordered pairs of numbers

$$f = \{(0, 10^{12}), (5, 10^{10}), (10, 10^{8}), (15, 10^{6}), (20, 10^{4}), (25, 10^{2}), (30, 1)\}$$

Here we say that N is a *function* of time t that is described by the set f. The two sets of numbers

$$t = \{0, 5, 10, 15, 20, 25, 30\}$$

and

$$N = \{10^{12}, 10^{10}, 10^{8}, 10^{6}, 10^{4}, 10^{2}, 1\}$$

are called the *domain* and the *range* of the function f, respectively.

Example 2.7   A mathematician working for the Fancy Shirt Company finds that the total cost c (in dollars) of producing x shirts is given by the equation

$$c = 0.001x^2 + 1.2x + 30, \qquad x \geq 1$$

Obviously, the company does not produce a fraction of a shirt. That is why we require that the domain of f be the set of positive integers.

In this case, the cost c is a function of the number x of shirts produced that is described by the set of ordered pairs

$$f = \{(x, c) \,|\, x \geq 1, \; c = 0.001x^2 + 1.2x + 30\}$$

where x belongs to the set of positive integers (the domain of f).

Example 2.8   The area A of a square is dependent on the length of a side. We have

$$A = S^2$$

where S is the length of the side of a square (see Figure 2.11). Thus

Figure 2.11

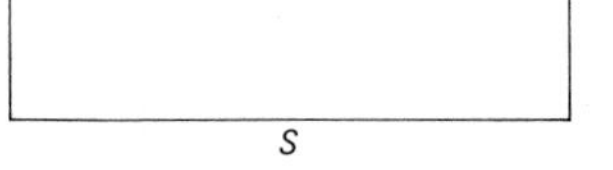

the area A is a function of S that is described by the set

$$f = \{(S, A) \,|\, S > 0, \; A = S^2\}$$

We observe that for this relation to be meaningful we require that the domain of f be the set $\{S \,|\, S > 0\}$.

In each of the preceding examples we see that we are dealing with
(a)   Two sets of numbers X and Y

(b)   A rule or formula that assigns to each element of X a *single* element of Y

We now give a precise definition of a function:

Definition 2.4   *A function f from a set X to a set Y is a relation f from X to Y such that for each $x \in X$, there is exactly one ordered pair (x, y) in f having x as its first component.*   ◄ *Function*

In other words, f is a function from X to Y, denoted by $f: X \to Y$, if the following conditions are satisfied.

(i) f is a relation from X to Y, that is, $f \subseteq X \times Y$

(ii)   Domain of $f = X$

and

(iii)   If $(x, y_1) \in f$ and $(x, y_2) \in f$, then $y_1 = y_2$

Since a function associates elements from a set X to the elements in a set Y, we have an alternative definition of a function:

Alternative Definition 2.4   *A function f from a set X to a set Y, denoted*   ◄ *by f: X → Y, is a correspondence (or a rule) that associates with each element $x \in X$ a unique $y \in Y$. The element y is called the image of x under f and is denoted by $f(x)$. The set X is called the* domain *of f. The set of all images of elements of X is called the* range *of f.*

We ask the reader to compare the two definitions of a function.

We usually write $y = f(x)$, read "y equal f of x," to denote that y is a function of x. The Swiss mathematician Leonhard Euler is responsible for this notation. We call x the *independent variable* because x can be selected arbitrarily from the set X. We call y the *dependent variable* since the value of y depends upon the value of x selected. We sometimes represent the correspondences pictorially as shown in Figure 2.12.

Figure 2.12

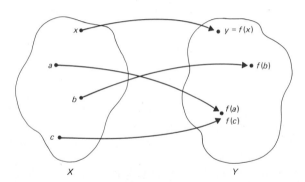

As in the case of a relation, the sets X and Y in the definition of a function need not be distinct. In fact, we shall be dealing with the sets X and Y, where X and Y are subsets of real numbers. The letters f, g, h, F, G, and H are customarily used to denote functions.

Let us repeat for emphasis: to each $x \in X$ there corresponds exactly one element $f(x) = y \in Y$. However, different elements such as a and c in Figure 2.12 may have the same image.

Let us agree that if a function is given by a formula and the domain is not specified, then we shall assume that the domain consists of all real numbers for which the formula is meaningful. For example, if

$$y = f(x) = \frac{x}{x - 3} \tag{2.6}$$

then the domain D of f, unless indicated, is assumed to be

$$D = \{x \,|\, x \text{ is a real number and } x \neq 3\}$$

Since division by zero is undefined, the real number 3 is excluded from the domain. By direct inspection of (2.6) it is not easy to find the range of f. One way to find the range of f is to solve Equation (2.6) for x in terms of y.

$$y = \frac{x}{x - 3}$$

$$y(x - 3) = x$$

$$xy - 3y = x$$

$$xy - x = 3y$$

$$x(y - 1) = 3y$$

or

$$x = \frac{3y}{y - 1} \tag{2.7}$$

From (2.7) we see that y can take on any real number except 1. Therefore, the range of f defined by (2.6) is given by

$$\text{range of } f = \{y \,|\, y \text{ is a real number and } y \neq 1\}$$

The domain of the function f defined by

$$f(x) = \sqrt{2 + x} \tag{2.8}$$

is understood to be

$$\{x \,|\, 2 + x \geq 0\} = \{x \,|\, x \geq -2\}$$

since for a real number b, $\sqrt{b}$ is real if and only if $b \geq 0$. However, if a certain set of values for the independent variable x is specified, such as

$$g(x) = \sqrt{2 + x}, \qquad 3 \leq x \leq 9 \tag{2.9}$$

then the domain of the function g is the closed interval $[3, 9]$. We note that the functions f and g defined by (2.8) and (2.9) are not equal, since they have different domains.

The *graph* of a function f is the graph of a relation f. It is easy to recognize from the graph of a relation whether the relation is a function or not. *If there are two or more points of the graph on the same vertical line, then the relation is not a function; otherwise, it is.* In Figure 2.13, (a) and (b) are the graphs of a function, whereas (c) is not the graph of a function.

Vertical line test

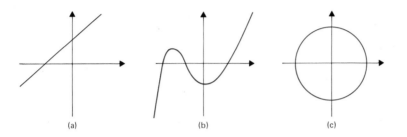

Figure 2.13

(a)   (b)   (c)

Let us consider the function f defined by

$$f(x) = x^2 - 2x + 1 \qquad (2.10)$$

The domain of f is assumed to be (by agreement stated earlier) the set of all real numbers. So any real number can be substituted for x. For example,

$$f(2) = (2)^2 - 2(2) + 1 = 4 - 4 + 1 = 1$$

$$f(-3) = (-3)^2 - 2(-3) + 1 = 9 + 6 + 1 = 16$$

$$f(0) = (0)^2 - 2(0) + 1 = 0 - 0 + 1 = 1$$

$$f(a) = a^2 - 2a + 1$$

$$f(a + h) = (a + h)^2 - 2(a + h) + 1$$

$$f(3c) = (3c)^2 - 2(3c) + 1$$

and so on.

Example 2.9 Let f be a function defined by

$$f(x) = x^2 - 2, \qquad x \in R \qquad (2.11)$$

If $a, h \in R$, where $h \neq 0$, find

(a) $\dfrac{f(a + h) - f(a)}{h}$

(b) Sketch the graph of f.

SOLUTION

(a) We have

$$f(a + h) = (a + h)^2 - 2 \text{ and } f(a) = a^2 - 2$$

Hence

$$\frac{f(a + h) - f(a)}{h} = \frac{(a + h)^2 - 2 - (a^2 - 2)}{h}$$

$$= \frac{a^2 + 2ah + h^2 - 2 - a^2 + 2}{h}$$

$$= \frac{2ah + h^2}{h} = 2a + h$$

(b) To sketch the graph of f, we list coordinates $(x, f(x))$ of some points on the graph of f in tabular form as follows.

x	-3	-2	-1	0	1	2	3
$f(x)$	7	2	-1	-2	-1	2	7

We then draw a smooth curve through these points obtaining the sketch shown in Figure 2.14.

Figure 2.14

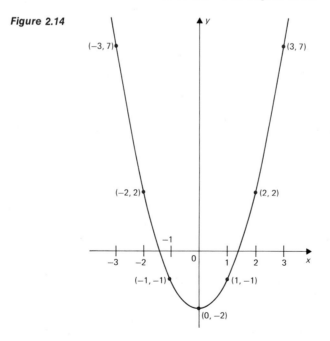

Identity function   **Example 2.10**   For the *identity function* defined by $f(x) = x$, find $f(-2)$, $f(-1)$, $f(0)$, $f(1)$, and $f(2)$. Describe the domain and range of f. Sketch the graph of f.

SOLUTION $f(-2) = -2$, $f(-1) = -1$, $f(0) = 0$, $f(1) = 1$, and $f(2) = 2$

domain $f = \{x \mid x \in R\}$ range $f = \{y \mid y \in R\}$

The graph of f is a straight line shown in Figure 2.15.

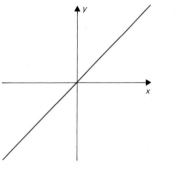

Figure 2.15

Example 2.11   The graphs of the *constant functions* $f(x) = 1$, $f(x) = 2$, and $f(x) = -1$ are shown in Figure 2.16. The domain of a constant

Constant function

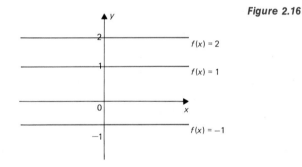

Figure 2.16

function $f(x) = c$, where c is a constant, is the set R of real numbers and its range is $\{c\}$. For example, if $f(x) = c$, then $f(3) = c$, $f(-2) = c$, $f(0) = c$, $f(-108) = c$, $f(139) = c$, etc.

Example 2.12   For the *absolute-value function*, $f(x) = |x|$, find $f(-2)$, $f(-1)$, $f(0)$, $f(1)$, and $f(2)$. Describe the domain and range of f. Sketch its graph.

Absolute-value function

SOLUTION   By definition of the absolute value, we have

$$f(x) = \begin{cases} x & \text{if } x \geq 0 \\ -x & \text{if } x < 0 \end{cases}$$

so that $f(-2) = -(-2) = 2$, $f(-1) = 1$, $f(0) = 0$, $f(1) = 1$, and $f(2) = 2$. The domain of f is R; the range of f is all nonnegative real numbers. The graph of f is shown in Figure 2.17.

Figure 2.17

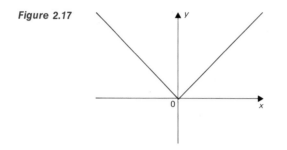

Greatest integer function

Example 2.13   For the *greatest integer function*

$$f(x) = [\![x]\!]$$

where $[\![x]\!]$ denotes the greatest integer less than or equal to x, find $f(3.5)$, $f(3.4)$, $f(3)$. $f(-2)$, $f(-2.5)$, $f(0.5)$, and $f(-0.5)$. Describe the domain and range of f. Sketch the graph of f.

  SOLUTION $f(3.5) = [\![3.5]\!] = 3$, $f(3.4) = 3$, $f(3) = 3$, $f(-2) = -2$, $f(-2.5) = -3$, $f(0.5) = 0$, and $f(-0.5) = -1$. The domain of f is R and its range is the set I of integers. The graph of f is shown in Figure 2.18.

Figure 2.18

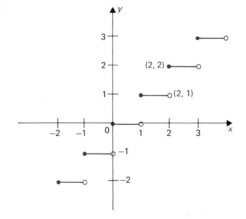

In the figure, the point $(2, 1)$ with empty circle means that this point is not included in the graph, whereas the point $(2, 2)$ with full circle means that this point is included in the graph.

In Problems 1 through 10, let $f(x) = 2x^2 - 2x + 3$ and $g(x) = 2x + 4$. Compute the following.

1. $f(0)$, $g(0)$

2. $f(-1)$, $g(-1)$

3. $f(\frac{1}{2})$, $g(\frac{1}{2})$

4. $f(2)$, $g(2)$

5. $f(1-2)$, $g(1-2)$

6. $f(x^2)$, $g(x^2)$

7. $f(a+h)$, $g(a+h)$

8. $\dfrac{f(a+h) - f(a)}{h}$, $h \neq 0$

9. $\dfrac{g(a+h) - g(a)}{h}$, $h \neq 0$

10. $f(g(x))$, $g(f(x))$

In Problems 11 through 18, identify which of the relations are functions and which are not. Identify the domain of the given relation and sketch its graph.

11. $\{(-1, 0), (2, 1), (3, 1)\}$

12. $\{(2, 0), (1, 1), (2, 2)\}$

13. $\{(x, y) | y = x + 2\}$

14. $\{(x, y) | y = |x|\}$

15. $\{(x, y) | |y| = x\}$

16. $\{(x, y) | y = x^2\}$

17. $\{(x, y) | y^2 = x^2\}$

18. $\{(x, y) | y < x - 2\}$

In Problems 19 through 32, determine the domain of the given function.

19. $f(x) = x + 4$

20. $f(x) = \dfrac{1}{x}$

21. $f(x) = \dfrac{1}{x^2 + 4}$

22. $f(t) = \dfrac{1}{t(t-5)}$

23. $f(s) = \dfrac{1}{s^2 - 4}$

24. $h(s) = \sqrt{s}$

25. $g(r) = \dfrac{r}{(r-1)(r+2)}$

26. $F(x) = \sqrt{2 - x}$

27. $H(x) = \sqrt[3]{x}$

28. $G(z) = \sqrt{4 - z^2}$

29. $F(x) = \sqrt{x^2 - 16}$

30. $H(t) = \sqrt[3]{\dfrac{1}{1-t}}$

31. $g(t) = \sqrt{t^2 + 3t + 2}$

32. $h(x) = \dfrac{2}{\sqrt{x^2 - x - 6}}$

In Problems 33 through 37, graph the given functions.

33. $h(x) = 2 + x$, domain $h = \{1, 2, 3, 4, 5\}$

34. $f(x) = x^2 - 3$, domain $f = \{-1, 0, 1, 2, 3, 4\}$

35. $g(s) = \sqrt{s + 1}$, domain $g = \{-1, 0, 1, 2, 3, 4\}$

36. $f(x) = 2 + \sqrt{x + 1}$, domain $f = \{-1, 0, 1, 2, 3, 4\}$

37. $H(u) = \dfrac{u + 1}{u - 1}$, domain $H = \{3, 5, 7, 9, 11\}$

38. Which of the following graphs represent functions?

Figure 2.19

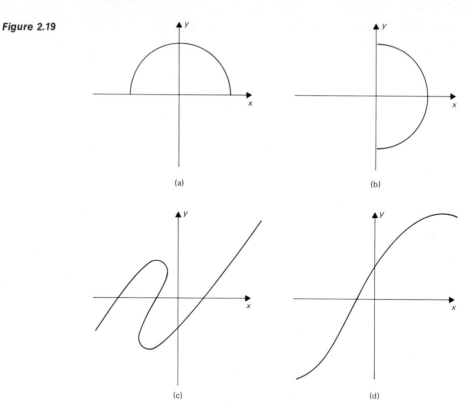

(a)

(b)

(c)

(d)

39. For the function f defined by $f(x) = x - |x|$, find $f(-2)$, $f(-1)$, $f(0)$, $f(1)$, and $f(2)$. What is the domain and range of f? Sketch the graph of f.

40. Repeat Problem 39 for the function g defined by $g(x) = x + |x|$.

In Problems 41 through 46, find the domain and range of the given function and sketch its graph.

41. $f(x) = 2|x|$ 42. $g(x) = 2[[x]]$

43. $f(x) = x - [[x]]$ 44. $g(x) = [[2x]]$

45. $f(x) = |x| - [[x]]$ 46. $f(x) = |x| + |x + 1|$

47. If $g(x) = 2x + 5$ defines a function g, find the number in the domain of g such that

 (a) $g(x) = 2$ (b) $g(x) = -1$ (c) $g(x) = c + 4$

48. If $f(x) = x^2 - 4$ defines a function f, find the numbers in the domain of f such that

 (a) $f(x) = 0$ (b) $f(x) = 3$ (c) $f(x) = -5$

49. Consider the function f defined by $f(x) = x^2$. Does $f(a) + f(b) = f(a + b)$? For what values of a and b will the equality hold?

Even and odd functions   50. **Definition 2.5** *A function f is said to be* even *if $f(x) = f(-x)$ for all x and*

−x in the domain of f. A function f is said to be odd *if f(−x)=−f(x) for all x and −x in the domain of f.*

Which of the following functions are even and which are odd?

(a) $f(x) = 2x^2$ (b) $f(x) = x^3$

(c) $f(x) = x^6 - x^4$ (d) $f(x) = x^2 - x$

51. Can a function be both odd and even?

2.3 Operations on Functions

The operations of addition, subtraction, multiplication, and division apply to functions as well as to numbers. If f and g are functions with domains A and B, respectively, then the sum $f + g$, the difference $f - g$, the product fg, and the quotient f/g are the functions defined as follows:

$$(f + g)(x) = f(x) + g(x) \qquad (2.12)$$

$$(f - g)(x) = f(x) - g(x) \qquad (2.13)$$

$$(fg)(x) = f(x)\, g(x) \qquad (2.14)$$

and

$$\left(\frac{f}{g}\right)(x) = \frac{f(x)}{g(x)} \qquad (2.15)$$

respectively. The domain of $f + g$, $f - g$, and fg is $A \cap B$. The domain of f/g is $\{x \mid x \in A \cap B,\ g(x) \neq 0\}$. If $A \cap B = \emptyset$, then $f \pm g$, fg, and f/g are not defined.

Example 2.14 Let the functions f and g be defined by the equations

$$f(x) = 2x^2 + 3, \qquad g(x) = \sqrt{x + 1}$$

Describe the functions $f + g$, $f - g$, fg, and f/g.

SOLUTION The domain of f is the set of all real numbers; the domain of g is $\{x \mid x \geq -1\}$. Therefore the functions $f + g$, $f - g$, and fg are defined by

How is the domain of g obtained?

$f(x) + g(x) = 2x^2 + 3 + \sqrt{x + 1},$ domain $f + g = \{x \mid x \geq -1\}$

$f(x) - g(x) = 2x^2 + 3 - \sqrt{x + 1},$ domain $f - g = \{x \mid x \geq -1\}$

$f(x)\, g(x) = (2x^2 + 3)(\sqrt{x + 1}),$ domain $fg = \{x \mid x \geq -1\}$

Since $g(x) = 0$ when $x = -1$,

$$\frac{f(x)}{g(x)} = \frac{2x^2 + 3}{\sqrt{x + 1}}, \qquad \text{domain } \frac{f}{g} = \{x \mid x > -1\} \qquad (2.16)$$

Note $g(x) = 0 \Rightarrow \sqrt{x + 1} = 0 \Rightarrow x = -1$

Example 2.15 Let the functions f and g be defined by the equations

$$f(x) = \sqrt{16 - x^2} \quad \text{and} \quad g(x) = 1 - x^2$$

Describe the functions f/g and g/f.

SOLUTION The domain of f is $\{x \mid -4 \leq x \leq 4\}$; the domain of g is the set of all real numbers. We note that $f(x) = 0$ for $x = -4, 4$ and $g(x) = 0$ for $x = 1, -1$. Thus f/g is defined by

$$\frac{f(x)}{g(x)} = \frac{\sqrt{16 - x^2}}{1 - x^2}, \qquad \text{domain } \frac{f}{g} = \{x \mid -4 \leq x \leq 4, \ x \neq 1, -1\}$$

and g/f is defined by

$$\frac{g(x)}{f(x)} = \frac{1 - x^2}{\sqrt{16 - x^2}}, \qquad \text{domain } \frac{g}{f} = \{x \mid -4 < x < 4\}$$

Let us now examine what happens when the two components of every ordered pair in a function f are interchanged. For example, if $f = \{(-1, 1), (0, 2), (1, 2), (2, 3)\}$, then interchanging the components of these ordered pairs yields the set $\{(1, -1), (2, 0), (2, 1), (3, 2)\}$. The resulting set is not a function, since the ordered pairs $(2, 0)$ and $(2, 1)$ have the same first component and different second components. Now suppose a function f has no two ordered pairs with the same second components. Such a function is said to be *one-to-one*, and interchanging the components of every ordered pair in f yields another function. The resulting function is called the *inverse function* of f and is denoted by f^{-1} (read "f inverse"). Clearly, the domain of f^{-1} is the range of f and the range of f^{-1} is the domain of f.

Example 2.16 Consider the function f defined by the set $\{(1, -1), (2, -2), (3, -3), (4, -4)\}$. Find f^{-1}.

SOLUTION We observe that f is one-to-one. Therefore by definition, we have

$$f^{-1} = \{(-1, 1), (-2, 2), (-3, 3), (-4, 4)\}$$

and we note that

$$\text{domain } f^{-1} = \{-1, -2, -3, -4\} = \text{range } f$$

$$\text{range } f^{-1} = \{1, 2, 3, 4\} \qquad = \text{domain } f$$

Example 2.17 Two commonly used temperature scales are the Celsius temperature scale and the Fahrenheit temperature scale. We have two sets of numbers representing the scales. The set of numbers for the Celsius scale is $X = \{x \mid x \geq -273.15\}$. The set of numbers for the Fahrenheit scale is $Y = \{y \mid y \geq -459.67\}$. If x is the number on the Celsius scale for a temperature, then the same temperature is represented by the number $y = \left(\frac{9}{5}\right)x + 32$ on the Fahrenheit scale. Thus we have a function f defined by $\{(x, y) \mid x \in X, \ y = \left(\frac{9}{5}\right)x + 32\}$. Find f^{-1}.

SOLUTION   The function f is defined by

$$y = \tfrac{9}{5}x + 32, \qquad x \in X \qquad\qquad (2.17)$$

To interchange the components of the ordered pairs in f, we interchange x and y in Equation (2.17) and define f^{-1} by

$$x = \tfrac{9}{5}y + 32 \qquad\qquad (2.18)$$

Solving for y in terms of x in Equation (2.18), we obtain

$$y = \tfrac{5}{9}(x - 32), \qquad x \in Y \qquad\qquad (2.19)$$

In other words,

$$f^{-1} = \left\{(x, y) \mid x \in Y, \ y = \tfrac{5}{9}(x - 32)\right\}$$

Here f is a function that converts Celsius degrees into Fahrenheit and f^{-1} is a function that converts Fahrenheit degrees into Celsius.

Note the interchange of the domain and range for f and f⁻¹.

Example 2.18   Consider the function f defined by the equation $y = x^2$. Discuss the existence of f^{-1}.

SOLUTION   If we take the domain of f to be all real x, then f is not one-to-one and so f^{-1} does not exist. However, if we restrict the domain to $\{x \mid x \geq 0\}$, then the resulting function g is one-to-one, and we find g^{-1} as follows (Figure 2.20). From the equation

f is not one-to-one because (x, x²) and (−x, x²) are both in f.

$$y = x^2$$

we obtain

$$x = y^2$$

by interchanging x and y. Solving for y in terms of x, we have

$$y = \sqrt{x} \quad \text{or} \quad y = -\sqrt{x}$$

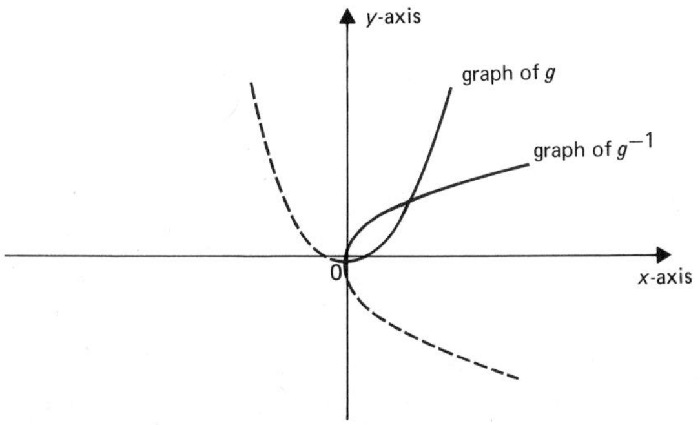

Figure 2.20

Since the domain of g is $\{x \mid x \geq 0\}$, this set is the range of g^{-1} defined by

$$y = \sqrt{x}$$

Therefore if $g = \{(x, y) \mid x \geq 0, y = x^2\}$, then $g^{-1} = \{(x, y) \mid x \geq 0, y = \sqrt{x}\}$. Now if we take the domain of f to be $\{x \mid x \leq 0\}$, then the resulting function h is again one-to-one. Using the equation

$$y = x^2$$

interchanging x and y, and solving for y, we obtain

$$y = \sqrt{x} \quad \text{or} \quad y = -\sqrt{x}$$

Since the domain of h is $\{x \mid x \leq 0\}$, this set is the range of h^{-1}. Therefore if $h = \{(x, y) \mid x \leq 0, y = x^2\}$, then $h^{-1} = \{(x, y) \mid x \geq 0, y = -\sqrt{x}\}$.

In later chapters the reader will encounter another useful operation on functions, called the composition of functions.

It is important to observe that the notation $(f \circ g)(x)$ indicates that we first apply g to x and then apply f to $g(x)$. See Figure 2.21.

▶ *Definition 2.6   If f and g are two functions, the* composite function, *denoted by $f \circ g$, has values given by*

$$(f \circ g)(x) = f(g(x)) \tag{2.20}$$

The domain of $f \circ g$ is $\{x \mid x \in \text{domain } g, g(x) \in \text{domain } f\}$ (see Figure 2.21).

Figure 2.21

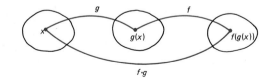

Example 2.19   Let the functions f and g be defined by the equations

$$f(x) = 3x^2 + 5 \qquad g(x) = \sqrt{x - 1}$$

Determine $f \circ g$ and $g \circ f$.

SOLUTION   The domain of f is all real numbers; the domain of g is $\{x \mid x \geq 1\}$. Using (2.20), we have for $f \circ g$:

$$(f \circ g)(x) = f(g(x))$$
$$= f(\sqrt{x - 1})$$
$$= 3(\sqrt{x - 1})^2 + 5$$
$$= 3(x - 1) + 5 = 3x + 2$$

and

$$\text{domain } f \circ g = \{x \,|\, x \in \text{domain } g,\ g(x) \in \text{domain } f\}$$
$$= \{x \,|\, x \geq 1,\ \sqrt{x-1} \text{ is real}\}$$
$$= \{x \,|\, x \geq 1\}$$

For $g \circ f$, we have

$$(g \circ f)(x) = g\big(f(x)\big)$$
$$= g(3x^2 + 5)$$
$$= \sqrt{(3x^2 + 5) - 1}$$
$$= \sqrt{3x^2 + 4}$$

and

$$\text{domain } g \circ f = \{x \,|\, x \in \text{domain } f,\ f(x) \in \text{domain } g\}$$
$$= \{x \,|\, x \text{ is real},\ 3x^2 + 5 \geq 1\}$$
$$= \{x \,|\, x \text{ is real}\}$$

The reader will note that the function $f \circ g$ is different from $g \circ f$. For example, we find that

$$(f \circ g)(2) = f\big(g(2)\big) = f(\sqrt{2} - 1) = f(1) = 8$$

and

$$(g \circ f)(2) = g\big(f(2)\big) = g(3 \cdot 2^2 + 5) = g(17) = 4$$

Example 2.20   Show that if the function f is defined by

$$f(x) = 2x + 4$$

then

(i) $(f \circ f^{-1})(x) = x$ (ii) $(f^{-1} \circ f)(x) = x$

SOLUTION   The reader can show that f^{-1} is defined by

$$f^{-1}(x) = \tfrac{1}{2}x - 2$$

Hence

(i) $(f \circ f^{-1})(x) = f\big(f^{-1}(x)\big)$
$$= f\big(\tfrac{1}{2}x - 2\big)$$
$$= 2\big(\tfrac{1}{2}x - 2\big) + 4$$
$$= x - 4 + 4 = x$$

Note   *The function $h(x) = 3x + 2$ is not the same as the function $(f \circ g)(x) = 3x + 2$ since the domain of h is the set of real numbers, whereas the domain of $f \circ g$ is $\{x \,|\, x \geq 1\}$.*

Note   *The inequality $3x^2 + 5 \geq 1$ holds for all x.*

Similarly,

(ii) $(f^{-1} \circ f)(x) = f^{-1}(f(x))$

$\qquad = f^{-1}(2x + 4)$

$\qquad = \frac{1}{2}(2x + 4) - 2$

$\qquad = x + 2 - 2 = x$

In general, if a function f has an inverse f^{-1}, then

$$(f \circ f^{-1})(x) = x \quad \text{and} \quad (f^{-1} \circ f)(x) = x$$

Indeed, this property can be used to check whether two given functions are such that one is the inverse of the other.

Exercises 2.3

In Problems 1 through 6, given the values of the functions f and g, describe the functions $f + g$, $f - g$, fg, and f/g.

1. $f(x) = 2x + 5, \quad g(x) = x^2$
2. $f(x) = \sqrt{x}, \quad g(x) = x^2 + 1$
3. $f(x) = \sqrt{x^2 - 4}, \quad g(x) = x^2 + 2x - 3$
4. $f(x) = \dfrac{1}{x + 1}, \quad g(x) = \dfrac{x}{2x - 1}$
5. $f(x) = \dfrac{1}{\sqrt{x}}, \quad g(x) = \dfrac{1}{x^2}$
6. $f(x) = \sqrt[3]{x}, \quad g(x) = \sqrt{x^2 - 1}$

In Problems 7 through 10, determine which of the given functions are one-to-one. If the function is one-to-one, find f^{-1}.

7. $f = \{(1, 4), (3, 2), (5, 6), (0, 7)\}$ 8. $f = \{(1, 1), (2, 2), (0, 0)\}$
9. $f = \{(2, -1), (1, 3), (3, -1), (4, 4)\}$
10. $f = \{(0, 2), (1, 4), (2, 6), (4, 1)\}$

In Problems 11 through 16, each of the equations defines a function f. Describe the inverse function f^{-1}. Show that $(f \circ f^{-1})(x) = (f^{-1} \circ f)(x) = x$.

11. $y = x - 1$
12. $y = 4x + 6$
13. $y + 2x = 5$
14. $2y + 3x = 4$
15. $y = x^3$
16. $y = 2x^3 - 1$
17. Let the function f be defined by the equation $y = 3x^2$. Discuss the existence of f^{-1}.

In Problems 18 through 25, describe $f \circ g$ and $g \circ f$ for the given pairs of functions f and g.

18. $f(x) = 2 + x, \quad g(x) = 3x$
19. $f(x) = 3x^2 - 4, \quad g(x) = 2 - x$
20. $f(x) = x^2 + 4, \quad g(x) = \sqrt{x}$
21. $f(x) = \sqrt[3]{x + 1}, \quad g(x) = 2x^3 + 1$
22. $f(x) = \dfrac{1}{x + 1}, \quad g(x) = x - 1$
23. $f(x) = \dfrac{x}{x^2 + 2}, \quad g(x) = \dfrac{1}{x}$
24. $f(x) = \sqrt{2x^2 + 3}, \quad g(x) = \dfrac{2}{x^2}$
25. $f(x) = x^{-1/2}, \quad g(x) = x - 2$
26. If $f(x) = 3x + 2$ and $g(x) = 2x^2$, find

(a) $(f+g)(2)$ (b) $(f-g)(6)$ (c) $(fg)(-1)$

(d) $(f/g)(\frac{1}{2})$ (e) $(f \circ g)(3)$ (f) $(g \circ f)(3)$

27. Repeat Problem 26 for $f(x) = \sqrt{x^2 - 16}$, $g(x) = \sqrt{x^2 + 4}$.

28. For each function f defined, determine $f \circ f$.

(a) $f(x) = \dfrac{1}{x}$ (b) $f(x) = 2 + 3x$

(c) $f(x) = \sqrt{4 + x}$ (d) $f(x) = \dfrac{1}{2 + x}$

(e) $f(x) = 2x + \dfrac{1}{x}$

29. If $f(x) = 3x - 1$, show that $f(3x) - 3f(x) = 2$.

30. If $f(x) = 2x^2$, show that $f(a + b) - f(a - b) = 8ab$.

31. If $f(x) = 2x - 1$ and $g(x) = 3x - 2$, show that $f \circ g = g \circ f$.

In Problems 32 through 35, let $f(x) = 2x + 1$, $g(x) = x + 4$, and $h(x) = x^2$; find the given function.

32. $f(g(h(x)))$ 33. $g(f(h(x)))$

34. $h(g(f(x)))$ 35. $h(f(g(x)))$

2.4 Linear Functions

In this section we shall study a class of simple and widely applied functions called *linear* functions. We shall begin by introducing the notion of the slope of a straight line in R^2.

Let $P(x_1, y_1)$ and $Q(x_2, y_2)$ be any two points in R^2 with $x_1 \neq x_2$. We define the *slope* of the line segment PQ to be the number

$$m = \frac{y_2 - y_1}{x_2 - x_1}, \qquad x_2 \neq x_1 \qquad (2.21)$$

In Figure 2.22 since $x_2 - x_1 > 0$ and $y_2 - y_1 > 0$, the line segment PQ

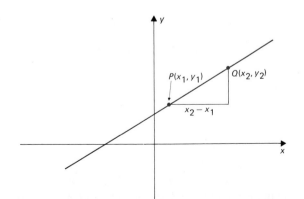

Figure 2.22

has slope $m > 0$. In Figure 2.23, however, since $x_2 - x_1 > 0$ and

Figure 2.23

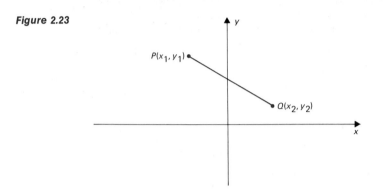

$y_2 - y_1 < 0$, the line segment PC has slope $m < 0$. For the line segment in Figure 2.24 which is parallel to the x-axis, the slope $m = 0$. For the

Figure 2.24

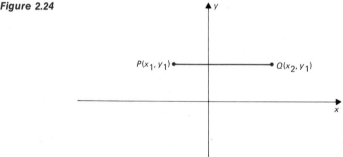

case where the line segment is parallel to y-axis (see Figure 2.25) and hence $x_2 = x_1$, the slope is undefined.

Figure 2.25

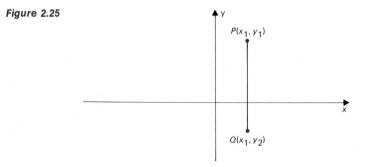

To help you remember the sign of the slope of a line segment: If a slippery bug placed at one endpoint of a line segment slides down from right to left, the slope is positive; if the bug slides down from left to right, the slope is negative; if the bug fails to slide, the slope is zero; and if the bug plunges vertically, the slope is undefined.

We note that any two distinct points in R^2 determine a straight line. Consider the two points A and B and the line L in Figure 2.26. Suppose

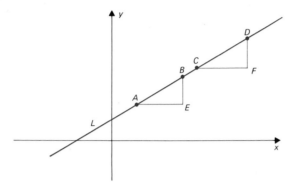

Figure 2.26

that C and D are two other points on L. Now the two right triangles ABE and CDF are associated with each pair of points A, B and C, D, respectively. The reader can see that the two triangles are similar. Therefore the slope of the line segment AB is equal to the slope of the line segment CD (Why?). Hence the slope of the line L is equal to the slope of any of the line segments belonging to L.

Example 2.21 A line L contains the points $P(-1, -2)$ and $Q(4, 6)$ (see Figure 2.27). Find the slope of L.

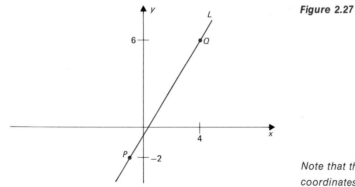

Figure 2.27

Note that the subscripts on the coordinates of the points P and Q may be interchanged. Thus if $P = (x_2, y_2)$ and $Q = (x_1, y_1)$, we again find that

$$m = \frac{-2 - 6}{-1 - 4} = \frac{-8}{-5} = \frac{8}{5}$$

SOLUTION Using (2.21), we have

$$m = \frac{y_2 - y_1}{x_2 - x_1} = \frac{6 - (-2)}{4 - (-1)} = \frac{8}{5}$$

Example 2.22   Given the line L with slope $m = -3$, find the number c if $P(2, 1)$ and $Q(c, 4)$ are two points on L (see Figure 2.28).

Figure 2.28

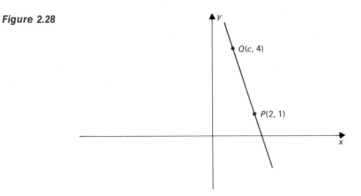

SOLUTION   Using (2.21), we have

$$m = \frac{y_2 - y_1}{x_2 - x_1} = \frac{4 - 1}{c - 2} = \frac{3}{c - 2}$$

Setting $m = -3$, we obtain

$$-3 = \frac{3}{c - 2} \quad \text{or} \quad c - 2 = -1; \quad c = 1$$

Now again let $A(x_1, y_1)$ and $B(x_2, y_2)$ be two points in R^2 with $x_1 \neq x_2$. If $P(x, y)$ is any point on the line L containing A and B, then the slope of the line segment AP must equal the slope of the line segment AB, that is,

$$\frac{y - y_1}{x - x_1} = \frac{y_2 - y_1}{x_2 - x_1} \tag{2.22}$$

Equation (2.22) describes the set of all points $P(x, y)$ belonging to the line L. If the slope L is m, Equation (2.22) may be written in the *point-slope form*

Point-slope form

$$y - y_1 = m(x - x_1) \tag{2.23}$$

Equation (2.23) describes a line with slope m and containing the point (x_1, y_1). If the point (x_1, y_1) is the point of intersection of L and the y-axis, then $x_1 = 0$ and Equation (2.23) takes the form

Slope-intercept form

$$y = mx + y_1 \tag{2.24}$$

This is called the *slope-intercept form* of the equation describing L. Here m is the slope of the line and y_1 is the y-intercept.

Linear function ►   *Definition 2.7   A function f is said to be* linear *if it is of the form*

$$f(x) = mx + b$$

where m and b are constants. The domain of a linear function is R.

The graph of f is a straight line (we shall not prove this here), and hence we have the name *linear function* (see Figure 2.29).

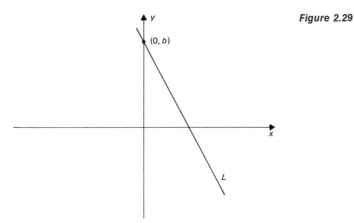

(0, b)

L

Figure 2.29

Example 2.23 Describe the linear function f whose graph contains the points $P(-2, 1)$ and $Q(3, -4)$.

SOLUTION Since f is a linear function, its graph in R^2 is a straight line. Using (2.21), we have

$$\frac{y - 1}{x - (-2)} = \frac{-4 - 1}{3 - (-2)}$$

or

$$\frac{y - 1}{x + 2} = \frac{-5}{5}$$

Simplifying, we obtain

$$y = -x - 1$$

Therefore f is defined by $f(x) = -x - 1$, and its graph is sketched in Figure 2.30.

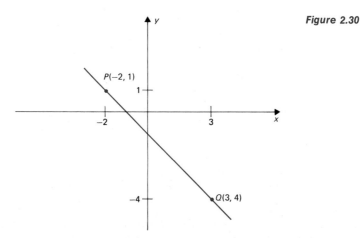

$P(-2, 1)$

$Q(3, 4)$

Figure 2.30

Example 2.24   What is the linear function f whose domain is the set of all real numbers such that the graph of f has slope 3 and contains the point $(-4, 1)$?

SOLUTION   Using the point-slope form of the equation of a line, we have

$$y - y_1 = m(x - x_1)$$

Setting $m = 3$ and $(x_1, y_1) = (-4, 1)$, we obtain

$$y - 1 = 3[x - (-4)] \quad \text{or} \quad y = 3x + 13$$

Therefore f is defined by $f(x) = 3x + 13$ (see Figure 2.31).

Figure 2.31

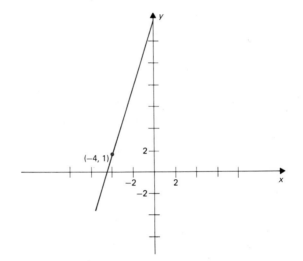

Example 2.25   Consider the linear function f defined by

$$\{(x, y)\,|\,2x - 3y + 4 = 0\}$$

Find the slope, y-intercept, and x-intercept of the graph of f.

SOLUTION   From the equation

$$2x - 3y + 4 = 0$$

we obtain

$$y = \tfrac{2}{3}x + \tfrac{4}{3}$$

which is the slope-intercept form. Therefore,

$$m = \tfrac{2}{3} \quad \text{(slope)}, \qquad b = \tfrac{4}{3} \quad (y\text{-intercept})$$

The zeros of a function are those values of x for which $f(x) = 0$. For example, the function $f(x) = 1$ has no zeros, $f(x) = x$ has one zero, while $f(x) = 0$ has an infinite number of zeros.

To determine the x-intercept, or the x-coordinate of the point where the line intersects the x-axis (also called the *zero* of the function f), we ask: For what value of x is y zero? In the equation $2x - 3y + 4 = 0$, set $y = 0$ and solve for x:

$$2x - 3(0) + 4 = 0 \quad \text{or} \quad x = -2$$

This is the x-intercept (Figure 2.32).

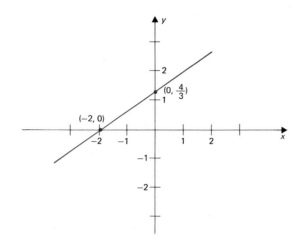

Figure 2.32

The equation of a line in *standard form* is

$$ax + by + c = 0 \qquad (2.25)$$

where a, b, and c are constants and a and b are not both zero. If $b=0$, the resulting equation does not define a function (Why?). If $a=0$ and $b \neq 0$, the resulting equation $y = -c/b$ defines a *constant function*.

A constant function is a function of the form $f(x) = k$.

Economics is one area in which linear functions have useful applications. Let us consider a manufacturer producing x units of an item. Suppose that the total cost y is a linear function of the number of units produced. We shall assume that the graph of the function relating y to x is given by Figure 2.33.

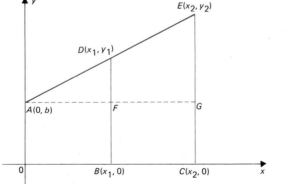

Figure 2.33

We interpret the graph as follows. If x_1 units are manufactured, then the total cost is y_1, which is the length of the line segment BD. Similarly, y_2, the length of the line segment CE, is the total cost of producing x_2 units of the item. If zero units of the item are produced, the cost is b; this is called the *fixed cost*.

Fixed cost may be due to such expenses as fire insurance or the purchase of equipment.

Variable cost is the change in cost as the number of units produced changes.

The fixed cost of producing x_2 units is given by the line segment CG whose length is b, and the line segment GE represents the *variable cost*. Thus the variable cost of making x_1 units is given by the line segment FD.

We now observe that the slope of the line segment AE is given by

$$m = \frac{y_2 - b}{x_2 - 0} = \text{change in total cost when one more unit is produced}$$

Slope = marginal cost

Economists refer to this as the *marginal cost*. When the total cost is linearly related to output, the marginal cost is constant.

Example 2.26 In Figure 2.34 the line L is the graph of a function re-

Figure 2.34

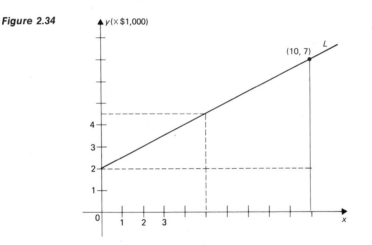

lating the total manufacturing cost y and the number of units x of a product manufactured. Find the following.

(a) The fixed cost

(b) The total cost when 10 units are produced

(c) The variable cost when 10 units are produced

(d) The marginal cost

(e) The variable cost when 5 units are produced

SOLUTION

(a) Fixed cost $= \$2000$.

(b) Total cost when 10 units are produced is $\$7000$

(c) Variable cost when 10 units are produced is total cost minus fixed cost:

$$\$7000 - \$2000 = \$5000$$

(d) Marginal cost = slope of $L = \dfrac{7-2}{10-0} = \dfrac{1}{2}$

(e) First we find the total cost for any x; that is, we find the equation of the line L. Using the slope-intercept form, we have

$$y = mx + b = \tfrac{1}{2}x + 2$$

Setting $x = 5$, we obtain

$$y = \tfrac{1}{2}(5) + 2 = 4.5$$

In other words, the total cost is $4.5 \times \$1000 = \$4500$. Therefore the variable cost when 5 units are produced is

$$(\text{total cost}) - (\text{fixed cost}) = \$4500 - \$2000 = \$2500$$

In Problems 1 through 9, find the equation of the line through each of the given pair of points and sketch the graph.

Exercises 2.4

1. $(3, 2),\ (-2, 4)$ 2. $(1, 1),\ (2, -2)$

3. $(-3, -3),\ (4, 9)$ 4. $(-1, -3),\ (-2, 5)$

5. $(0, 0),\ (3, 2)$ 6. $(5, 0),\ (0, 5)$

7. $(-2, -3),\ (5, -7)$ 8. $\left(\tfrac{1}{2}, 2\right),\ \left(\tfrac{5}{2}, -2\right)$

9. $\left(\tfrac{1}{3}, \tfrac{1}{4}\right),\ \left(\tfrac{2}{3}, \tfrac{1}{3}\right)$

In Problems 10 through 17, find the equation of the line with the given slope and containing the given point. Sketch the graph.

10. $\left(\tfrac{1}{3}, 3\right),\ m = \tfrac{1}{2}$ 11. $(-2, 5),\ m = -\tfrac{2}{3}$

12. $(-4, -3),\ m = 1$ 13. $(-3, -3),\ m = -1$

14. $(0, 3),\ m = -2$ 15. $(3, 0),\ m = 2$

16. $(-4, 3),\ m = 0$ 17. $(1, -3),\ m = 0$

In Problems 18 through 21, determine the slope, y-intercept, and x-intercept of the graph of the function defined.

18. $\{(x, y)\,|\,3x + 4y - 6 = 0\}$ 19. $\{(x, y)\,|\,x - 2y + 4 = 0\}$

20. $\{(x, y)\,|\,-4x + 5y + 12 = 0\}$ 21. $\{(x, y)\,|\,2x + y = 0\}$

22. (a) What is the equation of the line that passes through the points $(0, 3)$ and $(5, 0)$?

(b) Show that the equation of the line that passes through the points $(0, b)$ and $(a, 0)$ can be written in the following form (called the *intercept form*)

$$\frac{x}{a} + \frac{y}{b} = 1 \qquad (a \cdot b \neq 0)$$

23. The graph of a linear function has slope $m = 2$. If $P(-1, 3)$ and $Q(c, -2)$ belong to f, find the number c.

24. Repeat Problem 23 for $m = -5$ and the points $P(3, c)$ and $Q(-2, 4)$.

25. In Figure 2.35 the line L is the graph of a function that relates the total

Figure 2.35

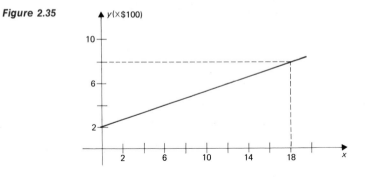

manufacturing cost y and the number of units x of a product manufactured. Find the following.

(a) The total cost when 20 units are produced
(b) The fixed cost
(c) The variable cost when 20 units are produced
(d) The marginal cost
(e) The variable cost when 7 units are produced

26. The total cost changes from $500 to $800 as the number of units manufactured increases from 300 to 900. Assuming that the total cost y is a linear function of x (the number of units manufactured), determine

(a) The fixed cost
(b) The marginal cost
(c) The total cost when 500 units are manufactured
(d) The variable cost when 500 units are manufactured

27. On an average weekday in the bustling city of Garbville, the pollution index p starts to rise at 8:00 A.M. At 10:00 A.M., $p = 30$ and at 2:00 P.M., $p = 50$. Assume that p is a linear function of time between 8:00 A.M. and 6:00 P.M.

(a) Find an equation relating p and t.
(b) Sketch the graph of the line.
(c) Find p at 8:00 A.M., 12:00 noon, and 6:00 P.M.
(d) What is the rate of increase of p with respect to time (that is, what is the slope of the line)?

28. Suppose that every night the pollution index at Garbville remains constant between 6:00 P.M. and 8:00 P.M. and then drops linearly to a fixed level at 8:00 A.M.

(a) Find the equation of the line relating p and t between 8:00 P.M. and 8:00 A.M.
(b) Calculate p at 12:00 midnight, at 2:00 A.M., and at 4:00 A.M.
(c) What is the rate of decrease (slope) of p with respect to t?

2.5 Intersection of Lines

Many geometrical problems involve the intersection of the graphs of functions. If the equations describing linear functions are given, it is a straightforward algebraic task to find a point of intersection, if any such point exists. The reader is aware of what we mean by the intersection of lines, for example, the point P in Figure 2.36. Analytically,

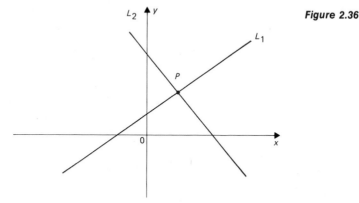

Figure 2.36

the intersection of the lines L_1 and L_2 ($L_1 \cap L_2$) is an ordered pair of numbers (x, y) such that $(x, y) \in L_1$ and $(x, y) \in L_2$. Here we shall outline the procedure for determining the point of intersection of two lines.

Let L_1 and L_2 be two lines given by

$$L_1: a_1 x + b_1 y + c_1 = 0, \qquad a_1^2 + b_1^2 \neq 0$$
$$L_2: a_2 x + b_2 y + c_2 = 0, \qquad a_2^2 + b_2^2 \neq 0$$

$a^2 + b^2 \neq 0$ *is another way of saying that not both a and b are zero.*

Then $L_1 \cap L_2$ is

$$\{(x, y) | a_1 x + b_1 y + c_1 = 0 \text{ and } a_2 x + b_2 y + c_2 = 0\}$$

We find the point of intersection by solving the simultaneous equations

$$a_1 x + b_1 y + c_1 = 0 \tag{2.26}$$
$$a_2 x + b_2 y + c_2 = 0 \tag{2.27}$$

In other words, we wish to find a single point (x, y) which satisfies (2.26) and (2.27) simultaneously. Multiplying (2.26) by b_2 and (2.27) by b_1, we obtain

$$a_1 b_2 x + b_1 b_2 y + c_1 b_2 = 0 \tag{2.28}$$
$$a_2 b_1 x + b_1 b_2 y + c_2 b_1 = 0 \tag{2.29}$$

Subtracting (2.29) from (2.28), we eliminate y to solve for x:

$$(a_1 b_2 - a_2 b_1) x + c_1 b_2 - c_2 b_1 = 0$$

or

$$x = \frac{b_1 c_2 - b_2 c_1}{a_1 b_2 - a_2 b_1} \qquad \text{provided that } a_1 b_2 \neq a_2 b_1 \qquad (2.30)$$

Remark The condition $a_1 b_2 \neq a_2 b_1$ is necessary and sufficient for L_1 and L_2 to have a point of intersection.

Similarly, we can eliminate x and solve for y to obtain

$$y = \frac{a_2 c_1 - a_1 c_2}{a_1 b_2 - a_2 b_1} \qquad \text{provided that } a_1 b_2 \neq a_2 b_1 \qquad (2.31)$$

The ordered pair of numbers (x, y) defined by (2.30) and (2.31) is the intersection of L_1 and L_2. There is no need to memorize (2.30) and (2.31); the procedure just outlined can be followed for each problem.

Example 2.27 Determine the point of intersection of the given lines:

$$L_1: x + y - 2 = 0, \qquad L_2: 2x + 3y - 1 = 0$$

SOLUTION We find the point of intersection of L_1 and L_2 by solving simultaneously the equations

$$x + y - 2 = 0 \qquad (2.32)$$
$$2x + 3y - 1 = 0 \qquad (2.33)$$

Multiply Equation (2.32) by -2 to obtain

$$-2x - 2y + 4 = 0 \qquad (2.34)$$

and then adding Equations (2.33) and (2.34), we have

$$y + 3 = 0 \qquad \text{or} \qquad y = -3$$

At this point we could follow the procedure just used and eliminate y to solve for x. Alternatively, putting $y = -3$ into either of the two equations (2.32) or (2.33), we can solve for x. Thus, from Equation (2.32), we obtain

Substituting $y = -3$ into Equation (2.32) yields $2x + 3(-3) - 1 = 0$ $\Rightarrow x = \frac{10}{2} = 5$.

$$x + (-3) - 2 = 0 \qquad \text{or} \qquad x = 5$$

Therefore the point of intersection is $(5, -3)$. As a check the reader should verify that $(5, -3) \in L_1$ and $(5, -3) \in L_2$ (see Figure 2.37).

Figure 2.37

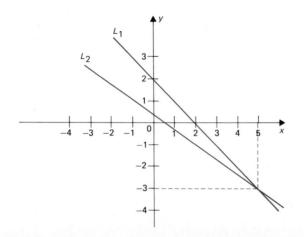

We now illustrate how the point of intersection of two lines is used in economic theory.

Example 2.28 A company expects the variable cost to be $25,000 at a sales volume (revenue) of $40,000 with a fixed cost of $12,000. Assuming that the linear model applies, find the following.

(a) An equation relating total cost and sales

(b) The *break-even point*, that is, the volume at which the company experiences neither loss nor profit   *Break-even point*

(c) The net profit if sales should turn out to be $50,000

SOLUTION

(a) First we determine the marginal cost:

$$\text{Marginal cost} = \frac{\text{variable cost}}{\text{sales volume}} = \frac{25,000}{40,000} = \frac{5}{8}$$

This result represents the slope of the line segment AB (see Figure 2.38). Using the slope-intercept form of the equation of the line L_1, we have the total cost

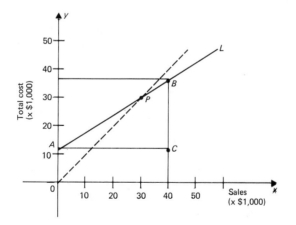

Figure 2.38
A break-even chart.

$$y = \tfrac{5}{8}x + b \tag{2.35}$$

where b represents the fixed cost (y-intercept). Hence

$$y = \tfrac{5}{8}x + 12,000 \tag{2.36}$$

is the equation relating total cost and sales.

(b) To determine the break-even point, we are asked to find a point on L where total cost equals sales, that is, where $y = x$. In other words, we are asked to solve simultaneously the equations

$$y = \tfrac{5}{8}x + 12{,}000 \qquad\qquad (2.37)$$

$$y = x \qquad\qquad (2.38)$$

Subtracting (2.37) from (2.38), we obtain

$$\tfrac{3}{8}x - 12{,}000 = 0$$

from which we can calculate that the break-even level of sales is

$$x = 32{,}000$$

So

$$y = 32{,}000$$

This result tells us that when the sales are at \$32,000, the total cost to the company is also \$32,000. The company's sales must be above \$32,000 to bring in a profit. If the sales are less than \$32,000, the company will have a loss.

(c)   Setting $x = 50{,}000$ in Equation (2.36), we obtain the total cost:

$$y = \tfrac{5}{8}(50{,}000) + 12{,}000 = 31{,}250 + 12{,}000 = 43{,}250$$

The net profit (before taxes) is

$$\text{net profit} = (\text{sales}) - (\text{cost}) = \$50{,}000 - \$43{,}250 = \$6750$$

Example 2.29   Consider the lines

$$L_1: x - y + 5 = 0 \quad \text{and} \quad L_2: -2x + 2y - 1 = 0$$

Find $L_1 \cap L_2$.

SOLUTION   Consider the pair of equations

$$x - y + 5 = 0 \qquad\qquad (2.39)$$

$$-2x + 2y - 1 = 0 \qquad\qquad (2.40)$$

We find that the method of elimination of variables employed in Example 2.27 fails to yield a solution for x and y. If we multiply the first equation by 2 and add the result to the second equation, we obtain $9 = 0$, a false statement. Since the operations leading to this statement are valid, we must conclude that (2.39) and (2.40) cannot both be true for the same point (x, y). Consequently, the two lines do not intersect (see Figure 2.39). This example leads us to the following definition.

Parallel lines   ▶   **Definition 2.8**   *Two distinct lines L_1 and L_2 are said to be* parallel *if they have no point of intersection.*

We state the following theorem concerning parallel lines.

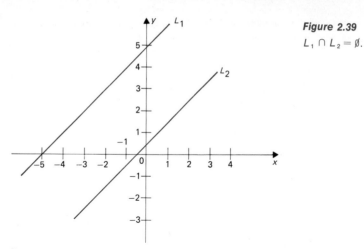

Figure 2.39
$L_1 \cap L_2 = \emptyset$.

Theorem 2.2   *The two lines*

$$L_1: a_1x + b_1y + c_1 = 0, \qquad a_1^2 + b_1^2 \neq 0$$

$$L_2: a_2x + b_2y + c_2 = 0, \qquad a_2^2 + b_2^2 \neq 0$$

are parallel (or identical) if and only if $a_1b_2 - a_2b_1 = 0$.

We could look at this from another point of view. The expression

$$a_1b_2 - a_2b_1 = 0$$

implies that

$$\frac{a_1}{b_1} = \frac{a_2}{b_2} \qquad \text{provided } b_1b_2 \neq 0$$

and so

$$-\frac{a_1}{b_1} = -\frac{a_2}{b_2}$$

Since $-a_1/b_1$ is the slope of L_1 and $-a_2/b_2$ is the slope of L_2 (see Section 2.4), the lines L_1 and L_2 are parallel if and only if their slopes are equal.

For the case where $b_1 = 0$, L_1 is the vertical line $x = -c_1/a_1$ and its   *Why is a_1 nonzero?*
slope is undefined. Here we find that b_2 must also be zero (Why?).
Hence L_2 is also a vertical line and is parallel to L_1 (Why?).

Example 2.30   Show that the lines

$$L_1: 2x - 3y + 4 = 0 \quad \text{and} \quad L_2: y = \tfrac{2}{3}x + 16$$

are parallel.

SOLUTION   Since

$$a_1b_2 - a_2b_1 = (2)(1) - \left(-\tfrac{2}{3}\right)(-3) = 0$$

by Theorem 2.2, the lines L_1 and L_2 are parallel.

Example 2.31 Determine the number c so that the line L_1 through the points $(c,\ 1)$ and $(2,\ -3)$ is parallel to the line

$$L_2\text{:}\ y = x + 5$$

Figure 2.40

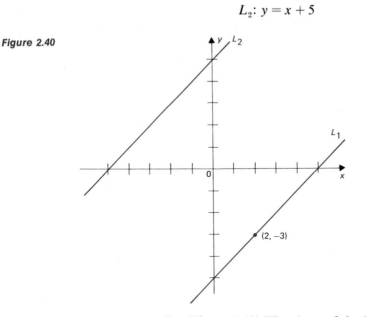

SOLUTION (See Figure 2.40) The slope of the line L_2 is

$$m_2 = 1$$

If m_1 is the slope of L_1 and L_1 is parallel to L_2, then $m_1 = 1$. But

$$m_1 = \frac{-3 - 1}{2 - c}$$

Setting $m_1 = 1$ and solving for c, we obtain

$$1 = \frac{-4}{2 - c}$$

$$2 - c = -4$$

$$c = 6$$

Next we shall consider *perpendicular* lines. If L_1 and L_2 are two lines neither of which is parallel to an axis, the following theorem (stated without proof) gives a test for perpendicularity in terms of the slopes of the lines.

Theorem 2.3 *The two lines L_1 and L_2 with slopes m_1 and m_2, respectively, are perpendicular if and only if $m_1 m_2 = -1$.*

Example 2.32 Show that the points $A(9,\ -1)$, $B(3,\ 1)$, and $C(4,\ 4)$ are the vertices of a right triangle (see Figure 2.41).

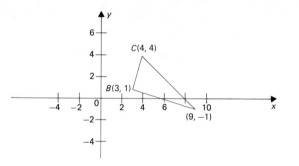

Figure 2.41

SOLUTION The slopes of the sides of the triangle ABC are as follows:

$$\text{slope of } AB = \frac{1 - (-1)}{3 - 9} = \frac{2}{-6} = -\frac{1}{3}$$

$$\text{slope of } AC = \frac{4 - (-1)}{4 - 9} = \frac{5}{-5} = -1$$

$$\text{slope of } BC = \frac{4 - 1}{4 - 3} = 3$$

Since

$$(\text{slope of } AB)(\text{slope of } BC) = \left(-\tfrac{1}{3}\right)(3) = -1$$

the line through A and B and the line through B and C are perpendicular. Therefore the triangle is a right triangle with right angle at B.

In Problems 1 through 10, determine the point of intersection of the given lines.   **Exercises 2.5**

1. $x - 2y = 3, \quad x + y + 4 = 0$

2. $2x + y - 5 = 0, \quad x - y + 16 = 0$

3. $3x - y + 2 = 0, \quad 2x - 3y + 1 = 0$

4. $4x - 2y + 7 = 0, \quad x + y + 6 = 0$

5. $-x + 5y + 2 = 0, \quad 3x + y + 1 = 0$

6. $y = 5x - 6, \quad y = \tfrac{2}{5}x + 4$

7. $y = 16x + 2, \quad 3y = 4x - 2$

8. $4x - 6y = 3, \quad x + 3y = 6$

9. $x = 4y - 3, \quad x + 4y = 7$

10. $2x + 7y = 8, \quad 3x - 2y = 4$

In Problems 11 through 18, determine which of the given pairs of lines are parallel and which are perpendicular.

11. $2x + y - 6 = 0, \quad 4x + 2y + 3 = 0$

12. $y + 2 = 0$, $x + 4 = 0$

13. $5x - y = 4$, $2x + 10y - 7 = 0$

14. $3x - 2y + 1 = 0$, $y = \frac{3}{2}x + 10$

15. $y + 4 = 3x$, $x + 16 = \frac{1}{3}y$

16. $x = 2y + 12$, $4y = 2x - 3$

17. $3x - y + 6 = 0$, $\frac{1}{3}x + y = 0$

18. $7x + 14y - 6 = 0$, $y = -\frac{1}{2}x + 4$

19. A line passes through the points $(2, c)$ and $(1, -3)$. Determine the number c so that the line
    (a) has slope 3
    (b) has slope zero
    (c) is parallel to the line $5x + y + 6 = 0$
    (d) is perpendicular to the line $2x + y = 0$

20. A line L_1 passes through the points $(c, 1)$ and $(5, -1)$. Determine the number c so that L_1
    (a) has slope -2
    (b) is parallel to the line L_2: $2x + 3y = 1$
    (c) is perpendicular to L_2
    Is it possible for L_1 to have slope zero? Why?

21. Using slopes, identify each of the figures with the given vertices.
    (a) $A(-4, 3)$, $B(-5, 0)$, $C(1, -2)$
    (b) $A(-1, 4)$, $B(-2, 1)$, $C(4, -1)$, $D(5, 2)$
    (c) $A(4, 3)$, $B(9, -2)$, $C(10, 1)$
    (d) $A(-9, 5)$, $B(-2, 13)$, $C(6, 6)$, $D(-1, -2)$

22. Consider the line segment with endpoints $(2, 1)$ and $(6, 5)$. Show that the point $(5, 2)$ is on the perpendicular bisector of the given line segment.

23. In Figure 2.42 ABC is a triangle and D and E are the midpoints of the line segments AB and AC, respectively.

Figure 2.42

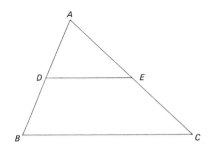

(a) Show that the line through D and E is parallel to the line through B and C.
(b) Show that the length of the line segment DE is one-half the length of the line segment BC.

24. A company expects the variable cost to be $44,000 on sales of $70,000 with a fixed cost of $22,000. Assume that the linear model applies.

(a) Write the equation relating sales and total cost.
(b) Sketch the break-even chart.
(c) Find the break-even point.
(d) What will the net profit (before taxes) be on sales of $80,000?

25. Suppose that the equation

$$y = \tfrac{1}{3}x + 24{,}000$$

relates the total cost y and the sales volume x. Find the following.
(a) The fixed cost
(b) The variable cost on sales of $50,000
(c) The break-even point
(d) Net profit (before taxes) on sales of $40,000

26. Given that the freezing point of water is measured at $0°C$ (Celsius) and
$32°F$ (Fahrenheit) and the boiling point of water is measured at $100°C$
and $212°F$.
(a) Find C as a linear function of F.
(b) Find F as a linear function of C.
(c) At what temperatures will $°C = °F$?
(d) Sketch a graph of each function and indicate the point of intersection.

27. Suppose that f is a linear function with nonzero slope m.
(a) Show that f^{-1} is a function.
(b) Show that f^{-1} is linear.
(c) What is the slope of f^{-1}?
(d) Can the graph of f^{-1} be perpendicular to the graph of f?
(e) Can the graph of f^{-1} be parallel to the graph of f?

2.6 Quadratic Functions

In many instances, functions whose graphs are curves rather than
straight lines provide better approximations to real-world situations.
One such class of functions which we shall study in this section, is the
class of quadratic functions.
   We now consider functions defined by

$$f(x) = ax^2 + bx + c \qquad (2.41)$$

where a, b, and c are constants and $a \neq 0$.

Definition 2.9 *A function f that can be described by (2.41) is called a*   ◀ Quadratic functions
quadratic function. *The graph of f is called a* parabola.

   Consider the special case of (2.41),

$$f(x) = x^2 \qquad (2.42)$$

Assigning values to x and computing the corresponding values $f(x)$,
we obtain some points belonging to f. For example, in tabular form, we
have

x	-3	-2	-1	0	1	2	3
$f(x) = x^2$	9	4	1	0	1	4	9

The graph of these points is shown in Figure 2.43. If we compute

Figure 2.43

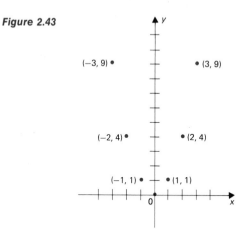

additional points belonging to f and connect them with a smooth curve as shown in Figure 2.44, we find that the graph of f forms a simple

Figure 2.44

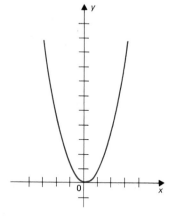

pattern. Obviously, the resulting curve is an approximation to the parabola describing the function. The graph is in the shape shown in Figure 2.44. The curve is said to be concave upward. We observe that the pairs of points

$$(1, 1) \quad \text{and} \quad (-1, 1) \quad \text{and} \quad (2, 4) \quad \text{and} \quad (-2, 4)$$

and in general

$$(x, x^2) \quad \text{and} \quad (-x, x^2)$$

Note To obtain an exact graph is an impossible task, since we are unable to locate every point belonging to f.

belong to f. Geometrically, this means that the portion of the graph of f to the left of the y-axis is the mirror image of the portion to the right of the y-axis. In other words, if we fold the paper along the y-axis, the curve to the left of the y-axis will coincide with the curve to the right of the y-axis. Such a graph is called *symmetric*, and the y-axis is called the *line of symmetry*. The point of intersection of the line of symmetry and the graph of f is called the *vertex* of the parabola. In this case, the vertex represents the lowest point of the parabola. We find that

Line of symmetry

Vertex of a parabola

$$f(0) = 0 \quad \text{and} \quad f(0) < f(x)$$

for every $x \neq 0$. We say that f has a *minimum* value at $x = 0$.

Example 2.33 Graph

(a) $f(x) = \frac{1}{4}x^2$ (b) $f(x) = 2x^2$ (c) $f(x) = 3x^2$

SOLUTION First we tabulate a few points belonging to each function.

(a)

x	-3	-2	$-\frac{3}{2}$	-1	$-\frac{1}{2}$	0	$\frac{1}{2}$	1	$\frac{3}{2}$	2	3
$f(x) = \frac{1}{4}x^2$	$\frac{9}{4}$	1	$\frac{9}{16}$	$\frac{1}{4}$	$\frac{1}{16}$	0	$\frac{1}{16}$	$\frac{1}{4}$	$\frac{9}{16}$	1	$\frac{9}{4}$

(b)

x	-2	$-\frac{3}{2}$	-1	$-\frac{1}{2}$	0	$\frac{1}{2}$	1	$\frac{3}{2}$	2
$f(x) = 2x^2$	8	$\frac{9}{2}$	2	$\frac{1}{2}$	0	$\frac{1}{2}$	2	$\frac{9}{2}$	8

(c)

x	-2	$-\frac{3}{2}$	-1	$-\frac{1}{2}$	0	$\frac{1}{2}$	1	$\frac{3}{2}$	2
$f(x) = 3x^2$	12	$\frac{27}{4}$	3	$\frac{3}{4}$	0	$\frac{3}{4}$	3	$\frac{27}{4}$	12

We illustrate the graphs of these functions in Figures 2.45, 2.46, and

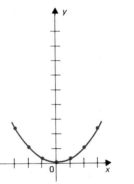

Figure 2.45
Graph of $f(x) = \frac{1}{4}x^2$.

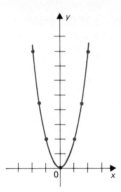

2.47, and note that as the coefficient of x^2 takes on larger values, the steepness of the curve increases. However, the vertex in each case is the point $(0, 0)$.

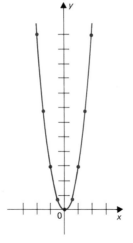

Next, we consider (2.41) for $a < 0$. Suppose that f is the function defined by

$$f(x) = -x^2 \qquad (2.43)$$

Computing some of the points belonging to the graph of f, we obtain

x	-3	-2	$-\frac{3}{2}$	-1	$-\frac{1}{2}$	0	$\frac{1}{2}$	1	$\frac{3}{2}$	2	3
$f(x) = -x^2$	-9	-4	$-\frac{9}{4}$	-1	$-\frac{1}{4}$	0	$-\frac{1}{4}$	-1	$-\frac{9}{4}$	-4	-9

The graph of these points and an approximate graph of f is shown in Figure 2.48.

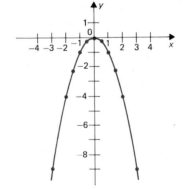

Figure 2.48

Here again the graph of the function is in the shape shown in Figure 2.48. The curve is concave downward. The graph is symmetric with respect to the y-axis, and the vertex is at the origin. This point is called the highest point of the parabola. We find that

$$f(0) = 0 \quad \text{and} \quad f(0) > f(x)$$

for every $x \neq 0$. We say that f attains a *maximum* value at $x = 0$.

In Figures 2.49 and 2.50, we illustrate the graphs of the functions

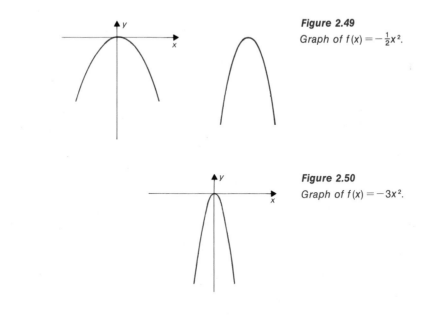

Figure 2.49

Graph of $f(x) = -\frac{1}{2}x^2$.

Figure 2.50

Graph of $f(x) = -3x^2$.

(a) $f(x) = -\frac{1}{2}x^2$ (b) $f(x) = -\frac{3}{2}x^2$ (c) $f(x) = -3x^2$

Note that the steepness of the curve increases as the absolute value of the coefficients of x^2 increases.

Returning to (2.41), we now can show that the graph of f has either a lowest or a highest point. We *complete the square* as follows.

Adding and subtracting $b^2/4a$ to complete the square using the first two terms

$$f(x) = ax^2 + bx + c, \qquad a \neq 0$$

$$= ax^2 + bx + c + \left(\frac{b^2}{4a} - \frac{b^2}{4a} \right)$$

$$= ax^2 + bx + \frac{b^2}{4a} + c - \frac{b^2}{4a}$$

Factoring out a in the first three terms and adding the last two

$$= a\left(x^2 + \frac{b}{a}x + \frac{b^2}{4a^2} \right) + \frac{-b^2 + 4ac}{4a}$$

Note $\left(x + \dfrac{b}{2a} \right)^2 = x^2 + \dfrac{b}{a}x + \dfrac{b^2}{4a^2}$

$$= a\left(x + \frac{b}{2a} \right)^2 + \frac{-b^2 + 4ac}{4a} \qquad (2.44)$$

Case 1 $a > 0$.

Now,

$$\left(x + \frac{b}{2a} \right)^2 = 0 \qquad \text{at} \quad x = \frac{-b}{2a}.$$

For all other values of x, if $a > 0$, then

$$a\left(x + \frac{b}{2a} \right)^2 > 0 \qquad \text{(Why?)}$$

Therefore $f(-b/2a) < f(x)$ for all $x \neq -b/2a$. Hence the graph of f has a minimum value at $x = -b/2a$; that is, the lowest point is

$$\left(-\frac{b}{2a}, \frac{-b^2 + 4ac}{4a} \right) \qquad (2.45)$$

Case 2 $a < 0$.

At $x = -b/2a$,

$$\left(x + \frac{b}{2a} \right)^2 = 0$$

For all other values of x, $a(x + b/2a)^2 < 0$ (Why?). Therefore $f(-b/2a) > f(x)$ for all $x \neq -b/2a$ in this case, and the function f attains a maximum value at $x = -b/2a$. In other words, the graph of f has a highest point and this point is given by (2.45).

We observe from (2.44) that if we set $f(x) = 0$, we obtain

$$a\left(x + \frac{b}{2a} \right)^2 = \frac{b^2 - 4ac}{4a}$$

$$\left(x + \frac{b}{2a} \right)^2 = \frac{b^2 - 4ac}{4a^2}$$

$$x + \frac{b}{2a} = \pm \frac{\sqrt{b^2 - 4ac}}{2a}$$

$$x = -\frac{b}{2a} \pm \frac{\sqrt{b^2 - 4ac}}{2a}$$

Hence

$$x = \frac{-b + \sqrt{b^2 - 4ac}}{2a} \quad \text{and} \quad x = \frac{-b - \sqrt{b^2 - 4ac}}{2a} \qquad (2.46)$$

These are the points where the graph of f intersects the x-axis provided that $b^2 - 4ac \geq 0$. If $b^2 - 4ac < 0$, the graph of f lies either entirely above or entirely below the x-axis.

Note *The student may recognize (2.46) as the solutions of the quadratic equation $ax^2 + bx + c = 0$.*

Example 2.34 Graph the function f, where

$$f(x) = x^2 - 6x + 5$$

SOLUTION   First we determine the highest or lowest point of the graph. We observe that since $a = 1 > 0$, the graph has a lowest point. Here,

$$a = 1, \qquad b = -6, \quad \text{and} \quad c = 5$$

Therefore

$$-\frac{b}{2a} = -\frac{(-6)}{2(1)} = 3$$

and

$$\frac{-b^2 + 4ac}{4a} = \frac{-(-6)^2 + 4(1)(5)}{4(1)} = \frac{-36 + 20}{4} = -4$$

Hence the lowest point is $(3, -4)$. Next we tabulate a few points belonging to f.

x	0	1	2	3	4	5
$f(x) = x^2 - 6x + 5$	5	0	-3	-4	-3	0

Figure 2.51 illustrates the graph of f. Note that the vertical line

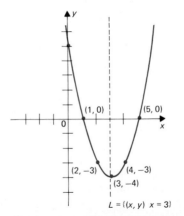

Figure 2.51

$L = \{(x, y) \mid x = 3\}$ is the line of symmetry. Using (2.46), we obtain the points of intersection of the graph of f with the x-axis: $x = 1$ and $x = 5$.

Example 2.35 Graph the function f, where

$$f(x) = -3x^2 - 8x + 1$$

SOLUTION Since $a = -3 < 0$, the graph of f has a highest point. Here

$$a = -3, \qquad b = -8, \quad \text{and} \quad c = 1$$

Therefore

$$-\frac{b}{2a} = -\frac{(-8)}{2(-3)} = -\frac{4}{3}$$

and

$$\frac{-b^2 + 4ac}{4a} = \frac{-(-8)^2 + 4(-3)(1)}{4(-3)} = \frac{76}{12} = \frac{19}{3}$$

The highest point on the graph is $\left(-\frac{4}{3}, \frac{19}{3}\right)$. A few points on the graph of f are

x	-2	$-\frac{3}{2}$	$-\frac{4}{3}$	-1	$-\frac{1}{2}$	0	$\frac{1}{2}$	1
$f(x) = -3x^2 - 8x + 1$	5	$\frac{25}{4}$	$\frac{19}{3}$	6	$\frac{17}{4}$	1	$-\frac{15}{4}$	-10

The graph of f is shown in Figure 2.52. The line of symmetry is the vertical line $x = -\frac{4}{3}$.

Figure 2.52

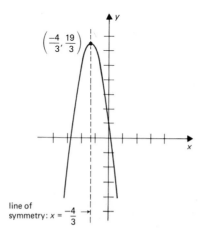

line of
symmetry: $x = \dfrac{-4}{3}$

We can summarize our results for the graph of the quadratic function $f(x) = ax^2 + bx + c$.

1. If $a > 0$, the parabola is concave upward.

2. If $a < 0$, the parabola is concave downward.

3. The vertex of the parabola is the point $\left(-\dfrac{b}{2a}, \dfrac{-b^2 + 4ac}{4a}\right)$

4. The line of symmetry is given by $x = -b/2a$.

5. The y-intercept is the point $(0, c)$.

6. The x-intercepts are $x = \dfrac{-b \pm \sqrt{b^2 - 4ac}}{2a}$

Now we shall illustrate how certain real-world situations give rise to quadratic functions.

Example 2.36 A farmer wishes to enclose a rectangular lot of maximum area with a fence 400 meters long (see Figure 2.53). Find the dimensions of the rectangle.

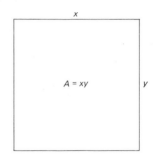

Figure 2.53

SOLUTION   Suppose that the length and the width of the rectangle are x and y meters, respectively, Then

$$2x + 2y = 400 \qquad (2.47)$$

or

$$x + y = 200 \qquad (2.48)$$

The area of the rectangle is

$$A = xy \qquad (2.49)$$

In Equation (2.49) A is dependent on two variables, x and y. We wish to express A as a function of one of the variables, say x. Solving for y in Equation (2.48) and substituting the result into (2.49), we get

$$A(x) = x(200 - x) = -x^2 + 200x \qquad (2.50)$$

Equation (2.50) describes a quadratic function. Since the coefficient of x^2 is negative, the graph of $A(x)$ has a highest point; that is, A attains a maximum value and this occurs at

$$x = -\frac{b}{2a} = -\frac{200}{2(-1)} = 100$$

Note The reader is encouraged
to try other values for x and find
the corresponding values of y and
A. It should be clear that A is
maximum when x = 100.

Setting $x = 100$ in Equation (2.48), we find that

$$y = 200 - x = 100$$

Therefore the maximum area is enclosed if the rectangle is a square having side length 100 meters.

Example 2.37 Sophia operates a pizza parlor. Having studied mathematics, she finds that the cost of operating her shop is given by

$$C(x) = \tfrac{3}{100}x^2 - 12x + 1800$$

where $C(x)$ is the daily cost (in dollars) to make x pizzas. Determine the number of pizzas Sophia should make daily to minimize the cost.

   SOLUTION   Here, the cost C is a quadratic function. Since the coefficient of x^2 is positive, C attains a minimum value when

$$x = -\frac{b}{2a} = -\frac{(-12)}{2\left(\tfrac{3}{100}\right)} = 200$$

In other words, when 200 pizzas are made, the cost is minimum and this minimum cost is

$$C(200) = \tfrac{3}{100}(200)^2 - 12(200) + 1800 = 600$$

This result tells Sophia that it will cost her more than \$600 if she prepares *more or less* than 200 pizzas. Assuming that she can sell all the pizzas she produces, Sophia will have to charge \$3 per pizza to break even.

Exercises 2.6

In Problems 1 through 10, sketch the graph of the given function and indicate the highest or the lowest point and the line of symmetry. Find the points of intersection (if any) of the graph of f with the x-axis.

1. $f(x) = x^2 - 4$

2. $f(x) = x^2 - 3x + 2$

3. $f(x) = x^2 - 5x + 4$

4. $f(x) = x^2 + x - 2$

5. $f(x) = -x^2 + 3x - 2$

6. $f(x) = -x^2 + 5x - 4$

7. $f(x) = -2x^2 - 3x + 2$

8. $f(x) = -\tfrac{1}{2}x^2 - x$

9. $f(x) = \tfrac{1}{4}x^2 + 3$

10. $f(x) = \tfrac{1}{4}x^2 + 3x$

11. Find two numbers whose sum is 10 and whose product is as large as possible.

12. A piece of wire 20 in. long is to be bent to form three sides of a rectangle. Find the dimensions of the rectangle thus formed that will have maximum area.

13. The profit p in the manufacture and sale of x units of a product is given by

$$p(x) = 200x - 0.001x^2$$

    (a) Find the number x that yields maximum profit.
    (b) Sketch a graph of the function p.

14. A window is to be constructed in the shape of a rectangle surmounted by a semicircle (see Figure 2.54). If the perimeter of the window is 18 ft, find its dimensions for maximum area.

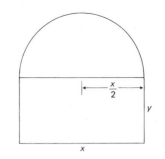

Figure 2.54

15. The Blue-Yonder Travel Club is organizing a South American trip for a group of university students. The trip will cost $300 per student if not more than 120 students make the trip; however, the cost per student will be reduced by $1.00 for each student in excess of 120.
    (a) If x is the number of students in excess of 120, show that the gross income r is given by

$$r = -x^2 + 180x + 36,000$$

    (b) For what value of x is r a maximum?
    (c) Find the maximum value of r.

16. The manager of an 80-unit apartment complex finds that at a rental of $110 per month all units will be occupied. However, for each $4.00 increase in rent, one unit will remain vacant. Assume that the maintenance cost is $10 per month for an occupied unit.

    Note   *The $10 maintenance fee is paid by the management.*

    (a) Show that the function relating gross profit p and number of unoccupied units x is given by $p(x) = -4x^2 + 220x + 8000$.
    (b) Find the number of vacant units for which gross profit is maximum.
    (c) What is the maximum gross profit?
    (d) What is the rental for maximum profit?

Part 1   *(Oral)*

Chapter 2 Review

Define or discuss each of the following:

    1. Cartesian product of two sets

2. The distance formula

3. Relation, domain, and range

4. Function

5. Graph

6. Composite function

7. Inverse function

8. Point-slope form of an equation of a line

9. Linear function

10. Quadratic formula

11. Parallel lines

12. Perpendicular lines

13. Vertex of a parabola

Part 2 (*Written*)

1. Let $A = \{1, 3, 5\}$ and $B = \{c, d\}$. Find $A \times B$ and $B \times A$.

In Problems 2 through 5, find the distance between the points with the given coordinates. Also, find the coordinates of the midpoint.

2. $A(3, 5)$, $B(7, 9)$ 3. $A(4, -2)$, $B(-3, 6)$

4. $A\left(\frac{1}{2}, 0\right)$, $B(2, -3)$ 5. $A(-4, 0)$, $B(1, -3)$

6. A *circle* is defined as the set of all points in the plane whose distance from a fixed point is constant. The fixed point is called the center of the circle and the fixed distance is called the radius of the circle. Show that the equation of a circle with center (h, k) and radius r is $(x - h)^2 + (y - k)^2 = r^2$. [*Hint:* Use the distance formula.]

In Problems 7 through 12, find the equation of a circle satisfying the given conditions.

7. Center at $(3, 4)$ and radius 2

8. Center at $(-1, 2)$ and radius 4

9. Center at $(3, -4)$ and containing the point $(-1, 5)$

10. Diameter the line segment with endpoints $(3, 1)$ and $(7, 5)$

11. Center at $(-1, 3)$ and touching the line $y = -1$

12. Center at $(2, 2)$ and touching the line $x = 5$

In Problems 13 through 19, find the equation of the line and sketch the graph:

13. Through $(2, 3)$ and $(5, 7)$

14. Through $(1, 0)$ and $(6, -3)$

15. Through $(1, -2)$ and slope $\frac{1}{2}$

16. Through $(2, -5)$ and parallel to x-axis

17. Through $(3, 7)$ and parallel to y-axis

18. Slope $-\frac{2}{3}$ and y-intercept 4

19. The x-intercept -2 and y-intercept 3

20. Find the slope, x-intercept, and y-intercept for the line $3x - 4y + 12 = 0$

21. Let the slope of a line L_1 be $-\frac{2}{3}$. What is the slope of the line L_2 if

(a) L_2 is parallel to L_1?   (b) L_2 is perpendicular to L_1?

22. The line $y = mx + b$ passes through the point of intersection of $y = 2x - 4$ and $y = 5x - 10$. Find the value of m

23. Find the equation of the line passing through $(2, 3)$ and parallel to the line $5x - 12y + 3 = 0$.

24. Find the equation of the line passing through $(1, -1)$ and perpendicular to the line $3x + 4y + 7 = 0$.

25. Find the equation of the line passing through the point of intersection of the lines $x + y - 1 = 0$ and $2x - y - 5 = 0$ and satisfying the following condition.

(a) Parallel to the line $4x + 3y + 1 = 0$

(b) Perpendicular to the line $4x - 3y + 2 = 0$

26. Let $A(-1, 0)$, $B(2, -2)$, and $C(3, -4)$ be a triangle. Find the following.

(a) The equation of the altitude from A to BC

(b) The equation of the altitude from B to AC

(c) The equation of the altitude from C to AB

(d) Show that the three altitudes meet at a point. Such a point is called the *orthocenter* of the triangle.

In Problems 27 through 30, graph the given quadratic function.

27. $f(x) = x^2 + 2$ 28. $f(x) = -x^2 + 4$

29. $f(x) = 2x^2 - 3x + 1$ 30. $f(x) = -x^2 - 6x + 2$

31. Find a linear cost function for a fixed cost of $5000 and $32 for each item produced.

32. A woman earned $8000 in the first year. She earned $9200 the second year. If her earnings increase at the same rate, write an equation relating her earnings to years. How much did she earn in the twelfth year?

33. Find a linear cost function from the following data: A cost of $425 occurs when 50 units are produced and a cost of $525 occurs when 250 units are produced. Also find the following.

(a) The total cost when 100 units are produced

(b) The fixed cost

(c) The variable cost when 80 units are produced

(d) The marginal cost

34. Find two positive numbers whose sum is 50 and whose product is as large as possible.

35. If the cost $C(x)$ of manufacturing x units of a product is given by

$$C(x) = \tfrac{1}{8}x^2 - 64x + 7$$

find the number x for which the cost is a minimum.

36. If the profit $P(x)$ from the manufacture and sale of x units of a product is given by

$$P(x) = -20 + 200x - \tfrac{1}{4}x^2$$

find the number x for which the profit is maximum.

37. The owner of an 100-unit Resort Inn finds it possible to rent all units nightly at $20 per night. However, for each dollar increase in the room rate, two units will be vacant. Find the rate (per night) the owner should charge to maximize income. Also find the maximum income.

Numerical Applications Using Calculators

Pocket calculators can be used to sketch the graph of a function. At this moment we can only sketch the graph of a function $y = f(x)$ by plotting points on the graph, i.e., by selecting different values $x_1, x_2, \ldots, x_n$ in the domain of the function and computing the function values $f(x_i)$ to obtain points $(x_i, f(x_i))$ on the graph. Connecting these points by a smooth curve will give us a rough sketch of the graph. However, the rough sketch might not give us the real picture of the graph if we fail to plot enough points. A pocket calculator can be used to locate the maxima and minima of the graph. With this information, we can sketch a more accurate graph.

Let us consider the example

$$y = x^3 - 6x^2 + 9x - 7$$

Using a calculator, we obtain the following table:

x	−3	−2	−1	0	1	2	3	4
y	−115	−57	−23	−7	−3	−5	−7	105

We observe from the table that the function values increase from $x = -3$ to $x = 1$, decrease from $x = 1$ to $x = 3$, then increase again from $x = 3$ to $x = 4$. Therefore we suspect that the maximum and minimum might occur around $x = 1$ and $x = 3$. How do we find out? We need the calculator to plot more points around $x = 1$ and $x = 3$.

Table 2.1

x	x^2	x^3	$-6x^2$	$9x$	$x^3 - 6x^2 + 9x - 7$
0.6	0.36	0.216	−2.16	5.4	−3.544
0.7	0.49	0.343	−2.94	6.3	−3.297
0.8	0.64	0.512	−3.84	7.2	−3.128
0.9	0.81	0.729	−4.86	8.1	−3.031
1.0	1.0	1.0	−6.0	9.0	−3
1.1	1.21	1.331	−7.26	9.9	−3.029
1.2	1.44	1.728	−8.64	10.8	−3.112
1.3	1.69	2.197	−10.14	11.7	−3.243
1.4	1.96	2.744	−11.76	12.6	−3.416

Using the data in Table 2.1, we can see that the point $(1, -3)$ is a maximum point. Let us check by calculator if a minimum point can be obtained near $x = 3$.

Table 2.2

x	x^2	x^3	$-6x^2$	$9x$	$x^3 - 6x^2 + 9x - 7$
2.6	6.76	17.576	−40.56	23.4	−6.584
2.7	7.29	19.683	−43.74	24.3	−6.757
2.8	7.84	21.952	−47.04	25.2	−6.888
2.9	8.41	24.389	−50.46	26.1	−6.971
3.0	9.0	27.0	−54	27	−7
3.1	9.61	29.791	−57.66	27.9	−6.969
3.2	10.24	32.768	−61.44	28.8	−6.872
3.3	10.89	35.937	−65.34	29.7	−6.703

We see that the point $(3, -7)$ is a minimum point. A sketch of the graph is given in Figure 2.55.

Figure 2.55

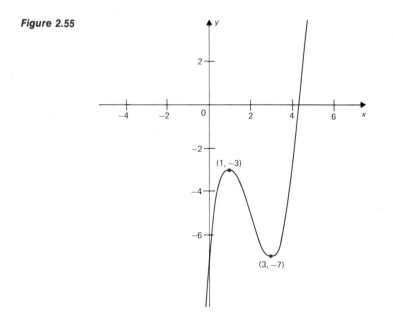

Exercises   Use a pocket calculator to locate as close as possible all the maxima and minima of each of the following functions and sketch the graph.

1. $f(x) = 2x^2 - 3x - 1$ 2. $f(x) = x^3 - 1$

3. $f(x) = x^3 + 3x^2 - 9x + 7$ 4. $f(x) = 2x^3 - 6x^2 + 9$

5. $f(x) = \frac{1}{4}x^4 + \frac{2}{3}x^3 - \frac{1}{2}x^2 - 2x + 1$

Limit and Continuity
CHAPTER 3

3.1 The Limit Concept

The concept of the limit of a function is fundamental in the development of calculus. Roughly speaking, limits examine how a function $f(x)$ behaves as the values x in the domain get closer and closer to (but do not take on) a certain value c.

We shall begin with an intuitive discussion of limits and then give a formal definition. (Historically, the limit concept was developed in just this way.) The concept of the limit of a function will then be used to introduce the reader to the notion of continuity of functions.

Consider the following examples:

Both Newton and Leibniz considered the idea of limit, but their notion of the concept was vague and lacking in mathematical precision. It was the great mathematician Weierstrass who finally gave the notion of limit a rigorous mathematical foundation.

Example 3.1 Let the function f be defined by

$$f(x) = x + 2$$

The reader can determine the graph of f, which is given in Figure 3.1. Let us see what happens to $f(x)$ as x gets closer and closer to 2 as shown in the following table.

x	1	1.5	1.75	1.80	1.90	1.97	1.99	1.9999
$f(x) = x + 2$	3	3.5	3.75	3.80	3.90	3.97	3.99	3.9999

Figure 3.1

Graph of $f(x) = x + 2$.

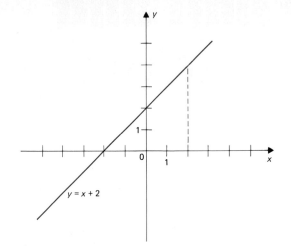

Obviously, as x approaches 2, $f(x)$ approaches 4. Of course, x could approach 2 through values greater than 2. Thus

x	3	2.50	2.40	2.30	2.20	2.10	2.05	2.001	2.0001
$f(x) = x + 2$	5	4.50	4.40	4.30	4.20	4.10	4.05	4.001	4.0001

Again $f(x)$ approaches 4 as x approaches 2 through values greater than 2. In the first case we say that $f(x)$ approaches 4 as x approaches 2 from the left, and we write

Left-hand limit

$$f(x) \to 4 \qquad \text{as} \quad x \to 2^-$$

In the second case we say that $f(x)$ approaches 4 as x approaches 2 from the right, and we write

Right-hand limit

$$f(x) \to 4 \qquad \text{as} \quad x \to 2^+$$

Therefore the function $f(x)$ approaches 4 as x approaches 2 from the left ($x \to 2^-$) or from the right ($x \to 2^+$). In such a situation we say that the limit of $f(x)$ as x approaches 2 is the number 4, and write

$$\lim_{x \to 2} f(x) = 4$$

Example 3.2   Consider the function f defined by Figure 3.2.

$$f(x) = x^2, \qquad x \neq -1$$

To illustrate a point, we have removed the number -1 from the domain of f; $f(-1)$ is not defined, but we have

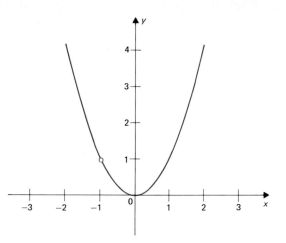

Figure 3.2
Graph of $f(x) = x^2$, $x \neq -1$.

x	-2	-1.50	-1.40	-1.30	-1.20	-1.10	-1.01	-1.001
$f(x) = x^2$	4	2.25	1.96	1.69	1.44	1.21	1.0201	1.002001

In other words

$$f(x) \to 1 \qquad \text{as} \quad x \to -1^-$$

Also we have the following:

x	0	-0.50	-0.60	-0.70	-0.80	-0.90	-0.99	-0.999
$f(x) = x^2$	0.00	0.25	0.36	0.49	0.64	0.81	0.9801	0.998001

or

$$f(x) \to 1 \qquad \text{as} \quad x \to -1^+$$

Thus we may write

$$\lim_{x \to -1} f(x) = 1$$

We note that even though the function is not defined at -1, the limit of $f(x)$ as x approaches -1 still exists and is equal to 1.

Example 3.3   Let f be the function defined by

$$f(x) = \begin{cases} x - 1, & x > 3 \\ x^2, & x \leq 3 \end{cases}$$

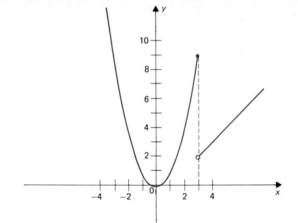

A graph of f is shown in Figure 3.3. The reader who moves a pencil on the curve just to the left of the line $x = 3$ will find that $f(x)$ approaches 9 as the pencil gets closer and closer to the line $x = 3$. Similarly, if the pencil is moved along the curve just to the right of the line $x = 3$, the reader will find that $f(x)$ approaches 2 as the pencil approaches the line $x = 3$. Thus

$$f(x) \to 9 \quad \text{as} \quad x \to 3^-$$

whereas

$$f(x) \to 2 \quad \text{as} \quad x \to 3^+$$

Since $f(x)$ approaches two different numbers depending upon whether x approaches 3 from the left or from the right, we say that

$$\lim_{x \to 3} f(x) \quad \text{does not exist.}$$

Example 3.4 Consider the function f defined by

$$f(x) = \sqrt{x + 4}$$

Here the domain of f is $\{x \mid x \geq -4\}$. The graph of f is shown in Figure 3.4. We find that

$$f(x) \to 0 \quad \text{as} \quad x \to -4^+$$

However, when $x < -4$, $f(x)$ is not defined. Consequently, $f(x)$ does not approach a real number as $x \to -4^-$. We say that

$$\lim_{x \to -4} f(x) \quad \text{does not exist.}$$

In view of the preceding examples, for the expression

$$\lim_{x \to c} f(x) = L \tag{3.1}$$

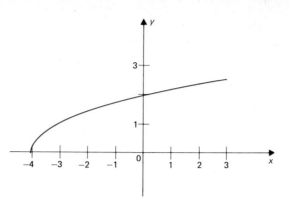

Figure 3.4
Graph of $f(x) = \sqrt{x+4}$.

(read "the limit of $f(x)$ as x approaches c is L"), we make the following observations.

(i)   The value of $f(x)$ approaches (gets closer to) L as x approaches c.

(ii) $f(x)$ approaches L whether x approaches c from the left or from the right.

(iii) The function need not even be defined at $x = c$.

In general, the expression

$$\lim_{x \to c} f(x) = L$$

intuitively means that the value of $f(x)$ can be made as close to L as we wish by making x close enough to c. To put it more formally, we have the following definition.

Definition 3.1 (*optional*)   *The $\lim_{x \to c} f(x) = L$ if and only if for every positive number ϵ (epsilon), there can be found a positive number δ (delta) such that $|f(x) - L| < \epsilon$, whereas $0 < |x - c| < \delta$. The notation*   ◀ *Limit*

$$\lim_{x \to c^+} f(x) = L$$

means that $f(x)$ approaches L as x approaches c from the right. Similarly,

$$\lim_{x \to c^-} f(x) = L$$

means that $f(x)$ approaches L as x approaches c from the left.

Definition 3.1 is not useful for finding the limit L—only for verifying that the given number L is the desired limit. To find the limit, we may make a table for the values of $f(x)$ as x gets closer and closer to c (from the left and from the right). This technique was used in Examples 3.1 and 3.2. In those examples, you may have noted that the limit could have been evaluated by substituting c for x in $f(x)$. For example, in Example 3.1,

$$\lim_{x \to 2} f(x) = \lim_{x \to 2} (x + 2) = 2 + 2 = 4$$

and in Example 3.2,

$$\lim_{x \to -1} f(x) = \lim_{x \to -1} x^2 = (-1)^2 = 1$$

The technique is quite valid and will work for this type of problem. Unfortunately, this is not the case always. Let us consider another example.

Example 3.5 Evaluate $\lim_{x \to 1} \dfrac{x^2 - 1}{x - 1}$

SOLUTION The function

$$f(x) = \frac{x^2 - 1}{x - 1}$$

is not defined when $x = 1$, because it causes division by zero. So we cannot take the shortcut of substituting for x. However, we note that

$$\frac{x^2 - 1}{x - 1} = \frac{(x - 1)(x + 1)}{x - 1} \tag{3.2}$$

Since x is only approaching 1 and is not equal to 1, in Equation (3.2), we have $x - 1 \neq 0$. Consequently, we can divide the numerator and the denominator of the right-hand side of (3.2) by $x - 1$. So when $x - 1 \neq 0$, we have

$$\frac{x^2 - 1}{x - 1} = \frac{(x - 1)(x + 1)}{x - 1} = x + 1 \tag{3.3}$$

and therefore

$$\lim_{x \to 1} \frac{x^2 - 1}{x - 1} = \lim_{x \to 1} \frac{(x - 1)(x + 1)}{x - 1} = \lim_{x \to 1} (x + 1) = 1 + 1 = 2$$

Exercises 3.1

1. Consider the function f whose graph is shown in Figure 3.5. Find each of the following:

   (a) $f(3)$ (b) $\lim_{x \to 1+} f(x)$

   (c) $\lim_{x \to 3-} f(x)$ (d) $\lim_{x \to 3+} f(x)$

   (e) $\lim_{x \to 4-} f(x)$ (f) $\lim_{x \to 4+} f(x)$

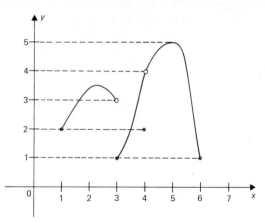

Figure 3.5

(g) $f(4)$

(h) $\lim\limits_{x \to 6+} f(x)$

(i) $\lim\limits_{x \to 6-} f(x)$

(j) $\lim\limits_{x \to 5} f(x)$

In Problems 2 through 14, sketch the graphs of the functions defined and use the graph to find the indicated limit, if it exists.

2. $\lim\limits_{x \to -1} (3x - 5)$

3. $\lim\limits_{x \to 2} (2 - 3x)$

4. $\lim\limits_{x \to 2} (4x - 3)$

5. $\lim\limits_{x \to 2} 3x^2$

6. $\lim\limits_{x \to 2} (2x^2 + 4)$

7. $\lim\limits_{x \to 3} (x - 1)^2$

8. $\lim\limits_{x \to 2} f(x), \quad f(x) = \begin{cases} 2x + 3, & x \le 2 \\ 3x - 1, & x > 2 \end{cases}$

9. $\lim\limits_{x \to -1} f(x), \quad f(x) = \dfrac{x^2 - 1}{x + 1}$

10. $\lim\limits_{x \to 2} f(x), \quad f(x) = \begin{cases} \dfrac{x^2 - 4}{x - 2}, & x \ne 2 \\ 5, & x = 2 \end{cases}$

11. $\lim\limits_{x \to 1} f(x), \quad f(x) = \dfrac{x^2 - 5x + 4}{x - 1}$

12. $\lim\limits_{x \to -3} f(x), \quad f(x) = x + 3$

13. $\lim\limits_{x \to 2} f(x), \quad f(x) = [\![x]\!]$

14. $\lim\limits_{x \to 3} f(x), \quad f(x) = x - [\![x]\!]$

In Problems 15 through 22, use a table of values of $f(x)$ for x close to the given number c to evaluate $\lim\limits_{x \to c} f(x)$. (You may use a calculator or computer to construct the table.)

15. $f(x) = 2 - 7x, \quad c = 3$

16. $f(x) = 3x - 1, \quad c = 2$

17. $f(x) = x^2 + x, \quad c = 1$

18. $f(x) = \dfrac{x^2 + 3x}{x}, \quad c = 0$

19. $f(x) = \dfrac{x^2 - 16}{x - 4}, \quad c = 4$

20. $f(x) = \dfrac{x^3 - x^2 + x - 1}{x - 1}, \quad c = 1$

21. $f(x) = \dfrac{3x^2 - 2x + 4}{2x + 3}, \quad c = -5$ 22. $f(x) = \dfrac{x^3 + 4}{3x^2 - 3}, \quad c = 3$

In Problems 23 through 26, use the method of Example 3.5 to evaluate the given limit (if it exists).

23. $\displaystyle\lim_{x \to 2} \frac{x^3 - 8}{x - 2}$ 24. $\displaystyle\lim_{x \to -2} \frac{x^3 + 8}{x + 2}$

25. $\displaystyle\lim_{x \to 1} \frac{x^4 - 1}{x^3 - 1}$ 26. $\displaystyle\lim_{x \to 2} \frac{x^2 - 4}{x^2 - 3x + 2}$

3.2 Theorems on Limits

In the previous section we introduced the methods of evaluating limits in an intuitive way. In this section we state several theorems which allow us to evaluate limits for the commonly used functions. Each of these theorems can be proved by using Definition 3.1. However, we shall state the theorems without proof and illustrate their use by examples.

►   *Theorem 3.1*   *If c and k are constants, then*

$$\lim_{x \to c} k = k \tag{3.4}$$

Theorem 3.1 states that the limit of a constant is that constant. For example,

$$\lim_{x \to 3} 6 = 6 \quad \text{and} \quad \lim_{x \to -5} 6 = 6$$

►   *Theorem 3.2*   *If a, b, and c are constants, then*

$$\lim_{x \to c} (ax + b) = ac + b \tag{3.5}$$

In other words, if x approaches c, then the limit of $ax + b$ is obtained by substituting c for x. For example,

$$\lim_{x \to 3} (2x - 5) = 2(3) - 5 = 6 - 5 = 1$$

$$\lim_{x \to 4} (7 - 3x) = 7 - 3(4) = 7 - 12 = -5$$

►   *Theorem 3.3*   *If f and g are functions and c is a constant, then*

$$\lim_{x \to c} [f(x) \pm g(x)] = \lim_{x \to c} f(x) \pm \lim_{x \to c} g(x) \tag{3.6}$$

provided that each limit exists.

In words, the limit of a sum or difference is equal to the sum or difference of the limits.

Theorem 3.4 *If f and g are functions and c is a constant, then*   ◀

$$\lim_{x \to c} [f(x) \cdot g(x)] = \lim_{x \to c} f(x) \cdot \lim_{x \to c} g(x) \qquad (3.7)$$

provided that each limit exists.

In words, the limit of the product is equal to the product of the limits.

Example 3.6

(a) $\displaystyle \lim_{x \to 5} (2 - x)(x + 3) = \lim_{x \to 5} (2 - x) \cdot \lim_{x \to 5} (x + 3)$

$$= (2 - 5)(5 + 3) = -3 \cdot 8 = -24$$

(b) $\displaystyle \lim_{x \to 4} (3x^2 + 7) = \lim_{x \to 4} 3 \cdot \lim_{x \to 4} x^2 + \lim_{x \to 4} 7$ (Theorems 3.3 and 3.4)

$$= 3 \lim_{x \to 4} (x \cdot x) + 7 \qquad \text{(Theorem 3.1)}$$

$$= 3 \lim_{x \to 4} x \cdot \lim_{x \to 4} x + 7 \qquad \text{(Theorem 3.4)}$$

$$= 3 \cdot 4 \cdot 4 + 7 \qquad \text{(Theorem 3.2)}$$

$$= 48 + 7 = 55$$

As a special case of Theorem 3.4, we observe that if $f(x) = k$, where k is a constant, then

$$\lim_{x \to c} kg(x) = \lim_{x \to c} k \cdot \lim_{x \to c} g(x) = k \lim_{x \to c} g(x) \qquad (3.7')$$

For example,

$$\lim_{x \to 4} 7(3x - 2) = 7 \lim_{x \to 4} (3x - 2) = 7(3 \cdot 4 - 2) = 70$$

Theorem 3.5 *If f and g are functions and c is a constant, then*   ◀

$$\lim_{x \to c} \left[\frac{f(x)}{g(x)} \right] = \frac{\lim_{x \to c} f(x)}{\lim_{x \to c} g(x)} \qquad (3.8)$$

provided that each limit exists and $\lim_{x \to c} g(x) \neq 0$.

In words, the limit of a quotient is the quotient of the limits.

Example 3.7 $\displaystyle \lim_{x \to 5} \frac{3x + 4}{2x - 3} = \frac{\lim_{x \to 5}(3x + 4)}{\lim_{x \to 5}(2x - 3)}$ (Theorem 3.5)

$$= \frac{3 \cdot 5 + 4}{2 \cdot 5 - 3} \qquad \text{(Theorem 3.2)}$$

$$= \frac{19}{7}$$

▶   **Theorem 3.6**   *If f is a function and c and n $\in$ I are constants,*

$$\lim_{x \to c} \left[f(x) \right]^{1/n} = \left[\lim_{x \to c} f(x) \right]^{1/n} \qquad (3.9)$$

provided the right side of (3.9) exists, and (i) $\lim_{x \to c} f(x) > 0$ if n is even; (ii) $\lim_{x \to c} f(x) \neq 0$ if $n < 0$.

Example 3.8

(a) $\lim_{x \to 2} x^{1/3} = \left(\lim_{x \to 2} x \right)^{1/3}$ (Theorem 3.6)

$\qquad\qquad = (2)^{1/3}$ (Theorem 3.2)

(b) $\lim_{x \to 2} \sqrt{x^2 + 5} = \lim_{x \to 2} (x^2 + 5)^{1/2} = \left[\lim_{x \to 2} (x^2 + 5) \right]^{1/2}$

$\qquad\qquad = (9)^{1/2} = 3$

Using the previous theorems, it is easy to see that the limits of polynomial functions can be obtained by substitution. For example, if

$$f(x) = a_0 x^n + a_1 x^{n-1} + \cdots + a_{n-1} x + a_n$$

where $a_0, a_1, \ldots, a_n$ are constants, then

$$\lim_{x \to c} f(x) = a_0 c^n + a_1 c^{n-1} + \cdots + a_{n-1} c + a_n$$

However, in calculus some important limits cannot be obtained by direct substitution. Let us consider some more examples.

Example 3.9   Evaluate $\lim_{x \to 2} \dfrac{x^2 - 5x + 6}{x - 2}$

SOLUTION   If 2 is substituted for x, we obtain

$$\frac{x^2 - 5x + 6}{x - 2} = \frac{(2)^2 - 5(2) + 6}{2 - 2} = \frac{4 - 10 + 6}{2 - 2} = \frac{0}{0}$$

Before we jump to the conclusion that the limit does not exist, we try factoring.

$$\frac{x^2 - 5x + 6}{x - 2} = \frac{(x - 2)(x - 3)}{x - 2} \qquad (3.10)$$

Now it is permissible to cancel the common factor $x - 2$, since $x \neq 2$. Therefore

$$\lim_{x \to 2} \frac{x^2 - 5x + 6}{x - 2} = \lim_{x \to 2} \frac{(x - 2)(x - 3)}{x - 2} = \lim_{x \to 2} (x - 3)$$

$$= 2 - 3 \qquad \text{(Theorem 3.2)}$$

$$= -1$$

Example 3.10 Evaluate $\lim\limits_{h \to 0} \dfrac{\sqrt{3+h} - \sqrt{3}}{h}$

SOLUTION If we substitute 0 for h, we obtain

$$\frac{\sqrt{3+0} - \sqrt{3}}{0} = \frac{\sqrt{3} - \sqrt{3}}{0} = \frac{0}{0}$$

Again we have the form 0/0. In this example the numerator cannot be factored as in the last example. So we try rationalizing the numerator by multiplying and dividing the given expression by $\sqrt{3+h} + \sqrt{3}$.

$$\frac{\sqrt{3+h} - \sqrt{3}}{h} = \frac{\sqrt{3+h} - \sqrt{3}}{h} \cdot \frac{\sqrt{3+h} + \sqrt{3}}{\sqrt{3+h} + \sqrt{3}} = \frac{(3+h) - 3}{h(\sqrt{3+h} + \sqrt{3})}$$

$$= \frac{h}{h(\sqrt{3+h} + \sqrt{3})} = \frac{1}{\sqrt{3+h} + \sqrt{3}} \qquad (\text{since } h \neq 0)$$

Therefore

$$\lim_{h \to 0} \frac{\sqrt{3+h} - \sqrt{3}}{h} = \lim_{h \to 0} \frac{1}{\sqrt{3+h} + \sqrt{3}}$$

$$= \frac{1}{\sqrt{3+0} + \sqrt{3}} = \frac{1}{2\sqrt{3}}$$

Example 3.11 Evaluate $\lim\limits_{x \to 2} \sqrt[3]{\dfrac{x^3 - 8}{x^2 - 4}}$

SOLUTION $\lim\limits_{x \to 2} \sqrt[3]{\dfrac{x^3 - 8}{x^2 - 4}} = \lim\limits_{x \to 2} \left(\dfrac{x^3 - 8}{x^2 - 4} \right)^{1/3} = \left(\lim\limits_{x \to 2} \dfrac{x^3 - 8}{x^2 - 4} \right)^{1/3}$

$$= \left[\lim_{x \to 2} \frac{(x - 2)(x^2 + 2x + 4)}{(x - 2)(x + 2)} \right]^{1/3}$$

$$= \left(\lim_{x \to 2} \frac{x^2 + 2x + 4}{x + 2} \right)^{1/3} = \left(\frac{12}{4} \right)^{1/3}$$

$$= (3)^{1/3} = \sqrt[3]{3}$$

Difference of two cubes:
$a^3 - b^3 = (a - b)(a^2 + ab + b^2)$

In the next example, we consider the illustration where the limit does not exist.

Example 3.12 Evaluate (a) $\lim\limits_{x \to 2} \dfrac{5}{x - 2}$ and

  (b) $\lim\limits_{x \to -3} \sqrt{x + 1}$

SOLUTION

(a) The limit does not exist, since no factoring process or rationalization can change the division by zero.

(b) The limit does not exist, since

$$\lim_{x \to -3} \sqrt{x + 1} = [\lim_{x \to -3} (x + 1)]^{1/2} = \sqrt{-3 + 1} = \sqrt{-2}$$

but $\sqrt{-2}$ is not a real number.

Example 3.13 Evaluate $\lim_{h \to 0} \dfrac{(x + h)^2 - x^2}{h}$

SOLUTION   Here the variable x is to be treated as a constant, since we are interested in $h \to 0$. Consequently,

$$\lim_{h \to 0} \frac{(x + h)^2 - x^2}{h} = \lim_{h \to 0} \frac{x^2 + 2xh + h^2 - x^2}{h} = \lim_{h \to 0} \frac{2xh + h^2}{h}$$

$$= \lim_{h \to 0} \frac{(2x + h)h}{h} = \lim_{h \to 0} (2x + h) \qquad (h \neq 0)$$

$$= 2x + 0 = 2x$$

Exercises 3.2   In Problems 1 through 24, evaluate the indicated limits (if the limit exists).

1. $\lim\limits_{x \to 2} (3 + 5x)$

2. $\lim\limits_{x \to 3} (-7x)$

3. $\lim\limits_{x \to -2} (2x - 3)$

4. $\lim\limits_{x \to -4} (3 - 8x)$

5. $\lim\limits_{x \to 2} (5x^2 + 4)$

6. $\lim\limits_{x \to 1} (3x^2 - 8x + 2)$

7. $\lim\limits_{x \to 2} (x - 2)(x^2 + 3)$

8. $\lim\limits_{x \to -3} (2x^3 - 3x^2 + 6x + 1)$

9. $\lim\limits_{x \to 4} \dfrac{x^2}{2x - 1}$

10. $\lim\limits_{x \to 2} \dfrac{x^3 + 5}{x^2 + 3x}$

11. $\lim\limits_{x \to 2} \dfrac{x^2 - 6}{3x + 1}$

12. $\lim\limits_{x \to 4} \dfrac{2x^2 - 7}{x^3 + 3}$

13. $\lim\limits_{x \to 0} \dfrac{-x^2 + 4x - 8}{x^3 + 3x + 2}$

14. $\lim\limits_{x \to 2} \dfrac{3x + 1}{2x - 2}$

15. $\lim\limits_{h \to 0} \dfrac{h + h^2}{h}$

16. $\lim\limits_{h \to 0} \dfrac{x^2 + 2xh + h^2}{x}$

17. $\lim\limits_{x \to 2} \dfrac{x^2 - 4}{x^2 - 3x + 2}$

18. $\lim\limits_{x \to -2} \dfrac{x^3 + 8}{2x^2 + x - 1}$

19. $\lim\limits_{x \to 1} \dfrac{x^3 - 1}{x^2 - 1}$

20. $\lim\limits_{x \to -1} \sqrt[3]{\dfrac{x + 1}{x^3 + 1}}$

21. $\lim\limits_{h \to 0} \dfrac{\sqrt{1 + h} - \sqrt{h}}{h}$

22. $\lim\limits_{h \to 0} \dfrac{\sqrt{x + h} - \sqrt{x}}{h}$

23. $\lim\limits_{x \to 0} \dfrac{\sqrt{x + 16} - 4}{x}$

24. $\lim\limits_{y \to 0} \dfrac{\sqrt{y + 36} - 6}{y}$

In Problems 25 through 38, form the new function F, where

$$F(h) = \frac{f(x+h) - f(x)}{h}, \qquad h \neq 0$$

and evaluate $\lim_{h \to 0} F(h)$.

25. $f(x) = x$ 26. $f(x) = x + 1$

27. $f(x) = 1 - 3x$ 28. $f(x) = x^2$

29. $f(x) = \sqrt{x}$ 30. $f(x) = \sqrt{x+1}$

31. $f(x) = \sqrt{2x+3}$ 32. $f(x) = \sqrt{3-4x}$

33. $f(x) = x^3$ 34. $f(x) = 3\sqrt{x}$

35. $f(x) = x^2 - 2x + 5$ 36. $f(x) = x^3 + 2x^2 - 3x + 7$

37. $f(x) = 2$ 38. $f(x) = -4$

3.3   Limits at Infinity and Infinite Limits

We shall now consider some other important limits: the limit at infinity and the infinite limit. Consider the function f, where

$$f(x) = \frac{1}{x}, \qquad x \neq 0$$

The graph of f is given in Figure 3.6. A few points belonging to the graph of f are given in the following table:

x	-10^{100}	-10^{10}	-10^2	-10	10	10^2	10^{10}	10^{100}
$f(x) = \dfrac{1}{x}$	-10^{-100}	-10^{-10}	-10^{-2}	-10^{-1}	10^{-1}	10^{-2}	10^{-10}	10^{-100}

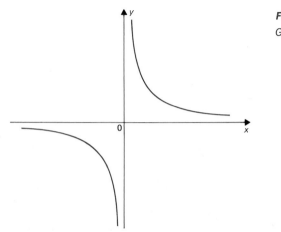

Figure 3.6
Graph of $f(x) = 1/x$.

As x gets larger and larger, $f(x)$ gets closer and closer to zero. The expression "x tends to infinity" is used to indicate that x increases without bound. The following symbols are used:

$$x \to +\infty \qquad \text{if } x \text{ increases without bound}$$

$$x \to -\infty \qquad \text{if } x \text{ decreases without bound}$$

From the preceding table, we see that $1/x$ approaches 0 as x increases without bound through positive values ($x \to +\infty$). Also, $1/x$ approaches 0 as x decreases without bound.

▶ **Theorem 3.7** *If f is the function defined by*

$$f(x) = \frac{1}{x^r}$$

where r is a positive integer, then

We write $\lim\limits_{x \to +\infty} \dfrac{1}{x^r} = 0$ *or*

$\dfrac{1}{x^r} \to 0 \quad$ *as* $\quad x \to +\infty$

and

$\lim\limits_{x \to -\infty} \dfrac{1}{x^r} = 0 \quad$ *or* $\quad \dfrac{1}{x^r} \to 0$

as $\quad x \to -\infty$

(i) $\quad \lim\limits_{x \to +\infty} f(x) = 0$

(ii) $\quad \lim\limits_{x \to -\infty} f(x) = 0$

The results of Theorems 3.1 through 3.6 still hold when the number c is replaced by the symbols $+\infty$ or $-\infty$. We apply these results in the following examples.

Example 3.14 Evaluate $\lim\limits_{x \to +\infty} \dfrac{2x+5}{7x-3}$

SOLUTION   For $x \neq 0$, we have

Dividing numerator and denominator by x

$$\frac{2x+5}{7x-3} = \frac{2 + \dfrac{5}{x}}{7 - \dfrac{3}{x}}$$

Now

Limit of sum = sum of limits

$$\lim_{x \to +\infty}\left(2 + \frac{5}{x}\right) = \lim_{x \to +\infty} 2 + \lim_{x \to +\infty} \frac{5}{x}$$

$\lim\limits_{x \to +\infty} b = b$, *limit of product =
product of limits*

$$= 2 + 5 \lim_{x \to +\infty} \frac{1}{x} = 2 + 0 = 2$$

and

$$\lim_{x \to +\infty}\left(7 - \frac{3}{x}\right) = \lim_{x \to +\infty} 7 + \lim_{x \to +\infty} -\frac{3}{x}$$

$$= 7 - 3 \lim_{x \to +\infty} \frac{1}{x} = 7 - 0 = 7$$

Then

$$\lim_{x \to +\infty} \frac{2x + 5}{7x - 5} = \lim_{x \to +\infty} \frac{2 + \dfrac{5}{x}}{7 - \dfrac{3}{x}}$$

$$= \frac{\displaystyle\lim_{x \to +\infty} \left(2 + \frac{5}{x}\right)}{\displaystyle\lim_{x \to +\infty} \left(7 - \frac{3}{x}\right)} = \frac{2}{7}$$ *Limit of quotient = quotient of limits*

Example 3.15 Evaluate $\displaystyle\lim_{x \to -\infty} \frac{x^2 - 3x + 5}{2x^3 - 3}$

SOLUTION We divide the numerator and denominator by the highest power of x occurring in either the numerator or denominator; in this case x^3 is the highest power of x involved. Thus we have

$$\lim_{x \to -\infty} \frac{x^2 - 3x + 5}{2x^3 - 3} = \lim_{x \to -\infty} \frac{\dfrac{1}{x} - \dfrac{3}{x^2} + \dfrac{5}{x^3}}{2 - \dfrac{3}{x^3}}$$

$$= \frac{\displaystyle\lim_{x \to -\infty} \left(\frac{1}{x} - \frac{3}{x^2} + \frac{5}{x^3}\right)}{\displaystyle\lim_{x \to -\infty} \left(2 - \frac{3}{x^3}\right)}$$ *Limit of quotient = quotient of limits*

$$= \frac{\displaystyle\lim_{x \to -\infty} \left(\frac{1}{x}\right) - 3 \lim_{x \to -\infty} \left(\frac{1}{x^2}\right) + 5 \lim_{x \to -\infty} \left(\frac{1}{x^3}\right)}{\displaystyle\lim_{x \to -\infty} (2) - 3 \lim_{x \to -\infty} \left(\frac{1}{x^3}\right)}$$ *Limit of sum = sum of limits*

$$= \frac{0 - (3 \cdot 0) + (5 \cdot 0)}{2 - (3 \cdot 0)} = \frac{0}{2} = 0$$

Example 3.16 Evaluate $\displaystyle\lim_{x \to +\infty} \frac{\sqrt[3]{x^3 + 1}}{3x - 5}, \qquad x \neq \frac{5}{3}$

SOLUTION We have

$$\frac{\sqrt[3]{x^3 + 1}}{3x - 5} = \frac{\dfrac{1}{x} \sqrt[3]{x^3 + 1}}{\dfrac{1}{x} (3x - 5)} = \frac{\sqrt[3]{1 + \dfrac{1}{x^3}}}{3 - \dfrac{5}{x}}$$

Now

$$\lim_{x \to +\infty} \left(1 + \frac{1}{x^3}\right) = \lim_{x \to +\infty} 1 + \lim_{x \to +\infty} \frac{1}{x^3} = 1 + 0 = 1$$

Also

$$\lim_{x \to +\infty} \left(3 - \frac{5}{x}\right) = \lim_{x \to +\infty} 3 + \lim_{x \to +\infty} \frac{-5}{x}$$

Give reasons!

$$= 3 - 5 \lim_{x \to +\infty} \frac{1}{x} = 3 - 0 = 3$$

Therefore

$$\lim_{x \to +\infty} \frac{\sqrt[3]{x^3 + 1}}{3x - 5} = \lim_{x \to +\infty} \frac{\sqrt[3]{1 + \frac{1}{x^3}}}{3 - \frac{5}{x}}$$

Application of Theorem 2(iii), (iv)

$$= \frac{\lim_{x \to +\infty} \sqrt[3]{1 + \frac{1}{x^3}}}{\lim_{x \to +\infty} \left(3 - \frac{5}{x}\right)}$$

$$= \frac{1}{3}$$

Example 3.17   Evaluate $\displaystyle\lim_{x \to -\infty} \frac{2x + 3}{\sqrt{3x^2 - 4}}$

Recall that $\sqrt{x^2} = |x| = -x$, $x < 0$.
Note   *$\sqrt{b}$ is a real number provided that $b \geq 0$.*

SOLUTION   Since we are considering negative values of x, we have $\sqrt{x^2} = -x$. Hence we divide numerator and denominator by $-\sqrt{x^2} = x$ to obtain

$$\lim_{x \to -\infty} \frac{2x + 3}{\sqrt{3x^2 - 4}} = \lim_{x \to -\infty} \frac{2 + \frac{3}{x}}{-\sqrt{\frac{3x^2 - 4}{x^2}}} = \lim_{x \to -\infty} \frac{2 + \frac{3}{x}}{-\sqrt{3 - \frac{4}{x^2}}}$$

$$= \frac{\lim_{x \to -\infty} \left(2 + \frac{3}{x}\right)}{\lim_{x \to -\infty} \left(-\sqrt{3 - \frac{4}{x^2}}\right)}$$

$$= \frac{\lim_{x \to -\infty} (2) + 3 \lim_{x \to -\infty} \left(\frac{1}{x}\right)}{-\sqrt{\lim_{x \to -\infty} \left(3 - \frac{4}{x^2}\right)}}$$

$$= \frac{2 + 3 \cdot 0}{-\sqrt{3} - 0} = -\frac{2}{\sqrt{3}}$$

ALTERNATIVE SOLUTION   If we substitute $x = -y$, then as $x \rightarrow -\infty$, we have $y \rightarrow +\infty$. Therefore

$$\lim_{x \to -\infty} \frac{2x + 3}{\sqrt{3x^2 - 4}} = \lim_{y \to +\infty} \frac{2(-y) + 3}{\sqrt{3(-y)^2 - 4}} = \lim_{y \to +\infty} \frac{-2y + 3}{\sqrt{3y^2 - 4}}$$

$$= \lim_{y \to +\infty} \frac{-2 + \dfrac{3}{y}}{\sqrt{3 - \dfrac{4}{y^2}}} = \frac{\lim\limits_{y \to +\infty} \left(-2 + \dfrac{3}{y} \right)}{\lim\limits_{y \to +\infty} \sqrt{3 - \dfrac{4}{y^2}}} = -\frac{2}{\sqrt{3}}$$

Returning to the function f defined by

$$f(x) = \frac{1}{x}$$

we find that (see Figure 3.6) $f(x)$ increases without bound as x approaches 0 from the right. We write

$$\lim_{x \to 0^+} \frac{1}{x} = +\infty$$

and as x approaches 0 from the left, $f(x)$ decreases without bound. We write

$$\lim_{x \to 0^-} \frac{1}{x} = -\infty$$

For the function whose graph is shown in Figure 3.7, we have

$$\lim_{x \to 1^+} f(x) = +\infty$$

and

$$\lim_{x \to 1^-} f(x) = +\infty$$

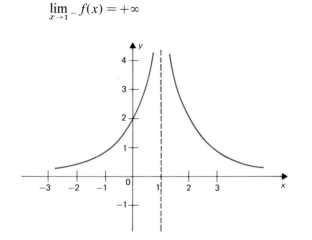

Figure 3.7
Graph of $f(x) = 2/(x - 1)^2$.

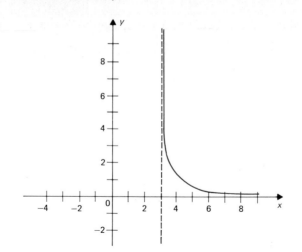

Figure 3.8

Graph of $f(x) = 1/\sqrt{x-3}$.

For the function whose graph is shown in Figure 3.8, we have

$$\lim_{x \to 3^+} f(x) = +\infty$$

and since $f(x)$ is not defined for $x \le 3$, we say that

$$\lim_{x \to 3^-} f(x) \text{ does not exist}$$

The reader should be careful when operating with limits involving the symbol ∞. It should be remembered that ∞ is *not* a number and cannot be treated as such. We have the following rules of operation.

Rule 1   If $\lim_{x \to c} f(x) = +\infty$ (or $-\infty$) and $\lim_{x \to c} g(x) = b$, then

$$\lim_{x \to c} [f(x) + g(x)] = +\infty \qquad (\text{or } -\infty)$$

If $\lim_{x \to c} g(x) = -\infty$ (or $+\infty$), nothing in general can be said about

$$\lim_{x \to c} [f(x) + g(x)]$$

Further investigation of the particular problem will be necessary.

Rule 2   If $\lim_{x \to c} f(x) = +\infty$ $(-\infty)$ and $\lim_{x \to c} g(x) = b$, then

(i) $\lim_{x \to c} f(x) \cdot g(x) = +\infty \, (-\infty) \qquad$ if $b > 0$

(ii) $\lim_{x \to c} f(x) \cdot g(x) = -\infty \, (+\infty) \qquad$ if $b < 0$

If $b = 0$, further investigation is needed.

Rule 3   If $\lim_{x \to c} f(x) = +\infty$ $(-\infty)$ and $\lim_{x \to c} g(x) = b$, then

$$\lim_{x \to c} \frac{g(x)}{f(x)} = 0$$

Rule 4   If $\lim_{x \to c} f(x) = 0$ and $\lim_{x \to c} g(x) = b$, where $b \ne 0$, then

(i) $\lim_{x \to c} \dfrac{g(x)}{f(x)} = \begin{cases} +\infty \text{ if } b > 0 \text{ and } f(x) \to 0 \text{ through positive values} \\ -\infty \text{ if } b > 0 \text{ and } f(x) \to 0 \text{ through negative values} \end{cases}$

(ii) $\displaystyle \lim_{x \to c} \frac{g(x)}{f(x)} = \begin{cases} -\infty & \text{if } b < 0 \text{ and } f(x) \to 0 \text{ through positive values} \\ +\infty & \text{if } b < 0 \text{ and } f(x) \to 0 \text{ through negative values} \end{cases}$

The above rules are valid when one-sided limits are used.

Example 3.18   Evaluate each of the following limits:

(a) $\displaystyle \lim_{x \to 1^+} \frac{2x^2 + x + 2}{x^2 + 2x - 3}$ (b) $\displaystyle \lim_{x \to 1^-} \frac{2x^2 + x + 2}{x^2 + 2x - 3}$

(c) $\displaystyle \lim_{x \to +\infty} \frac{x^2}{x + 2}$ (d) $\displaystyle \lim_{x \to +\infty} \frac{3x - x^2}{2x + 7}$

SOLUTION

(a) $\displaystyle \lim_{x \to 1^+} \frac{2x^2 + x + 2}{x^2 + 2x - 3} = \frac{\displaystyle \lim_{x \to 1^+} (2x^2 + x + 2)}{\displaystyle \lim_{x \to 1^+} (x + 3)(x - 1)}$

The limit of the numerator is 5 and the denominator is approaching 0 through positive values. Then, by Rule 4(i), we obtain

$$\lim_{x \to 1^+} \frac{2x^2 + x + 2}{x^2 + 2x - 3} = +\infty$$

(b) In this case the limit of the numerator is again 5 and the denominator is approaching 0 through negative values. Then, by Rule 4(i),

$$\lim_{x \to 1^-} \frac{2x^2 + x + 2}{x^2 + 2x - 3} = -\infty$$

(c) Dividing numerator and denominator by x^2, we get

$$\lim_{x \to +\infty} \frac{x^2}{x + 2} = \lim_{x \to +\infty} \frac{1}{\dfrac{1}{x} + \dfrac{2}{x^2}}$$

The limit of the denominator is

$$\lim_{x \to +\infty} \left(\frac{1}{x} + \frac{2}{x^2} \right) = \lim_{x \to +\infty} \frac{1}{x} + 2 \lim_{x \to +\infty} \frac{1}{x^2} = 0 + 2 \cdot 0 = 0$$

and the denominator is approaching zero through positive values. Thus

$$\lim_{x \to +\infty} \frac{x^2}{x + 2} = \lim_{x \to +\infty} \frac{1}{\dfrac{1}{x} + \dfrac{2}{x^2}} = +\infty$$

(d) We have

$$\lim_{x \to +\infty} \frac{3x - x^2}{2x + 7} = \lim_{x \to +\infty} \frac{\dfrac{3}{x} - 1}{\dfrac{2}{x} + \dfrac{7}{x^2}}$$

Considering the limits of the numerator and denominator separately, we have

$$\lim_{x \to +\infty} \left(\frac{3}{x} - 1 \right) = 3 \lim_{x \to +\infty} \frac{1}{x} - \lim_{x \to +\infty} 1 = 0 - 1 = -1$$

and

$$\lim_{x \to +\infty} \left(\frac{2}{x} + \frac{7}{x^2} \right) = 2 \lim_{x \to +\infty} \frac{1}{x} + 7 \lim_{x \to +\infty} \frac{1}{x^2} = 2 \cdot 0 + 7 \cdot 0 = 0$$

where the denominator is approaching zero through positive values. Then, by **Rule 4(ii)**, we have

$$\lim_{x \to +\infty} \frac{3x - x^2}{2x + 7} = -\infty$$

Exercises 3.3   In Problems 1 through 22, evaluate the limit.

1. $\displaystyle \lim_{x \to +\infty} \frac{5x - 1}{-x + 4}$

2. $\displaystyle \lim_{x \to +\infty} \frac{3x + 1}{4x - 2}$

3. $\displaystyle \lim_{x \to +\infty} \frac{x^2 + 3x + 7}{3x^2 - 2x - 1}$

4. $\displaystyle \lim_{x \to +\infty} \frac{2x + 5}{x^2 + 1}$

5. $\displaystyle \lim_{x \to +\infty} \frac{2x^2 + 4x - 1}{x^3 + 4}$

6. $\displaystyle \lim_{x \to -\infty} \frac{\sqrt{x^2 + 6}}{x + 6}$

7. $\displaystyle \lim_{x \to -\infty} \frac{x^2 + 6}{x + 6}$

8. $\displaystyle \lim_{x \to +\infty} \frac{2x^3 + 3x^2 - 4}{5x^3 + x + 2}$

9. $\displaystyle \lim_{x \to +\infty} \frac{2 - x^2}{3x + 5}$

10. $\displaystyle \lim_{x \to +\infty} \frac{2x^2 - 3x + 4}{x + 4}$

11. $\displaystyle \lim_{x \to 2} \frac{x}{x - 2}$

12. $\displaystyle \lim_{x \to 3} \frac{x^2}{9 - x^2}$

13. $\displaystyle \lim_{x \to 1^+} \frac{x + 1}{x^2 - 1}$

14. $\displaystyle \lim_{x \to 1^-} \frac{x + 1}{x^2 - 1}$

15. $\displaystyle \lim_{x \to 2} \frac{-2}{(x - 2)^2}$

16. $\displaystyle \lim_{x \to 2^+} \frac{x + 2}{x^2 - 4}$

17. $\displaystyle \lim_{x \to 0^+} \frac{\sqrt{1 + x^2}}{x}$

18. $\displaystyle \lim_{x \to 0^-} \frac{\sqrt{1 + x^2}}{x}$

19. $\displaystyle\lim_{x\to 3^-}\frac{\sqrt{9-x^2}}{x-3}$

20. $\displaystyle\lim_{x\to 3^+}\frac{\sqrt{9-x^2}}{x-3}$

21. $\displaystyle\lim_{x\to 0}\left(\frac{1}{x}-\frac{1}{x^2}\right)$

22. $\displaystyle\lim_{x\to 3}\left(\frac{1}{x-3}-\frac{2}{x^2-9}\right)$

3.4   Continuity of a Function

Let us consider the functions whose graphs are illustrated in Figures
3.9 through 3.12. In each case we are interested in what happens to
the graph at the point $x=c$. In the first three graphs we note that there
is a break in the curve at c. In other words, if you move your pencil
along the curve, you will have to lift the pencil at $x=c$. We say that the
functions involved are *discontinuous* at $x=c$. On the other hand, you
could move your pencil along the curve in Figure 3.12 near c without
lifting the pencil at c. We say that the function whose graph is shown
in Figure 3.12 is *continuous* at $x=c$.

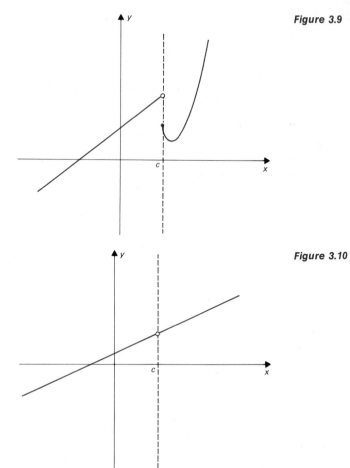

Figure 3.9

Figure 3.10

Figure 3.11

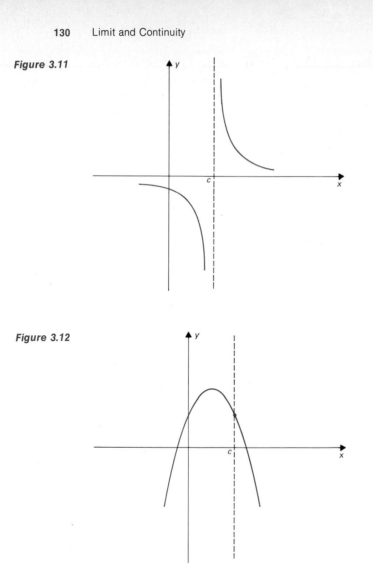

Figure 3.12

The mathematicians Bolzano and Weierstrass offered definitions of the concept of continuity, which as we have seen is connected with the notion of unbroken curves. Bolzano's definition was imprecise. Weierstrass offered the more rigorous definition, which is essentially the one that is used today.

We now see that continuity of a function at c is closely related to the limit of a function at $x = c$.

Bernhard Bolzano

Bernhard Bolzano (1781–1848), an Austrian mathematician and theologian, was a priest and a professor of religious philosophy at the University of Prague. In 1820 he was dismissed from his post by the Austrian government because his sermons were considered subversive. His mathematical work *Paradoxes of Infinity* was published after his death.

Definition 3.2 *A function f is said to be* continuous *at the number c if and only if*

(*i*)   *f(c) exists; that is, f is defined at c*

(*ii*) $\lim_{x \to c} f(x)$ *exists, and*

(*iii*) $\lim_{x \to c} f(x) = f(c)$

When a function f is not continuous at c, we say that f is discontinuous at c. If a function is continuous at every point in an open interval I, it is said to be continuous *on I.*

Using Definition 3.2, we shall consider some examples.

Example 3.19   Let *f* be the function defined by

$$f(x) = \begin{cases} 1, & x \geq 2 \\ -1, & x < 2 \end{cases}$$

Is *f* continuous at $x = 2$?

SOLUTION   Figure 3.13 illustrates the graph of *f*. Here $f(2) = 1$ and so condition (i) is satisfied. Now

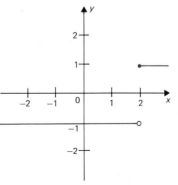

Continuous and discontinuous functions

Figure 3.13

Karl W. T. Weierstrass (1815–1897) was born in Ostenfelde, Germany, the eldest in a family of two brothers and two sisters. At 19 he entered the University of Bonn, where he indulged in fencing and beer drinking; he returned home after four years without a degree. At the age of 26, he took his teacher's certificate examinations and then taught in secondary schools until 1853.

Weierstrass is called the father of modern analysis. His early publications in an unknown high school paper went unnoticed. In 1854 he published a memoir on Abelian functions which created a sensation and won him an honorary doctorate from the University of Königsberg. He was elected to the Berlin Academy and was given a professorship at the University of Berlin, where he stayed until his death. His fame as an outstanding mathematician and an excellent teacher spread all over Europe and to America. He emphasized rigor and logic.

Karl W. T. Weierstrass

$$\lim_{x \to 2^+} f(x) = 1 \quad \text{and} \quad \lim_{x \to 2^-} f(x) = -1$$

Hence $\lim_{x \to 2} f(x)$ does not exist and so condition (ii) is not satisfied. Therefore f is discontinuous at $x = 2$.

Example 3.20 Let f be the function defined by

$$f(x) = \begin{cases} \dfrac{(x+2)(x-1)}{x-1}, & x \neq 1 \\ 2, & x = 1 \end{cases}$$

Is f continuous at $x = 1$?

SOLUTION   A graph of f is shown in Figure 3.14. Here $f(1) = 2$. Furthermore,

Figure 3.14

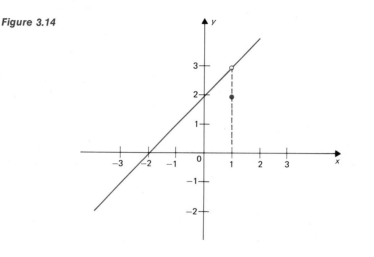

$$\lim_{x \to 1^+} f(x) = 3 \quad \text{and} \quad \lim_{x \to 1^-} f(x) = 3$$

Hence

$$\lim_{x \to 1} f(x) = 3$$

Thus conditions (i) and (ii) are satisfied. However,

$$\lim_{x \to 1} f(x) \neq f(1)$$

Therefore f is discontinuous at $x = 1$.

Example 3.21 Let the function f be defined by

$$f(x) = \frac{1}{x}$$

Is f continuous at $x = 0$?

SOLUTION   Since $f(0)$ is not defined, condition (i) is not satisfied and so f is discontinuous at $x = 0$ (see Figure 3.15).

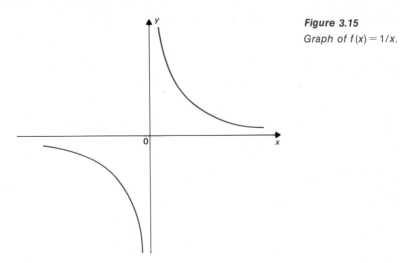

Figure 3.15
Graph of $f(x) = 1/x$.

Example 3.22   Let f be defined by

$$f(x) = \begin{cases} \dfrac{1}{x}, & x \neq 0 \\ 1, & x = 0 \end{cases}$$

Is f continuous at $x = 0$?

SOLUTION   Here $f(0) = 1$ and so condition (i) is satisfied. Now

$$\lim_{x \to 0^+} f(x) = +\infty \quad \text{and} \quad \lim_{x \to 0^-} f(x) = -\infty$$

Therefore condition (ii) is not satisfied and so f is discontinuous at $x = 0$ (see Figure 3.16).

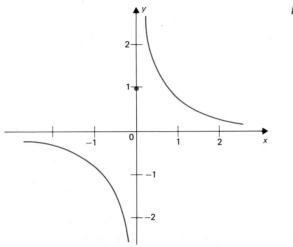

Figure 3.16

Example 3.23   Let f be the function defined by

$$f(x) = \sqrt{x + 1}$$

Is f continuous at $x = -1$?

SOLUTION   A graph of f is shown in Figure 3.17. We find that $f(-1) = 0$. Now

Figure 3.17

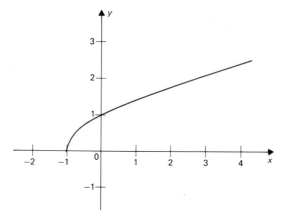

$$\lim_{x \to -1^+} f(x) = 0 \text{ but } \lim_{x \to -1^-} f(x) \text{ does not exist.}$$

Hence $\lim_{x \to -1} f(x)$ does not exist, and so condition (ii) is not satisfied. Therefore f is discontinuous at $x = -1$.

Example 3.24   Let f be the function defined by

$$f(x) = \frac{1}{x + 1}$$

(a)   Is f continuous on $I_1 = \{x | 0 < x < 4\}$?

(b)   Is f continuous on $I_2 = \{x | -3 < x < 4\}$?

SOLUTION
(a)   Let c be any number in I_1. Then $f(c) = 1/(c + 1)$. Also

$$\lim_{x \to c} f(x) = \lim_{x \to c} \frac{1}{x + 1} = \frac{1}{c + 1}, \qquad c \neq -1$$

Hence

$$\lim_{x \to c} f(x) = f(c)$$

Conditions (i), (ii), and (iii) are satisfied for every c in I_1. Therefore f is continuous on I_1.

(b) Here -1 is in I_2, and $f(-1)$ is not defined. Thus f is not continuous at -1. Therefore, f is not continuous on I_2. We have the following theorem.

Note that f is continuous at every other point in I_2. However f is still not continuous on I_2.

◄

Theorem 3.8 *Suppose that f and g are functions both continuous at c. Then*

(*i*) *$f + g$ is continuous at c*
(*ii*) *fg is continuous at c*
(*iii*) *f/g is continuous at c provided $g(c) \neq 0$.*

PROOF   We shall prove part (i) and leave parts (ii) and (iii) as exercises for the student.

By Definition 3.2, we have

$$\lim_{x \to c} f(x) = f(c) \quad \text{and} \quad \lim_{x \to c} g(x) = g(c)$$

Applying Theorem 3.3, we have

$$\lim_{x \to c} [f(x) + g(x)] = \lim_{x \to c} f(x) + \lim_{x \to c} g(x) = f(c) + g(c)$$

Thus Definition 3.2 is satisfied, and the function $f + g$ is continuous at c.

Example 3.25   Define $f(2)$ so that the function

$$f(x) = \frac{x^2 - x - 2}{x - 2}, \quad x \neq 2$$

is continuous at $x = 2$.

SOLUTION   We have

$$f(x) = \frac{x^2 - x - 2}{x - 2} = \frac{(x - 2)(x + 1)}{x - 2} = x + 1, \quad x \neq 2$$

Now,

$$\lim_{x \to 2} f(x) = \lim_{x \to 2} (x + 1) = 3$$

Therefore if we define $f(2) = 3$, then the three conditions of Definition 3.2 will be satisfied. Thus the function

$$f(x) = \begin{cases} \dfrac{x^2 - x - 2}{x - 2}, & x \neq 2 \\ 3, & x = 2 \end{cases}$$

is continuous at $x = 2$.

Exercises 3.4   In Problems 1 through 14, sketch the graphs of each of the functions defined and determine if each is continuous at the given value c.

1. $f(x) = \dfrac{3}{x-2}$ at $c = 2$

2. $f(x) = \begin{cases} \dfrac{1}{x+4}, & x \neq -4 \\ 0, & x = -4 \end{cases}$ at $c = -4$

3. $f(x) = \begin{cases} \dfrac{x^2 - 3x + 2}{x-1}, & x \neq 1 \\ -5, & x = 1 \end{cases}$ at $c = 1$

4. $f(x) = \sqrt{3x - 1}$ at $c = \frac{1}{3}$

5. $f(x) = \begin{cases} 2x + 1, & x \leq 1 \\ -x + 4, & x > 1 \end{cases}$ at $c = 1$

6. $f(x) = \begin{cases} 2x + 1, & x \leq 2 \\ -x + 4, & x > 2 \end{cases}$ at $c = 2$

7. $f(x) = |x - 2|$ at $c = 2$

8. $f(x) = \begin{cases} |x - 2|, & x \neq 2 \\ 1, & x = 2 \end{cases}$ at $c = 2$

9. $f(x) = \begin{cases} \dfrac{x^2 - 9}{x+3}, & x \neq -3 \\ 3, & x = -3 \end{cases}$ at $c = -3$

10. $f(x) = \begin{cases} \dfrac{x^2 - 4}{x-2}, & x \neq 2 \\ 4, & x = 2 \end{cases}$ at $c = 2$

11. $f(x) = \sqrt[3]{x}$ at $c = -1$

12. $f(x) = \sqrt[3]{x} - 1$ at $c = 1$

13. $f(x) = \begin{cases} \dfrac{x-4}{x^2 - 16}, & x \neq \pm 4 \\ \frac{1}{8}, & x = 4 \\ 0, & x = -4 \end{cases}$ at $c = 4, -4$.

14. $f(x) = \begin{cases} x + 2, & x \leq -1 \\ x^2, & -1 < x < 1 \\ 3x + 1, & x \geq 1 \end{cases}$ at $c = -1, 1$

In Problems 15 through 22, a function f is defined on an interval I. Determine whether f is continuous on I.

15. $f(x) = \dfrac{1}{x^2 + 1}$, $I = \{x | -1 < x < 5\}$

16. $f(x) = \dfrac{x}{x-1}$, $I = \{x | -3 < x < 1\}$

17. $f(x) = \dfrac{x}{x-1}$, $I = \{x | -1 < x < 4\}$

18. $f(x) = \dfrac{x^2 + 3x - 1}{2x^2 + 3}$, $I = \{x \mid -2 < x < 2\}$

19. $f(x) = \begin{cases} \dfrac{x^2 + x - 2}{x + 2}, & x \neq -2 \\ -3, & x = -2 \end{cases}$ $I = \{x \mid -3 < x < 3\}$

20. $f(x) = \begin{cases} \dfrac{x^3 - 1}{x - 1}, & x \neq 1 \\ 2, & x = 1 \end{cases}$ $I = \left\{ x \mid 0 < x < \frac{3}{2} \right\}$

21. $f(x) = \begin{cases} \dfrac{4x^2 - 1}{2x + 1}, & x \neq -\frac{1}{2} \\ -2, & x = -\frac{1}{2} \end{cases}$ $I = \{x \mid -1 < x < 1\}$

22. $f(x) = \dfrac{x + 2}{x^2 - 4}$, $I = \{x \mid 0 < x < 2\}$

In Problems 23 through 26, define $f(c)$ so that the function f is continuous at c.

23. $f(x) = \dfrac{(x - 1)(x + 2)}{x - 1}$, $x \neq 1$, $c = 1$

24. $f(x) = \dfrac{x^2 - 3x + 2}{x^2 - 4}$, $x \neq \pm 2$, $c = 2$

25. $f(x) = \dfrac{x^2 - 1}{x + 1}$, $x \neq -1$, $c = -1$

26. $f(x) = \dfrac{x^3 + 8}{x + 2}$, $x \neq -2$, $c = -2$

27. Can $f(-2)$ in Problem 24 be defined so that the function is continuous at -2?

*28. Suppose f is defined on the closed interval $I = \{x \mid a \leq x \leq b\}$. We say that the function f is continuous from the right at a if and only if $f(a)$ is defined and $\lim_{x \to a^+} f(x) = f(a)$.
   (i)   Define continuity of the function f from the left at b.
   (ii) Define continuity of the function f on I.

*29. Suppose f is continuous at c and $f(c) > 0$ [$f(c) < 0$]. Show that there exists an open interval I containing c such that $f(x) > 0$ [$f(x) < 0$] for every x in I.

Part 1   (Oral)   Chapter 3 Review

1. What is meant by the limit of $f(x)$ as x approaches c from the left?

2. What is meant by the limit of $f(x)$ as x approaches c from the right?

3. Define $\lim_{x \to c} f(x)$.

4. What is $\lim_{x \to c} (ax + b)$, where a and b are constants?

5. State a theorem for the limit of a sum of two functions.

6. State a theorem for the limit of a product of two functions.

7. State a theorem for the limit of a quotient of two functions.

8. Under what conditions is $\lim_{x \to c} [f(x)]^n = [\lim_{x \to c} f(x)]^n$?

9. Define continuity of a function at a point.

10. Define continuity of a function over an interval.

Part 2   (Written)

In Problems 1 through 20, find the limit (if it exists):

1. $\lim\limits_{x \to 2} 5x$

2. $\lim\limits_{x \to -5} -3x$

3. $\lim\limits_{x \to 2} (3x^2 - 2x + 5)$

4. $\lim\limits_{x \to 10} 98$

5. $\lim\limits_{x \to 1} (x - 1)^3$

6. $\lim\limits_{x \to -1} \dfrac{3x + 2}{x + 5}$

7. $\lim\limits_{x \to 7} \sqrt{x - 3}$

8. $\lim\limits_{x \to -2} (-2x^2)(3x + 4)$

9. $\lim\limits_{x \to 1} \dfrac{x^2 - 1}{x - 1}$

10. $\lim\limits_{x \to 9} \dfrac{\sqrt{x} - 3}{x - 9}$

11. $\lim\limits_{x \to 1} f(x)$, where $f(x) = \begin{cases} x + 3, & x \le 1 \\ x, & x > 1 \end{cases}$

12. $\lim\limits_{x \to 0} f(x)$, where $f(x) = \begin{cases} 2x^2, & x \ge 0 \\ x, & x < 0 \end{cases}$

13. $\lim\limits_{x \to 3} f(x)$, where $f(x) = \begin{cases} |x - 2|, & x \le 3 \\ \frac{1}{3}x, & x > 3 \end{cases}$

14. $\lim\limits_{x \to +\infty} \dfrac{5x}{3x - 1}$

15. $\lim\limits_{x \to -\infty} \dfrac{7x - 3}{1 - 4x}$

16. $\lim\limits_{x \to +\infty} \dfrac{2x^2 - x + 10}{3x^2 + 4}$

17. $\lim\limits_{x \to -\infty} \dfrac{x^3 + x - 1}{2x^4 + 4}$

18. $\lim\limits_{x \to +\infty} \dfrac{x^3 + 2x}{x^2 + 1}$

19. $\lim\limits_{x \to -\infty} \dfrac{x^4 + 3x + 1}{x^2 + 5}$

20. $\lim\limits_{x \to +\infty} [\sqrt{x} - \sqrt{x - 1}]$

In Problems 21 through 30, find the values of x at which the given function is discontinuous.

21. $f(x) = \dfrac{3}{x - 2}$

22. $f(x) = \dfrac{4}{x + 1}$

23. $f(x) = \dfrac{1}{x^2 - 3x + 2}$

24. $f(x) = \dfrac{3}{x^2 + 3}$

25. $f(x) = \begin{cases} 1, & x \ge 3 \\ -1, & x < 3 \end{cases}$

26. $f(x) = \begin{cases} x^2, & x \ne 2 \\ 6, & x = 2 \end{cases}$

27. $f(x) = \begin{cases} 1, & x > 2 \\ \dfrac{1}{x}, & -2 \le x \le 2 \\ -1, & x < -2 \end{cases}$ 28. $f(x) = [\![x]\!]$

29. $f(x) = [\![x - 1]\!]$ 30. $f(x) = \begin{cases} \dfrac{2}{x}, & x \ne 0 \\ 3, & x = 0 \end{cases}$

Numerical Applications Using Calculators

A pocket calculator can be used to find the limit of a function. Consider the following example:

Example 3.26 $\lim\limits_{x \to 2} (5x - 3)$

We shall use the calculator to compute $\lim\limits_{x \to 2-} (5x - 3)$.

Table 3.1

x	$5x$	$5x - 3$
1.9	9.5	6.5
1.99	9.95	6.95
1.999	9.995	6.995
1.9999	9.9995	6.9995
1.99999	9.99995	6.99995
1.999999	9.999995	6.999995
1.9999999	9.9999995	6.9999995

From Table 3.1 we can ascertain that $\lim\limits_{x \to 2-} (5x - 3) = 7$. We can also compute $\lim\limits_{x \to 2+} (5x - 3)$ by using the calculator.

Table 3.2

x	$5x$	$5x - 3$
2.1	10.5	7.5
2.01	10.05	7.05
2.001	10.005	7.005
2.0001	10.0005	7.0005
2.00001	10.00005	7.00005
2.000001	10.000005	7.000005
2.0000001	10.0000005	7.0000005

Table 3.2 informs us that $\lim\limits_{x \to 2+} (5x - 3) = 7$. Therefore

$\lim\limits_{x \to 2} 5x - 3 = 7$.

Example 3.27 Find $\lim\limits_{x \to 3} \dfrac{x^2 - 9}{x - 3}$

First we use the calculator to find

$$\lim_{X \to 3^-} \frac{x^2 - 9}{x - 3}$$

Table 3.3

x	x^2	$x^2 - 9$	$x - 3$	$\dfrac{x^2 - 9}{x - 3}$
2.9	8.41	−0.59	−0.1	5.9
2.99	8.9401	−0.0599	−0.01	5.99
2.999	8.994001	−0.005999	−0.001	5.999
2.9999	8.99940001	−0.00059999	−0.0001	5.9999
2.99999	8.99994	−0.00006	−0.00001	5.99999

Hence

$$\lim_{x \to 3^-} \frac{x^2 - 9}{x - 3} = 6$$

and

$$\lim_{x \to 3^+} \frac{x^2 - 9}{x - 3}$$

can be found from Table 3.4.

Table 3.4

x	x^2	$x^2 - 9$	$x - 3$	$\dfrac{x^2 - 9}{x - 3}$
3.1	9.61	0.61	0.1	6.1
3.01	9.0601	0.0601	0.01	6.01
3.001	9.006001	0.006001	0.001	6.001
3.0001	9.00060001	0.00060001	0.0001	6.0001
3.00001	9.00006	0.00006	0.00001	6.00001

Hence

$$\lim_{x \to 3^+} \frac{x^2 - 9}{x - 3} = 6$$

Therefore

$$\lim_{x \to 3} \frac{x^2 - 9}{x - 3} = 6$$

Example 3.28 Find $\lim\limits_{x \to \infty} \dfrac{x^2 + 5}{2x^2 - 3x + 1}$

From Table 3.5, it seems reasonable to conclude that

$$\lim_{x \to \infty} \frac{x^2 + 5}{2x^2 - 3x + 1} = 0.5$$

Table 3.5

x	$3x$	x^2	$2x^2$	$x^2 + 5$	$2x^2 - 3x + 1$	$\dfrac{x^2 + 5}{2x^2 - 3x + 1}$
10	30	100	200	105	171	0.614035
100	300	10000	20000	10005	19701	0.507842
1000	3000	1000000	2000000	1000005	1997001	0.500753
10000	30000	100000000	200000000	100000005	199970001	0.500075

In Problems 1 through 12 use a pocket calculator to find the indicated limits, if they exist.

Exercises

1. $\lim\limits_{x \to 0} 7x + 10$

2. $\lim\limits_{x \to 1} 2x^2 - x + 7$

3. $\lim\limits_{x \to 2} \dfrac{x^2 - 5x + 6}{x - 2}$

4. $\lim\limits_{x \to 3} x^3 - 27$

5. $\lim\limits_{x \to +\infty} \dfrac{x^2 + 3x + 7}{3x^2 - x + 1}$

6. $\lim\limits_{x \to -4} \sqrt{x^2 + 2x - 5}$

7. $\lim\limits_{x \to 3} \sqrt{\dfrac{x - 1}{x + 2}}$

8. $\lim\limits_{x \to 4} f(x)$, where $f(x) = \begin{cases} x - 2 & \text{if } x \le 4 \\ x^2 & \text{if } x > 4 \end{cases}$

9. $\lim\limits_{x \to 1} \dfrac{x - 1}{x^2 - 1}$

10. $\lim\limits_{x \to -3} f(x)$, where $f(x) = \begin{cases} 2x - 1 & \text{if } x > -3 \\ x + 3 & \text{if } \quad < -3 \\ 5 & \text{if } x = -3 \end{cases}$

11. $\lim\limits_{x \to \infty} \dfrac{-2x + 11}{5x^2 + 7x - 1}$

12. $\lim\limits_{x \to \infty} (\sqrt{x^2 + x + 1} - \sqrt{x^2 + 1})$

In Problems 13 through 16 use the calculator to determine if each of the following functions is continuous at the indicated values:

13. $f(x) = x^5 - 3x^2 + 7$ at $c = -2$

14. $f(x) = \begin{cases} \dfrac{2x^2 + 5x - 3}{2x - 1}, & x \ne \frac{1}{2} \\ 0, & x = \frac{1}{2} \end{cases}$ at $c = \frac{1}{2}$

15. $f(x) = \begin{cases} \dfrac{(x - 1)(x + 1)(x + 2)}{(x + 1)(x + 2)} & \text{if } x \ne -1, -2 \\ -2 & \text{if } x = -1 \\ 0 & \text{if } x = -2 \end{cases}$ at $c = -1, -2$

16. $f(x) = \begin{cases} \sqrt{x - 1} & \text{if } x \ge 1 \\ 0 & \text{if } x < 1 \end{cases}$ at $c = 0, 1$

The Derivative

4.1 The Derivative

In this chapter we shall study one of the most basic concepts in the development of calculus: the *derivative* of a function. Invented by Newton, the derivative has become a very valuable tool in both the physical and social sciences. Before we formulate a definition of this concept, we shall illustrate a few concepts needed to build up the definition. In later sections we shall indicate how the derivative is applied in various disciplines and develop some of its basic properties.

Consider the function f defined by

$$f(x) = x^2$$

Isaac Newton

Isaac Newton (1642–1727) was born on Christmas Day at Woolsthorpe, England. Newton's father died before the frail child's birth. Newton received his A.B. in 1664 at Trinity College in Cambridge. The Great Plague (1664–1665) forced Newton to return to Woolsthorpe to meditate for two years, during which time he invented calculus, proved that light is composed of different colors, and discovered the universal law of gravitation. In 1669 he became professor of mathematics at Trinity. He constructed a telescope and observed the satellites of Jupiter, and in 1686 he published (at Halley's expense) his masterpiece, *The Mathematical Principles of Natural Philosophy*.

Newton had a strong interest in theology and in alchemy. In addition, he spent some time in Parliament and was made Master of the Mint in 1699. Four years later he was elected President of the Royal Society and was knighted by Queen Anne.

When $x = 10$, we get $f(10) = 10^2 = 100$. Now suppose that we increase the value of x from 10 to 10.1. Then $f(10.1) = (10.1)^2 = 102.01$. The change in the value of x is

$$h = 10.1 - 10 = 0.1$$

and the corresponding change in $f(x)$ is

$$f(10.1) - f(10) = 102.01 - 100 = 2.01$$

The ratio of the change in $f(x)$ to the change in x is called the *difference quotient*:

Difference quotient

$$\text{difference quotient} = \frac{\text{change in } f(x)}{\text{change in } x} \qquad (4.1)$$

$$= \frac{f(x+h) - f(x)}{(x+h) - x} = \frac{f(x+h) - f(x)}{h}$$

This result is the *average rate of change* of f from x to $x + h$. In the preceding case, we have

$$\text{difference quotient} = \frac{f(10.1) - f(10)}{0.1} = 20.1$$

If the value of x changes from 10 to 9.9, the change in x is $h = 9.9 - 10 = -0.1$, a negative number. The corresponding change in $f(x)$ is

$$f(9.9) - f(10) = 98.01 - 100 = -1.99$$

so that

$$\text{different quotient} = \frac{f(9.9) - f(10)}{-0.1} = \frac{-1.99}{-0.1} = 19.9$$

Table 4.1 contains more difference quotients corresponding to different changes in the values of x. It is clear from Table 4.1 that as the change in x approaches zero (through positive or negative values), the difference quotient approaches the number 20.

Table 4.1

h(change in x)	$f(10 + h) - f(10)$	$\frac{f(10+h) - f(10)}{h}$
$h = 10.01 - 10 = 0.01$	0.2001	20.01
$h = 10.001 - 10 = 0.001$	0.020001	20.001
$h = 10.0001 - 10 = 0.0001$	0.00200001	20.0001
$h = 9.99 - 10 = -0.01$	-0.1999	19.99
$h = 9.999 - 10 = -0.001$	-0.019999	19.999
$h = 9.9999 - 10 = -0.0001$	-0.00199999	19.9999

We observe that

$$\lim_{h \to 0} \frac{f(10+h) - f(10)}{h} = \lim_{h \to 0} \frac{(10+h)^2 - 10^2}{h}$$

$$= \lim_{h \to 0} \frac{100 + 20h + h^2 - 100}{h}$$

$$= \lim_{h \to 0} (20 + h) = 20$$

Let f be a function and x be a fixed number in an open interval I in the domain of f. If we choose h such that $x + h$ is in I, then we have a new function F, where

F depends on x as well.

$$F(h) = \frac{f(x + h) - f(x)}{h}$$

Here F is not defined for $h = 0$. Furthermore $\lim_{h \to 0} F(h)$ may not exist. In general, this limit, if it exists, will depend upon x, and consequently will be a function of x. We call this new function the *derived* function of f or simply the *derivative* of f.

We now give a formal definition of the derivative of a function.

▶

Derivative.

Definition 4.1 *Let f be a function and x be a point in an open interval in the domain of f. The* derivative *of f at the point x, denoted by $f'(x)$, is defined by*

Note that h ≠ 0.

$$f'(x) = \lim_{h \to 0} \frac{f(x + h) - f(x)}{h} \qquad (4.2)$$

f′ is a new function obtained from f by the process of differentiation.

provided that the limit exists.

If the limit (4.2) exists, the function f is said to be differentiable at x.

Notation   The derivative notation f' was introduced by Lagrange. For a function f defined by the equation $y = f(x)$, Leibniz denoted the derivative of f by

$$\frac{dy}{dx} \quad \text{and} \quad \frac{df}{dx}$$

read "the derivative of y (or f) with respect to x." Other notations used are

$$Dy, \quad Df, \quad y'$$

Joseph Louis Lagrange

Joseph Louis Lagrange (1736–1813) was born in Turin, Italy. He did important mathematical work in partial differential equations, and he contributed immensely to the development of the calculus of variations.

At the age of 19, Lagrange became professor of mathematics at the Royal Artillery School of Turin. At 28 he won the coveted prize of mathematics of the Academy of Sciences of Paris, a prize he was to win several times in the succeeding years. In 1766 at the invitation of Frederick II of Prussia, he replaced Euler as director of the mathematics department of the Academy of Berlin. In 1787 at the invitation of Louis XVI, Lagrange went to Paris to work at the French Academy. Napoleon treated Lagrange as a friend, giving him the title of count and making him a senator and a high officer of the Legion of Honor.

When we wish to evaluate the derivative at a point c, we write

$$f'(c) = \frac{dy}{dx}\bigg|_{x=c} = \frac{df}{dx}\bigg|_{x=c} = y'(c) = Dy\big|_{x=c} = Df\big|_{x=c}$$

To find the derivative $f'(x)$ of a function $f(x)$, we perform the following steps:

Step 1   Find $f(x + h)$.

Step 2   Find the difference: $f(x + h) - f(x)$.

Step 3   Find the difference quotient: $\dfrac{f(x + h) - f(x)}{h}$

Step 4   If this limit exists, $f'(x) = \lim\limits_{h \to 0} \dfrac{f(x + h) - f(x)}{h}$

Example 4.1   Let f be the function defined by

$$f(x) = x^3$$

Find $f'(x)$, $f'(2)$, $f'(-4)$, and $f'(c)$.

SOLUTION

Step 1   Find $f(x + h)$. In this example,

$$f(x + h) = (x + h)^3 = x^3 + 3x^2h + 3xh^2 + h^3$$

Note $(a + b)^3 = a^3 + 3a^2b + 3ab^2 + b^3$

Step 2   Find the difference $f(x + h) - f(x)$. Here

$$f(x + h) - f(x) = (x^3 + 3x^2h + 3xh^2 + h^3) - x^3$$
$$= 3x^2h + 3xh^2 + h^3$$

Step 3   Find the difference quotient.

$$\frac{f(x + h) - f(x)}{h} = \frac{3x^2h + 3xh^2 + h^3}{h} = 3x^2 + 3xh + h^2$$

Step 4   See if $\lim\limits_{h \to 0} \dfrac{f(x + h) - f(x)}{h}$ exists. Here

$$f'(x) = \lim_{h \to 0} \frac{f(x + h) - f(x)}{h} = \lim_{h \to 0}(3x^2 + 3xh + h^2) = 3x^2$$

Hence if $f(x) = x^3$, then the derivative $f'(x) = 3x^2$. Therefore for $x = 2$, we have

$$f'(2) = 3(2)^2 = 12$$

Similarly,

$$f'(-4) = 3(-4)^2 = 48 \quad \text{and} \quad f'(c) = 3(c)^2 = 3c^2$$

To avoid excessive writing, we may in fact go directly to Step 4. We do this in the remaining examples.

Example 4.2 Let f be the function defined by

$$f(x) = \frac{1}{x}, \qquad x > 0$$

Find $f'(x)$.

SOLUTION Using Definition 4.1, we have for $h \neq 0$,

Since $x + h$ must be in the domain of f, then $x + h \neq 0$. (Why?)

$$f'(x) = \lim_{h \to 0} \frac{f(x+h) - f(x)}{h}$$

$$= \lim_{h \to 0} \frac{1}{h} \left(\frac{1}{x+h} - \frac{1}{x} \right) = \lim_{h \to 0} \frac{-1}{x(x+h)} = -\frac{1}{x^2}$$

Note
$$\frac{1}{x+h} - \frac{1}{x} = \frac{x - (x+h)}{x(x+h)}$$
$$= \frac{-h}{x(x+h)}$$

Example 4.3 Let f be the function defined by $f(x) = \sqrt{x}$. Find $f'(x)$.

SOLUTION We have

$$f'(x) = \lim_{h \to 0} \frac{f(x+h) - f(x)}{h} = \lim_{h \to 0} \frac{1}{h} (\sqrt{x+h} - \sqrt{x})$$

We observe that the indicated limit yields the indeterminate form 0/0. A useful trick at this point is to rationalize the numerator:

$$\sqrt{x+h} - \sqrt{x} =$$

$$(\sqrt{x+h} - \sqrt{x}) \left(\frac{\sqrt{x+h} + \sqrt{x}}{\sqrt{x+h} + \sqrt{x}} \right)$$

$$= \frac{(\sqrt{x+h})^2 - (\sqrt{x})^2}{\sqrt{x+h} + \sqrt{x}}$$

$$= \frac{h}{\sqrt{x+h} + \sqrt{x}}$$

$$= \lim_{h \to 0} \frac{1}{h} \left(\frac{h}{\sqrt{x+h} + \sqrt{x}} \right) = \lim_{h \to 0} \frac{1}{\sqrt{x+h} + \sqrt{x}} = \frac{1}{2\sqrt{x}}$$

Example 4.4 Find $f'(x)$ if $f(x) = 3x^2 + 4x + 8$.

SOLUTION We have

$$f'(x) = \lim_{h \to 0} \frac{f(x+h) - f(x)}{h}, \qquad h \neq 0$$

$$= \lim_{h \to 0} \frac{1}{h} \{ [3(x+h)^2 + 4(x+h) + 8] - [3x^2 + 4x + 8] \}$$

$$= \lim_{h \to 0} \frac{1}{h} [3x^2 + 6xh + 3h^2 + 4x + 4h + 8 - 3x^2 - 4x - 8]$$

$$= \lim_{h \to 0} \frac{1}{h} (6xh + 3h^2 + 4h) = \lim_{h \to 0} (6x + 3h + 4) = 6x + 4$$

Example 4.5 Let $f(x) = |x|$. Show that $f'(0)$ does not exist.

SOLUTION We point out that the function f is defined for all values of x. Now for $h \neq 0$, we have

$$f'(0) = \lim_{h \to 0} \frac{f(0+h) - f(0)}{h} = \lim_{h \to 0} \frac{|h|}{h}$$

We observe from the definition of the absolute value that

$$\lim_{h \to 0^+} \frac{|h|}{h} = \lim_{h \to 0^+} \frac{h}{h} = 1 \quad \text{and} \quad \lim_{h \to 0^-} \frac{|h|}{h} = \lim_{h \to 0^-} \frac{-h}{h} = -1$$

(See Figure 4.1.) Hence the $\lim_{h \to 0} |h|/h$ does not exist. Consequently, $f'(0)$ does not exist.

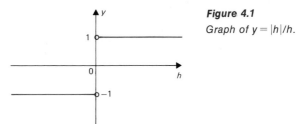

Figure 4.1
Graph of $y = |h|/h$.

In Example 4.5, we note that the function $f(x) = |x|$ is continuous at $x = 0$ (see Figure 4.2), but is not differentiable at $x = 0$. However, the converse of this statement is always true. We have the following theorem which we shall use in later sections.

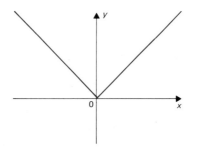

Figure 4.2
Graph of $y = |x|$.

Theorem 4.1   *If a function f is differentiable at $x = a$, then it is continuous at $x = a$.*   ◀

PROOF   We are given that

$$f'(a) = \lim_{h \to 0} \frac{f(a + h) - f(a)}{h} \tag{4.3}$$

exists, which means that $f'(a)$ is a fixed real number.

Now if we can show that $\lim_{h \to 0} f(a + h) = f(a)$, then by the definition of continuity, we can conclude that f is continuous at a. We have

$$f(a + h) = \frac{f(a + h) - f(a)}{h} h + f(a)$$

Thus

$$\lim_{h \to 0} f(a + h) = \lim_{h \to 0} \left[\frac{f(a + h) - f(a)}{h} h + f(a) \right]$$

$$= \lim_{h \to 0} \frac{f(a + h) - f(a)}{h} \cdot \lim_{h \to 0} h + f(a) \qquad \text{(Why?)}$$

$$= f'(a) \cdot 0 + f(a) = f(a)$$

This proves the theorem.

From the preceding examples and Theorem 4.1, we have the following information:

(i) If $f'(a)$ exists, then f is continuous at $x = a$.

(ii) If f is discontinuous at $x = a$, then $f'(a)$ does not exist.

(iii) If f is continuous at $x = a$, then $f'(a)$ may exist (Example 4.1) or may not exist (Example 4.5).

Exercises 4.1   In Problems 1 through 14, find $f'(x)$ and indicate the domain in which the function f is differentiable.

1. $f(x) = x$

2. $f(x) = x^2$

3. $f(x) = 5x^2$

4. $f(x) = 7$

5. $f(x) = 3x^2 + 2$

6. $f(x) = x^2 + 3x + 6$

7. $f(x) = \dfrac{2}{x}$

8. $f(x) = \dfrac{1}{x + 3}$

9. $f(x) = x + \dfrac{1}{x}$

10. $f(x) = \dfrac{x}{x^2 + 1}$

11. $f(x) = \dfrac{1}{\sqrt{x}}$

12. $f(x) = x - \sqrt{x}$

13. $f(x) = \sqrt{x^2 + 1}$

14. $f(x) = x^3 + \dfrac{1}{\sqrt{x}}$

In Problems 15 through 20, find the values of x for which $f'(x) = 0$, if any. Sketch a graph of f in each case.

15. $f(x) = 3x^2$

16. $f(x) = 2x^2 - 4x$

17. $f(x) = 2x^2 + 4x$

18. $f(x) = \sqrt{x + 1}$

19. $f(x) = x^3$

20. $f(x) = \dfrac{1}{x}$

21. Let $f(x) = |x - 2|$. Show that $f'(2)$ does not exist.

22. Let $f(x) = \sqrt[3]{x}$. Show that $f'(0)$ does not exist.

4.2 Some Applications

The notion of the derivative leads to the definition of the rate of change of a function. There are numerous applications of this notion in a wide variety of disciplines including the biological, social, and physical sciences, as well as economic theory. Here we shall illustrate some of these applications.

Example 4.6 The derivative in ecology Suppose a lake is being polluted by a liquid chemical. At any time t (minutes), the quantity of chemical in the lake is $Q(t)$ (gal). Hence Q is a function of time. As the time changes from t to $t + h$ ($h > 0$), the quantity of chemical in the lake changes by an amount

We assume that no liquid is leaving the lake. We disregard loss of water due to evaporation, or the addition of water due to rain, etc.

$$Q(t + h) - Q(t)$$

and the average rate of change of Q from t to $t + h$ is then

$$\frac{Q(t + h) - Q(t)}{(t + h) - t} = \frac{Q(t + h) - Q(t)}{h}, \qquad h \neq 0$$

The student should recognize this result as the difference quotient. Now we define the *instantaneous rate of change* (or simply the rate of change) of Q at time t to be

$$Q'(t) = \lim_{h \to 0} \frac{Q(t + h) - Q(t)}{h}$$

which is simply the derivative of Q with respect to t. At a given time t^*, the number $Q'(t^*)$ represents the rate at which the amount of undesirable chemical in the lake is increasing.

Example 4.7 The derivative in economic theory The total cost C of producing a commodity is often found to be dependent on the number x of items produced; that is, C is a function of x. If the number of items produced is changed from x to $x + h$, then the average rate of change of cost is given by

In reality $x \in N$. However, in analysis of this type, the function C is approximated by a continuous function of x.

$$\frac{C(x + h) - C(x)}{(x + h) - x} = \frac{C(x + h) - C(x)}{h}, \qquad h \neq 0$$

and the rate of change of the cost is given by

$$C'(x) = \lim_{h \to 0} \frac{C(x + h) - C(x)}{h}$$

This is called the *marginal cost*, which is simply the derivative of the cost C with respect to x.

We shall assume that the manufacturer is interested in revenue and profit. To sell the output of x items, the manufacturer charges a price p (dollars) per item. It is obvious that the price will depend on the number of items that can be sold. The *total revenue* is given by

marginal cost $= C'(x) =$
$\lim_{h \to 0} \dfrac{C(x + h) - C(x)}{h}$, $\quad h \neq 0$

Note that in general the higher the price, the smaller the number of items sold.

$$R(x) = xp(x)$$

and the *marginal revenue* (or the increase of revenue per unit increase in output) is given by

$$\text{marginal revenue} = \lim_{h \to 0} \frac{R(x + h) - R(x)}{h}, \qquad h \neq 0$$

$$= R'(x)$$

It is easy to see that the profit P is given by

$$P(x) = R(x) - C(x)$$

In the next chapter the reader will be shown how the manufacturer who has had an elementary course in calculus can then find out how to maximize profit by using certain properties of the derivative.

Example 4.8   The derivative in the physical sciences   Suppose the distance s traveled by an object moving along a straight line is a function of time, say,

$$s = f(t)$$

where t is time measured in seconds and s is the distance traveled in meters. The average rate of change of s from t to $t + h$ is

$$V_{av}(h) = \frac{f(t + h) - f(t)}{h}, \qquad h \neq 0$$

and is called the *average velocity* V_{av}. The instantaneous velocity (or the rate of change of s at time t) is

$$V = \lim_{h \to 0} V_{av}(h) = \lim_{h \to 0} \frac{f(t + h) - f(t)}{h}$$

For example, in an experiment conducted above the moon's surface, it is found that an object thrown from a spacecraft travels according to the law

$$s = kt^2 + t$$

where k is a constant, s (meters) is the distance of the object from the spacecraft, and t (seconds) is the time. As the time changes from t to $t + h$, we find that

$$V_{av} = \frac{f(t + h) - f(t)}{h} = \frac{[k(t + h)^2 + (t + h)] - [kt^2 + t]}{h}$$

$$= 2kt + kh + 1$$

Starting at $t = 0$ with $h = 0.5$, we obtain

$$V_{av} = 2k(0) + k(0.5) + 1 = 0.5k + 1$$

Table 4.2 illustrates the changes in the average velocity over shorter and shorter time intervals.

Table 4.2

$t + h$	h	$V_{av} = 2kt + kh + 1$
$0 + 0.5$	0.5	$k(0.5) + 1$
$0 + 0.1$	0.1	$k(0.1) + 1$
$0 + 0.05$	0.05	$k(0.05) + 1$
$0 + 0.01$	0.01	$k(0.01) + 1$
$0 + 0.001$	0.001	$k(0.001) + 1$

It is clear that as h approaches zero, the average velocity approaches 1. Thus, 1 is the instantaneous velocity at $t = 0$. That is,

$$V(t) = \lim_{h \to 0} V_{\text{av}}(h) = \lim_{h \to 0} (2kt + kh + 1) = 2kt + 1$$

At $t = 0$, we have

$$V(0) = 2k(0) + 1 = 1$$

This result implies that the object left the spacecraft with an initial instantaneous velocity of 1 m/sec.

Example 4.9   The derivative in the biological sciences   Consider a colony of bacteria introduced into a food supply in a jar. At the beginning there is no change in the number of bacteria (period of adaptation). Then there occurs a rapid increase in the number of bacteria (period of reproduction). Once the food supply is exhausted, the bacteria begin to die. In such a model, the number N of live bacteria is dependent on time t.

Now, the average rate of change of N from t to $t + h$ is

$$\frac{N(t + h) - N(t)}{h}, \qquad h \neq 0$$

and the instantaneous rate of change of N at the point t is

$$N'(t) = \lim_{h \to 0} \frac{N(t + h) - N(t)}{h}$$

In practice, it is often deduced from experimental data that the instantaneous rate of increase at time t is directly proportional to the number present. Thus we have the equation

$$N'(t) = kN(t)$$

where k is a constant of proportionality. Equations of this type, called differential equations, will be discussed in Chapter 8.

For studies of this type, the change in N is assumed to be continuous. In reality, the change in N occurs in jumps, and the function describing the model is usually approximated by a continuous function.

Exercises 4.2

1. A factory at the east end of the city of Polluteville starts operation at 7:00 A.M. Smoke emitted from the factory travels westward according to the law

    $$s(t) = 100t$$

    where s (ft) is the distance from the edge of the smoke cloud to the factory at t min after 7:00 A.M.

    (a)   Find the average velocity of the edge of the smoke cloud during the time interval h.

Figure 4.3

A

⟵— 8 miles —⟶

    (b)   Find the instantaneous velocity at $t = 0$.

    (c)   At what time will the smoke cloud reach the west end of town?

2.   If in Problem 1 we have

$$s(t) = 3t^2 + 50t$$

answer the following questions.

    (a)   At what velocity is the smoke moving when $t = 0$? $t = 10$ min? $t = 20$ min?

    (b)   When the cloud reaches point A, the instantaneous velocity is 650 ft/min. At what time will this occur?

    (c)   How far is A from the factory?

3.   Consider Example 4.8 and let

$$s(t) = 3t^2 + t$$

    (a)   Starting at $t = 1$ sec, construct a table similar to Table 4.1.

    (b)   Find the instantaneous velocity of the object at $t = 1$ sec.

4.   Given that the total cost C (dollars) of producing x units of a product is $C(x) = 2x + 160$.

    (a)   What is the average cost per unit if 80 units are made?

    (b)   What is the average rate of change of cost?

    (c)   What is the marginal cost of the 50th unit? 80th unit?

5.   Repeat Problem 4 with $C(x) = 2x^2$.

6.   Given that the total cost C of making x gal of oil is $C(x) = 20\sqrt{x} + 5$.

    (a)   Find the marginal cost at 160 gal output.

    (b)   At what level of output is the marginal cost $0.20? $0.40? [*Hint:* Solve for x such that $C'(x) = \$0.20$.]

7.   A manufacturer sells x units of a product at a price of p dollars per unit, where $p(x) = 100 + (x/50)$.

    (a)   Find the total revenue if the total output is 2500 units.

    (b)   Find the marginal revenue when the output is 500 units, 1000 units, 2500 units.

    (c)   At what level of output is the marginal revenue 150 dollars per unit?

8.   A colony of bacteria is sprayed with a bactericidal agent. Suppose that N represents the number of viable bacteria per milliliter remaining t min after spraying, and $N(t) = 10^6(40 - t)^2$.

    (a)   What is the average rate of decrease in the number of viable bacteria during the first 10 min? 20 min?

    (b)   What is the instantaneous rate of decrease in N at the end of 10 min? 20 min?

    (c)   At what time t^* would all the bacteria be dead?

    (d)   Find $N'(t^*)$.

4.3 Rules of Differentiation

We have introduced the notion of the derivative of a function and indicated some of its applications. We have also calculated the derivative of some simple functions by using the definition. Evidently, finding the derivative of a function by direct use of the definition can be tedious. We shall, therefore, develop standard formulas to simplify the task of finding derivatives.

Theorem 4.2 *The derivative of a constant function is the constant function 0.*

◀ *If $f(x) = c$ for all x, then $f'(x) = 0$ for all x.*

PROOF   A constant function is defined by

$$f(x) = c$$

for all $x \in R$, where c is a constant. Application of Definition 4.1 yields

$$f'(x) = \lim_{h \to 0} \frac{f(x+h) - f(x)}{h}, \qquad h \neq 0$$

Note   Since $f(x) = c$ for every x, $f(x+h) = c$.

$$= \lim_{h \to 0} \frac{c - c}{h} = \lim_{h \to 0} \frac{0}{h} = 0$$

This proves Theorem 4.2.

Theorem 4.3 *If f is the function defined by*

$$f(x) = x^n \tag{4.4}$$

◀

where n is a positive integer, then

$$f'(x) = nx^{n-1} \tag{4.5}$$

PROOF   First we shall consider specific values of n. For $n = 1$, we have $f(x) = x$ and, by Definition 4.1,

$$f'(x) = \lim_{h \to 0} \frac{(x+h) - x}{h} = \lim_{h \to 0} \frac{h}{h} = \lim_{h \to 0} 1 = 1$$

Here we have $f(x) = x^1$ and $f'(x) = 1 \cdot x^{1-1} = x^0 = 1$.

For $n = 2$, we have $f(x) = x^2$ and, by Definition 4.1,

$$f'(x) = \lim_{h \to 0} \frac{(x+h)^2 - x^2}{h} = \lim_{h \to 0} \frac{x^2 + 2xh + h^2 - x^2}{h}$$

$$= \lim_{h \to 0} (2x + h) = 2x$$

For $n = 3$, we have $f(x) = x^3$ and, by Definition 4.1,

$$f'(x) = \lim_{h \to 0} \frac{(x+h)^3 - x^3}{h} = \lim_{h \to 0} \frac{x^3 + 3x^2h + 3xh^2 + h^3 - x^3}{h}$$

$$= \lim_{h \to 0} (3x^2 + 3xh + h^2) = 3x^2$$

For n, we have

This is an application of the binomial theorem (see App. 1).

$(x + h)^n = x^n + nx^{n-1}h$

$+ \dfrac{n(n-1)}{2} x^{n-2}h^2 + \cdots$

$+ \dfrac{n!}{(n-k)!\,k!} x^{n-k}h^k + \cdots + h^n$

where for any positive integer k, k! (read k factorial) is defined as

$k! = k \cdot (k-1) \cdot (k-2) \cdots 3 \cdot 2 \cdot 1$

Thus

$5! = 5 \cdot 4 \cdot 3 \cdot 2 \cdot 1 = 120$

The binomial theorem was first proved by Isaac Newton.

$$f'(x) = \lim_{h \to 0} \frac{(x+h)^n - x^n}{h}$$

$$= \lim_{h \to 0} \frac{1}{h}\left[x^n + nx^{n-1}h + \frac{n(n-1)}{2}x^{n-2}h^2 + \cdots + h^n - x^n \right]$$

$$= \lim_{h \to 0} \left[nx^{n-1} + \frac{n(n-1)}{2}x^{n-2}h + \cdots + h^{n-1} \right]$$

Since all terms but the first contain a power of h, we obtain

$$f'(x) = nx^{n-1}$$

This result proves Theorem 4.3.

It can be shown that Theorem 4.3 holds for any real number n. We shall assume this result. The following generalization of Theorem 4.2 can be obtained:

▶ **Theorem 4.4** *If f is the function defined by*

$$f(x) = cx^n \tag{4.6}$$

where n is a real number and c is a constant, then

$$f'(x) = cnx^{n-1} \tag{4.7}$$

Example 4.10 Find $f'(x)$ for each of the following functions:

(a) $f(x) = x^{13}$ (b) $f(x) = 2x^{-4}$

(c) $f(x) = x^{2/3}$ (d) $f(x) = 5^3$

SOLUTION   By direct application of the preceding theorems, we have

(a) $f'(x) = 13x^{13-1} = 13x^{12}$

(b) $f'(x) = 2(-4)\,x^{-4-1} = -8x^{-5}$

(c) $f'(x) = \frac{2}{3}x^{2/3-1} = \frac{2}{3}x^{-1/3}$

(d) $f'(x) = 0$

Note that in part (d) f is a constant function, and so $f'(x) = 0$ (not $3 \cdot 5^2$).

Now we give another useful result.

▶ **Theorem 4.5** *If u and v are functions differentiable at x and if the function f is defined by*

The derivative of the sum is the sum of the derivatives.

$$f(x) = u(x) + v(x) \tag{4.8}$$

then f is differentiable at x and

$$f'(x) = u'(x) + v'(x) \tag{4.9}$$

PROOF   We have

$$f'(x) = \lim_{h \to 0} \frac{f(x+h) - f(x)}{h}$$

$$= \lim_{h \to 0} \frac{1}{h} \left[u(x+h) + v(x+h) - u(x) - v(x) \right]$$

$$= \lim_{h \to 0} \frac{1}{h} \left[u(x+h) - u(x) + v(x+h) - v(x) \right]$$

$$= \lim_{h \to 0} \left[\frac{u(x+h) - u(x)}{h} + \frac{v(x+h) - v(x)}{h} \right] \qquad \textit{Here we are using Theorem 3.4.}$$

$$= u'(x) + v'(x)$$

Example 4.11   Find the derivative of $f(x) = 5x^2 + 4x^7$.

SOLUTION   Application of Theorems 4.4 and 4.5 yields

$$f'(x) = (5)(2)\, x + (4)(7)\, x^6 = 10x + 28x^6$$

By an extension of Theorem 4.5, the derivative of the sum of any finite number of differentiable functions at the point x is the sum of the derivatives of the functions. In other words, if

$$f(x) = u_1(x) + u_2(x) + \cdots + u_n(x) \qquad (4.10)$$

then

$$f'(x) = u'_1(x) + u'_2(x) + \cdots + u'_n(x) \qquad (4.11)$$

Example 4.12   Suppose that the price per unit of a product with demand x is $p(x) = 3x^2 + 5x + 7$. Find the total revenue and the marginal revenue.

SOLUTION   We recall that the total revenue is given by

$$R(x) = xp(x) = x(3x^2 + 5x + 7) = 3x^3 + 5x^2 + 7x$$

and the marginal revenue is

$$R'(x) = (3)(3)\, x^2 + (5)(2)\, x + 7 = 9x^2 + 10x + 7 \qquad \textbf{Note}\;\; \textit{If } f(x) = 7x, \textit{ then } f'(x) =$$
$$(7)(1)\, x^0 = 7.$$

Example 4.13   A particle moves along a straight line according to the law

$$s(t) = -\tfrac{1}{4}t^2 + 5t \qquad (4.12)$$

where s is the distance measured in feet and t is the time measured in seconds.

(a)   Find the expression for the velocity of the particle at any instant t.

(b) At what distance from the starting point does the particle come to instantaneous rest?

SOLUTION

(a) We recall that $s'(t)$ gives the velocity of the particle at any instant t. From Equation (4.12), we have

$$s'(t) = -\tfrac{1}{2}t + 5 \qquad (4.13)$$

(b) Now the particle comes to instantaneous rest when the velocity is zero. Thus setting

$$-\tfrac{1}{2}t + 5 = 0$$

we obtain

$$t = 10$$

Hence the particle comes to rest 10 sec after the start and then reverses its direction. The distance traveled from the starting position in 10 sec may be obtained from Equation (4.12):

$$s(10) = -\tfrac{1}{4}(10)^2 + 5(10) = -25 + 50 = 25 \text{ ft}$$

Exercises 4.3   Use the theorems of this section to find the derivatives of each of the given functions in Problems 1 through 36.

1. $f(x) = 7x$

2. $f(x) = 3x^4$

3. $f(x) = -5x^2$

4. $f(x) = 2x^2 - 3x + 7$

5. $f(x) = \tfrac{1}{3}x^3 + \tfrac{3}{2}x^2$

6. $f(x) = x^3 + 7x^2 - 9x$

7. $f(x) = -4x^4 - 3x^3 + 1$

8. $f(x) = 5x^4 - 3x^3 - 10x^2 + 8x + 4$

9. $f(x) = 2x^{3/2}$

10. $f(x) = -\tfrac{3}{2}x^{1/2}$

11. $f(x) = \sqrt{x}$

12. $f(x) = 2\sqrt[3]{x}$

13. $f(x) = 2x^{-4}$

14. $f(x) = -3x^{-2}.$

15. $f(x) = -3x^{-1} + 1000$

16. $f(x) = -2\sqrt{x} + x^{3/2}$

17. $f(x) = 5x^{101} + 17x - 8$

18. $f(x) = (x + 3)^2$

19. $f(x) = \dfrac{1}{x}$

20. $f(x) = \dfrac{4}{x^3}$

21. $f(x) = \dfrac{1}{x} + \sqrt{x}$

22. $f(x) = \dfrac{1}{\sqrt{x}} - x^3$

23. $f(x) = \dfrac{1}{x^2} - 2$

24. $f(x) = \dfrac{x+1}{x}$ $\left[Hint: \dfrac{x+1}{x} = 1 + \dfrac{1}{x} \right]$

25. $f(t) = 3t^{1/2} - 9t^3$

26. $s(t) = \dfrac{t^2 - 10t - 3}{5}$

27. $w(s) = \left(s + \dfrac{1}{s}\right)^2$

28. $h(u) = \left(u - \dfrac{1}{u}\right)^2$

29. $f(x) = \left(x + \dfrac{1}{\sqrt{x}}\right)^2$

30. $w(s) = \left(s - \dfrac{1}{\sqrt{s}}\right)^3$

31. $f(x) = (x + a)(x + b)(x + c)$, where a, b, and c are constants

32. $f(x) = \dfrac{5x^3 + 3x^2 + 4x - 2}{x}$

33. $f(x) = \dfrac{5\sqrt{x} - 3x^2}{\sqrt{x}}$

34. $f(x) = \dfrac{x^{1/2} - x^{1/4}}{x^{1/3}}$

35. $f(x) = \dfrac{(x^2 + 3)^2}{2x^{3/2}}$

36. $f(x) = (x^2 + 3x + 5)(3x^2 - 2x + 4)$

37. Using the method of proof of Theorem 4.3, prove Theorem 4.4.

38. A brush fire in a valley spreads along a straight line according to the law

$$s(t) = \tfrac{3}{2}t - \tfrac{1}{4}t^2$$

where the distance s is measured in miles from a starting point and t is the time measured in hours.

(a) What is the instantaneous rate of spreading (velocity) of the fire at
     $t = 1$ hr? $t = 1.5$ hr? $t = 2$ hr?
(b) At what distance from the starting point does the fire stop spreading?
(c) If the fire spreads radially at the given rate, find the area of the circular region devastated by the fire.

39. Suppose that the total cost of making x units of a product is C dollars, where

$$C(x) = 20 + 18x^{2/3}$$

For what value of x will the marginal cost be $2?

40. If the total cost C of making x units of a product is

$$C(x) = 6x - 0.03x^2 + 0.00005x^3$$

construct a table showing the marginal cost at 50, 100, 150, 200, 250, and 300 units.

41. For what output will the marginal cost be $22 if the cost of making x units is C dollars, where

$$C(x) = 12x - 0.045x^2 + 0.0001\,\frac{x^3}{3}?$$

42. Suppose that W is the number of gallons of water in a pool being drained for cleaning, and t min after the draining starts

$$W(t) = 300(40 - t)^2$$

(a) Find the instantaneous rate at which the water is running out 5 min, 10 min, 20 min, and 30 min after the pool has started to drain.
(b) How many minutes after the pool has started to drain will the pool be empty?

43. Applying Newton's law of universal gravitation, we find that the force F exerted by the Earth on a spacecraft at a distance r from the center of the Earth is given by

$$F(r) = \frac{-k}{r^2}$$

where k is a positive number. Construct a table to show the instantaneous rate of change of F with respect to r when r is 4000 mi, 5000 mi, 10,000 mi, and 20,000 mi. (An approximate value for the radius of the Earth is 3959 mi.)

4.4 The Derivative of the Product and Quotient

In this section we develop formulas for the derivative of the product of two functions and for the quotient of two functions.

Derivative of a product ▶   **Theorem 4.6**   *Let $f = uv$ and suppose u and v are functions differentiable at x. Then f is differentiable at x and its derivative at x is given by*

$$f'(x) = u(x)\, v'(x) + u'(x)\, v(x)$$

PROOF   Setting up the difference quotient, we have

$$\frac{f(x+h) - f(x)}{h} = \frac{u(x+h)\, v(x+h) - u(x)\, v(x)}{h}$$

Here we have added and subtracted the term $\dfrac{v(x)\, u(x+h)}{h}$.

$$= \frac{u(x+h)\, v(x+h) - v(x)\, u(x+h) + v(x)\, u(x+h) - u(x)\, v(x)}{h}$$

$$= u(x+h)\left[\frac{v(x+h) - v(x)}{h}\right] + \left[\frac{u(x+h) - u(x)}{h}\right] v(x)$$

Now we are given that

$$\lim_{h \to 0} \frac{v(x+h) - v(x)}{h} = v'(x) \quad \text{and} \quad \lim_{h \to 0} \frac{u(x+h) - u(x)}{h} = u'(x)$$

Also, it follows from Theorem 4.1 that

$$\lim_{h \to 0} u(x+h) = u(x)$$

Since u is differentiable at x, u is continuous at x (see Theorem 4.1). Therefore

$$\lim_{z \to 0} u(z) = u(0)$$

Therefore

$$f'(x) = \lim_{h \to 0} \frac{f(x+h) - f(x)}{h}$$

$$= \lim_{h \to 0} \left\{ u(x+h) \left[\frac{v(x+h) - v(x)}{h}\right] + \left[\frac{u(x+h) - u(x)}{h}\right] v(x) \right\}$$

Here we have applied Theorem 4.1.

$$= u(x)\, v'(x) + u'(x)\, v(x)$$

Theorem 4.6 can be used to find the derivative of the product of three or more functions. For example, let

$$f = uvw$$

and suppose that u, v, and w are functions differentiable at x. Then

$$f'(x) = u(x)\, v(x)\, w'(x) + u(x)\, v'(x)\, w(x) + u'(x)\, v(x)\, w(x)$$

Example 4.14   It is found experimentally that the change in the temperature T of the body due to a dose of a drug is a function of the amount x of the drug administered. For a given drug, suppose that

$$T(x) = x^2(k_1 - k_2 x)$$

where k_1 and k_2 are positive constants.

(a)   Find the rate of change of T with respect to the amount of the drug administered.

(b)   For what values of x is the rate of change in T zero?

Analysis of this type is used to determine the exact dosage of a drug to be administered to give the greatest change in the reaction of the body with respect to a change in dosage.

SOLUTION

(a)   Let $u(x) = x^2$ and $v(x) = k_1 - k_2 x$. Then

$$T(x) = u(x)\, v(x)$$

Application of Theorem 4.6 yields

$$T'(x) = u(x)\, v'(x) + u'(x)\, v(x) = (x^2)(-k_2) + (2x)(k_1 - k_2 x)$$

$$= 2k_1 x - 3k_2 x^2$$

(b)   The rate of change of T with respect to x is zero when

$$2k_1 x - 3k_2 x^2 = 0$$

This occurs when $x = 0$ or $x = 2k_1/3k_2$.

Note   Another method for finding $T'(x)$ is to write $T(x)$ as

$$T(x) = k_1 x^2 - k_2 x^3$$

Applying the methods of Section 4.3, we get

$$T'(x) = 2k_1 x - 3k_2 x^2$$

Example 4.15   If $f(x) = (3x^3 + 5x^2 + 7)(2x^5 - 8x^3 + 6x)$, find $f'(x)$.

SOLUTION   Let $u(x) = 3x^3 + 5x^2 + 7$ and $v(x) = 2x^5 - 8x^3 + 6x$. Then

$$u'(x) = 9x^2 + 10x \quad \text{and} \quad v'(x) = 10x^4 - 24x^2 + 6$$

Now

$$f = uv$$

and, by Theorem 4.6, we have

$$f'(x) = u(x)\, v'(x) + u'(x)\, v(x)$$

Substitution then yields

$$f'(x) = (3x^3 + 5x^2 + 7)(10x^4 - 24x^2 + 6) + (9x^2 + 10x)(2x^5 - 8x^3 + 6x)$$

Derivative of a quotient ▶

Theorem 4.7 *Let $f = u/v$ and suppose that u and v are differentiable at x. Then f is differentiable at x and*

$$f'(x) = \frac{u'(x)\ v(x) - u(x)\ v'(x)}{[v(x)]^2} \tag{4.14}$$

provided that $v(x) \neq 0$.

PROOF   Setting up the difference quotient, we have

$$\frac{f(x+h) - f(x)}{h} = \frac{1}{h}\left[\frac{u(x+h)}{v(x+h)} - \frac{u(x)}{v(x)}\right]$$

Here we have added and subtracted the term $[u(x)\,v(x)]/h$

$$= \frac{1}{h}\left[\frac{u(x+h)\ v(x) - u(x)\ v(x+h)}{v(x+h)\ v(x)}\right]$$

$$= \left[\frac{u(x+h)\ v(x) - u(x)\ v(x) + u(x)\ v(x) - u(x)\ v(x+h)}{h}\right]$$

$$\times \frac{1}{v(x+h)\ v(x)}$$

$$= \left\{\left[\frac{u(x+h) - u(x)}{h}\right]v(x) - u(x)\left[\frac{v(x+h) - v(x)}{h}\right]\right\}$$

$$\times \frac{1}{v(x+h)\ v(x)}$$

And so we have

$$f'(x) = \lim_{h \to 0} \frac{f(x+h) - f(x)}{h}$$

Again we have used the property that $\lim_{h \to 0} v(x+h) = v(x)$.

$$= \lim_{h \to 0}\left\{\left[\left(\frac{u(x+h) - u(x)}{h}\right)v(x) - u(x)\left(\frac{v(x+h) - v(x)}{h}\right)\right]\right.$$

$$\left. \times \frac{1}{v(x+h)\ v(x)}\right\}$$

$$= \frac{u'(x)\ v(x) - u(x)\ v'(x)}{[v(x)]^2}$$

which proves Theorem 4.7.

Example 4.16 If $f(x) = (x^2 + 1)/(2x + 1)$, find $f'(x)$.

SOLUTION   Let $u(x) = x^2 + 1$ and $v(x) = 2x + 1$. Then

$$u'(x) = 2x \quad \text{and} \quad v'(x) = 2$$

and for $f = u/v$, we have

$$f'(x) = \frac{u'(x)\,v(x) - u(x)\,v'(x)}{[v(x)]^2}$$

Application of Theorem 4.7.

$$= \frac{(2x)(2x+1) - (x^2+1)(2)}{(2x+1)^2} = \frac{4x^2 + 2x - 2x^2 - 2}{(2x+1)^2}$$

$$= \frac{2x^2 + 2x - 2}{(2x+1)^2} = \frac{2(x^2 + x - 1)}{(2x+1)^2}$$

This holds provided that $x \neq -\frac{1}{2}$.

Example 4.17   If $f(x) = (x^2+2)(x^3+1)/(x^2-2)$ and $x^2 \neq 2$, find $f'(x)$.

SOLUTION   Set $f = u/v$, where

$$u(x) = (x^2+2)(x^3+1) \quad \text{and} \quad v(x) = x^2 - 2$$

We find that $v'(x) = 2x$, and application of Theorem 4.6 yields

$$u'(x) = (x^2+2)(3x^2) + (x^3+1)(2x) = 5x^4 + 6x^2 + 2x$$

Now

$$f'(x) = \frac{u'(x)\,v(x) - u(x)\,v'(x)}{[v(x)]^2}$$

Hence

$$f'(x) = \frac{(5x^4 + 6x^2 + 2x)(x^2-2) - (x^2+2)(x^3+1)(2x)}{(x^2-2)^2}$$

$$= \frac{3x^6 - 8x^4 - 12x^2 - 8x}{(x^2-2)^2}$$

In Problems 1 through 32, find $f'(x)$ and simplify the result as much as possible. (In Problems 1 through 14, verify your answer by using the methods of Section 4.3).

Exercises 4.4

1. $f(x) = x(x+2)$

2. $f(x) = (2x-1)(4x+3)$

3. $f(x) = (2-3x)(x+1)$

4. $f(x) = (6-5x)(2+3x)$

5. $f(x) = x(3x^2 + 4x - 1)$

6. $f(x) = (x^2-1)(3x-4)$

7. $f(x) = (2x^2+3)(4x^2-1)$

8. $f(x) = (x^3-1)(-2x+6x^2)$

9. $f(x) = (x+2x^3)(x^4-2x^5)$

10. $f(x) = (x^2-1)^2$

11. $f(x) = (2x^3 - x^2)^2$

12. $f(x) = \sqrt{x}\,(2x-2)$

13. $f(x) = 2\sqrt{x}\,(x^3 - 2x^2)$

14. $f(x) = (3x^{1/2} - 4)(2x^{-1/2})$

15. $f(x) = \dfrac{x+2}{x-1}$

16. $f(x) = \dfrac{1}{x+1}$

17. $f(x) = \dfrac{3x}{4+x}$

18. $f(x) = \dfrac{2+3x}{1-4x}$

19. $f(x) = \dfrac{2-3x}{2+5x}$

20. $f(x) = \dfrac{2x-1}{3-2x}$

21. $f(x) = \dfrac{9x^{-1} + 5x^{-2}}{x^3}$

22. $f(x) = \dfrac{5x^2 + 7x^3}{x^{-4}}$

23. $f(x) = \dfrac{1}{x^2 - 5}$

24. $f(x) = \dfrac{x^2 + 4x + 10}{x + 7}$

25. $f(x) = \dfrac{4x^2 - x}{2x^2 + 3}$

26. $f(x) = \dfrac{4}{2x^3 + 3x^2 + 5}$

27. $f(x) = \dfrac{2x^2 + 3x}{1 + x + x^{-1}}$

28. $f(x) = \dfrac{x^3 + x^{-3}}{x^3 - x^{-3}}$

29. $f(x) = \dfrac{x^2 + 5}{x(x + 7)}$

30. $f(x) = \dfrac{(2x + 3)^2}{(3x + 1)^3}$

31. $f(x) = \dfrac{3x^4 + 5x^2 + 9x - 7}{x^3 - 5x + 6}$

32. $f(x) = \dfrac{\sqrt{x}(2x^2 - x)}{3x + 4}$

[Hint: Apply Theorem 4.6.]   33. If u, v, and w are differentiable functions of x and if $f(x) = u(x)\,v(x)\,w(x)$, show that

$$f'(x) = u(x)\,v(x)\,w'(x) + u(x)\,v'(x)\,w(x) + u'(x)\,v(x)\,w(x)$$

In Problems 34 through 36, find $f'(x)$.

34. $f(x) = (2x - 1)(x^2 + 1)(4 - x)$

35. $f(x) = (x^5 - 3x^2)(4 - x^3)(5x^2 - x)$

36. $f(x) = (\sqrt{x} + 4x^{-1})(2x^{-1} - 3x^2)(4x^2 - x^{-2})$

4.5 The Derivative: A Geometric Interpretation

The problem of computing the velocity of a moving object led to the development of the derivative concept. A second problem that led to the development of the differential calculus was that of finding the direction of the tangent line at a specified point on a given curve. This problem was successfully solved by the French mathematician Fermat.

Here we shall illustrate how the derivative concept is used to define the tangent line to the graph of a given function at a given point. Consider a function f whose graph is given in Figure 4.4. Let P be a

Figure 4.4

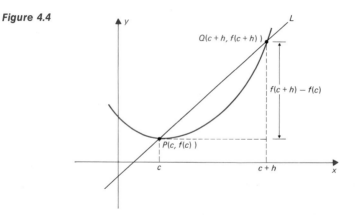

point on the graph of f with coordinates $(c, f(c))$. Let h be a number $(h \neq 0)$ such that $c + h$ is in the domain of f. Then there is a point Q on the graph of f with coordinates $(c + h, f(c + h))$. The line L containing points P and Q is called a *secant line* (in geometry, a line which crosses a circle at two points is called a secant line). Now the slope of L (see Chapter 2) is given by

$$\text{slope of } PQ = \frac{f(c + h) - f(c)}{h}, \qquad h \neq 0 \qquad (4.15)$$

We shall assume that the function f is continuous on an interval containing the points c and $c + h$. Now let the point Q approach P along the curve. Clearly, as Q approaches P, h approaches zero. Taking the limit in Equation (4.15), we have

$$\lim_{Q \to P} (\text{slope of } PQ) = \lim_{h \to 0} \frac{f(c + h) - f(c)}{h} \qquad (4.16)$$

If f has a derivative at the point P, then from Equation (4.16), we have

$$\lim_{Q \to P} (\text{slope of } PQ) = f'(c) \qquad (4.17)$$

We define the *tangent line* to the graph of f at the point P as the line L_{PT} having slope $f'(c)$ and containing the point P (Figure 4.5). Thus using the point slope form of the equation of a line, we obtain

$$L_{PT} \colon y - f(c) = f'(c)(x - c) \qquad (4.18)$$

Note that h is a positive number in Figure 4.4. However, h approaches zero through positive and negative values.

Figure 4.5

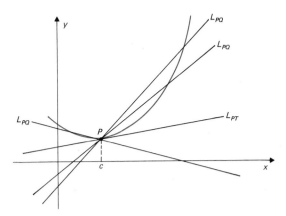

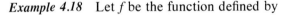

Example 4.18   Let f be the function defined by

$$f(x) = 2x - x^2$$

(a)   Find the equation of the tangent line to the graph of f at the point $\left(\frac{1}{2}, \frac{3}{4}\right)$.

(b)   At what point on the curve is the tangent line parallel to the
x-axis?

SOLUTION

(a)   Since $f(x) = 2x - x^2$, application of the rules of differentiation
yields

$$f'(x) = 2 - 2x \qquad (4.19)$$

Thus the slope of the tangent line to the graph of f at a point
(x, y) is $2 - 2x$. Hence the slope of the tangent line at the point
$\left(\frac{1}{2}, \frac{3}{4}\right)$ is

$$f'\left(\tfrac{1}{2}\right) = 2 - 2\left(\tfrac{1}{2}\right) = 1 \qquad (4.20)$$

The tangent line is then given by

$$L_{PT}: y - \tfrac{3}{4} = 1\left(x - \tfrac{1}{2}\right) \quad \text{or} \quad y = x + \tfrac{1}{4}$$

(b)   To find the point at which the tangent line is horizontal, or
parallel to the x-axis, we write

$$f'(x) = 0$$

and so

$$2 - 2x = 0 \quad \text{or} \quad x = 1$$

To find the corresponding value of f, we have

$$f(1) = 2(1) - (1)^2 = 1$$

Therefore at the point $P_2(1, 1)$, the tangent line is parallel to the
x-axis (see Figure 4.6).

Pierre de Fermat

Pierre de Fermat (1601–1665) was born at Beaumont-de-Lomagne,
France. His father, Dominique Fermat, was a leather merchant. Fermat
led an uneventful life. He received his education in his home town and
later at Toulouse. At the age of 30, he married and fathered three children, a boy and two girls. The latter became nuns. In 1648 he became
Counselor of the Parliament of Toulouse, a position which gave him
ample time to do mathematics.

Fermat is considered one of the creators of calculus (along with
Newton and Leibniz). He solved the problem of finding the extrema of
the graph of a function and deduced the laws of reflection and refraction.
He extended Descartes' analytic geometry to three dimensions, and was
a cofounder (with Pascal) of the theory of probability. His greatest work,
however, was done in the theory of numbers. He stated and claimed to
have proved a famous problem (Fermat's Last Theorem): If n is an
integer, $n > 2$, then there exist no rational numbers x, y, and z ($xyz \neq 0$)
such that $x^n + y^n = z^n$. This remains an unproved theorem today.

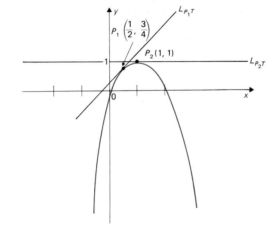

Figure 4.6

Example 4.19   Find the point at which the slope of the tangent line to the curve

$$y = 3x - x^3 + 5 \qquad\qquad (4.21)$$

is 3.

SOLUTION   From Equation (4.21), we obtain

$$\frac{dy}{dx} = 3 - 3x^2$$

To find the point at which the tangent line has slope 3, we set

$$\frac{dy}{dx} = 3$$

Hence

$$3 - 3x^2 = 3,$$

which implies that

$$x = 0$$

To find the corresponding value of y, we substitute x into Equation (4.21) to obtain

$$y = 3(0) - (0)^3 + 5 = 5$$

Therefore the required point is $(0, 5)$.

In Example 4.5, we proved that the function $f(x) = |x|$ does not have a derivative at $x = 0$; that is, $f'(0)$ does not exist. This conclusion could have been suspected if we look at the graph of $f(x) = |x|$ (see Figure 4.7).

We observe in Figure 4.7 that the graph of $f(x) = |x|$ has a sharp point at $x = 0$ and no unique tangent line. So, in general, if a graph has a sharp point, then we should strongly suspect that the function (although continuous) may not have a derivative at that point. For ex-

Figure 4.7

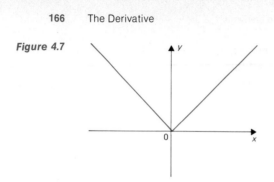

ample, in the graph of Figure 4.8, we would have reason to suspect that the function f defined by $y = f(x)$ is continuous at $x = a$, but $f'(a)$ does not exist.

Figure 4.8

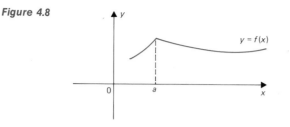

We shall now consider some important results that are crucial to the next chapter with its applications. First, for many curves which cross the x-axis at two points, there is at least one point between the crossings where the tangent line is parallel to the x-axis (see Figure 4.9).

Figure 4.9

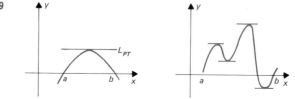

Before we state the next important theorem, the reader is advised to refer to Problem 28 in Exercises 3.4 for the definition of continuity on a closed interval $[a, b]$.

Rolle's theorem ▶   *Theorem 4.8   Suppose that*

(*i*) *f is a function continuous on $[a, b]$.*

(*ii*) *$f'(x)$ exists for each x on (a, b).*

(*iii*) *$f(a) = f(b) = 0$*

Then there is at least one point c between a and b such that

$$f'(c) = 0$$

It is important to note that hypothesis (ii) is essential. It ensures that the graph has no sharp points. If a function has no derivative at some point between a and b, Rolle's theorem does not necessarily apply.

For example, the function f, where

$$f(x) = 1 - x^{2/3}$$

is continuous on $[-1, 1]$ and $f(-1) = f(1) = 0$. However,

$$f'(x) = -\tfrac{2}{3}x^{-1/3}$$

is zero nowhere. Here we find that f' is not defined at $x = 0$, and so Rolle's theorem does not apply.

Example 4.20   If $f(x) = x^3 - x$, find all numbers c between 0 and 1 such that $f'(c) = 0$.

   SOLUTION   We find that

$$f(0) = (0)^3 - 0 = 0 \quad \text{and} \quad f(1) = 1^3 - 1 = 0$$

Computing the derivative, we have

$$f'(x) = 3x^2 - 1$$

Setting $f'(x)$ equal to zero for some value $x = c$, we obtain

$$3c^2 - 1 = 0 \quad \text{or} \quad c = \sqrt{\tfrac{1}{3}} \quad \text{or} \quad c = -\sqrt{\tfrac{1}{3}}$$

Now $-\sqrt{\tfrac{1}{3}}$ is not between 0 and 1. Therefore the only answer is $c = \sqrt{\tfrac{1}{3}}$ (see Figure 4.10). This is the point called for by Rolle's theorem.

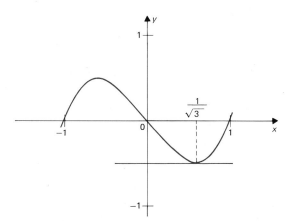

Figure 4.10
Graph of $f(x) = x^3 - x$.

A generalization of Rolle's theorem is

Theorem 4.9   *Suppose that*   ◀ *The mean value theorem*

   (i)   *f is a continuous function on $[a, b]$.*

(ii)   f'(x) exists for each x on (a, b).

Then there is a number c between a and b such that

$$f'(c) = \frac{f(b) - f(a)}{b - a} \qquad (4.22)$$

The proof of this theorem is left for the student as an exercise. Here we shall give a geometrical interpretation of the theorem. In Figure 4.11 we have the graph of a smooth function f between the points a and b. The points P and Q have coordinates $(a, f(a))$ and $(b, f(b))$, respectively. We find that the slope of the line segment PQ is

Figure 4.11

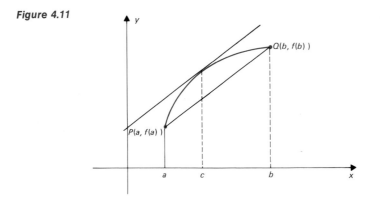

$$\frac{f(b) - f(a)}{b - a}$$

which is the right-hand side of Equation (4.22). The theorem says that the tangent line to the graph of f at c is parallel to the line segment PQ, where c is some point in the open interval (a, b).

Example 4.21   If $f(x) = x^2 - x - 2$, find all numbers c between 1 and 3 which satisfy the equation

$$f'(c) = \frac{f(3) - f(1)}{3 - 1}$$

Michel Rolle

Michel Rolle (1652–1719), a French mathematician, was born on April 21, 1652 at Ambert, and died on October 17, 1719 in Paris. Rolle first attracted the attention of the scientific community when he solved an algebraic problem proposed by Ozanam. His remarkable work in algebra and geometry gained him membership in the French Academy of Sciences. His works include *A Study of Descartes' Geometry* and *A Treatise of Algebra.*

SOLUTION   Computing the derivative of f, we find that

$$f'(x) = 2x - 1$$

Now

$$\frac{f(3) - f(1)}{3 - 1} = \frac{(3^2 - 3 - 2) - (1^2 - 1 - 2)}{2} = 3$$

Thus for some value c, we have by Theorem 4.9

$$f'(c) = 3 \quad \text{or} \quad 2c - 1 = 3$$

that is,

$$c = 2$$

The student can check that $f'(2) = 3$. In Figure 4.12 the slope of the line segment PQ is equal to the slope of the tangent line to the graph of f at $(2, 0)$.

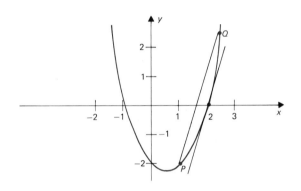

Figure 4.12

In Problems 1 through 5, find the equation of the tangent line to the graph of the given function at the given point.

Exercises 4.5

1. $f(x) = x^3$ at $(2, 8)$ 2. $f(x) = \sqrt{x}$ at $(4, 2)$

3. $f(x) = \dfrac{1}{x}$ at $\left(2, \frac{1}{2}\right)$ 4. $f(x) = \frac{1}{4}x^2 + \frac{3}{4}x + \frac{1}{2}$ at $(2, 3)$

5. $f(x) = x^2 + 2x - 3$ at points where the x-coordinates are -1 and 1.

6. At what point on the curve $y = 2x - x^2$ is the slope of the tangent line 5?

7. If $f(x) = 3x^2$, show that the slope of the tangent line to the graph of f at $(4, 48)$ is twice the slope at $(2, 12)$.

8. At what point on the curve $y = x + (4/x)$ is the tangent line parallel to the x-axis?

9. Let $y = 6x - x^2$. Find the point at which the slope is -4.

10. Find the coordinates of the point on the curve $y = 5x^2 + (80/x)$ at which the tangent line is parallel to the x-axis.

11. $A(1, -3)$ and $B(4, 3)$ are two points on the curve $y = x - (4/x)$. Find the coordinates of the points on the arc AB of the curve at which the tangent line to the curve is parallel to the line through A and B.

In Problems 12 through 16, find all possible values c between a and b which satisfy the equation

$$f'(c) = \frac{f(b) - f(a)}{b - a}$$

12. $f(x) = -3x + 1, \quad a = 0, b = 2$

13. $f(x) = x^2 - 3x - 4, \quad a = -1, b = 3$

14. $f(x) = x^3 - 2x^2 + 3x - 2, \quad a = 0, b = 2$

15. $f(x) = x^2 + 1, \quad a = 1, b = 2$

16. $f(x) = \dfrac{x - 2}{x + 2}, \quad a = 0, b = 3$

17. Given that $f(x) = (x + 3)/(2x + 1)$, $a = -1$ and $b = 2$, discuss the validity of the mean value theorem.

18. Given that $f(x) = (x^2 + 4x + 3)/(x - 1)$, $a = 0$, $b = 2$, show that there is no number c between a and b which satisfies the mean value theorem.

*19. Prove Theorem 4.9. [*Hint*: Use the function

$$g(x) = f(x) - \frac{f(b) - f(a)}{b - a}(x - a) - f(a)$$

and apply Rolle's theorem.]

4.6 The Chain Rule

Suppose we wish to find the derivative of f, where

$$f(x) = (3x^3 + 7x^2 + 8)^{70} \qquad (4.23)$$

From what we have learned so far, the only way to find $f'(x)$ is to compute the seventieth power of $3x^3 + 7x^2 + 8$ and then differentiate the resulting polynomial. To avoid this unpleasant task, we can fortunately use what is called the *chain rule*. By using the chain rule, we shall see that from (4.23) we obtain

$$f'(x) = 70(3x^3 + 7x^2 + 8)^{69}(9x^2 + 14x) \qquad (4.24)$$

Before studying the chain rule, the reader should review Section 2.3 on the composition of functions. We recall that if g and u are functions, then the composition of g with u is a function f defined by

$$f(x) = g\big(u(x)\big)$$

We now state the chain rule (without proof).

Theorem 4.10   *Suppose the function u is differentiable at x and the function g is differentiable at u(x). Then f = g ∘ u is differentiable at x and*

$$f'(x) = (g \circ u)'(x) = g'(u(x)) \cdot u'(x)$$

◀ The chain rule

Suppose that $f(x) = g(u(x))$.
Then, using Leibniz' notation, we have

$$\frac{df}{dx} = \frac{dg}{du} \cdot \frac{du}{dx}$$

Example 4.22   Let us return to the function f defined by

$$f(x) = (3x^3 + 7x^2 + 8)^{70}$$

Find $f'(x)$.

SOLUTION   We can write $f(x)$ as

$$f(x) = g(u(x))$$

where

$$u(x) = 3x^3 + 7x^2 + 8 \quad \text{and} \quad g(u) = u^{70}$$

Now

$$g'(u) = 70u^{69} \quad \text{and} \quad u'(x) = 9x^2 + 14x$$

Therefore application of Theorem 4.10 yields

$$f'(x) = g'(u(x)) \cdot u'(x)$$
$$= 70[u(x)]^{69} (9x^2 + 14x) = 70(3x^3 + 7x^2 + 8)^{69} (9x^2 + 14x)$$

Example 4.23   A rocket is fired vertically upward. The equation of motion is given by

$$s(t) = R\left(1 + \frac{3}{2} \cdot \frac{v_0}{R} t\right)^{2/3} \tag{4.25}$$

where R is the radius of the Earth, s is the distance of the rocket from the Earth's center, v_0 is a constant $= \sqrt{2gR}$ called the escape velocity, and t is time (see Figure 4.13). Find the instantaneous velocity of the rocket.

SOLUTION   Let

$$u(t) = 1 + \frac{3}{2} \cdot \frac{v_0}{R} t$$

Then

$$u'(t) = \frac{3}{2} \cdot \frac{v_0}{R}$$

Figure 4.13

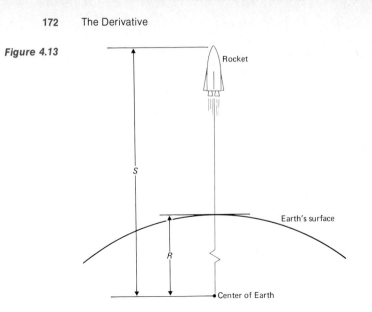

Rocket

S

Earth's surface

R

Center of Earth

Now from Equation (4.25), we have

$$s(t) = Ru^{2/3}$$

and by the chain rule

$$s'(t) = \left(R\,\frac{2}{3}\,u^{-1/3} \right) u'(t) = \left(\frac{2}{3}\,Ru^{-1/3} \right)\left(\frac{3}{2} \cdot \frac{v_0}{R} \right) = \frac{v_0}{\left[1 + \frac{3}{2} \cdot \frac{v_0}{R}\,t \right]^{1/3}},$$

which is the instantaneous velocity of the rocket.

Example 4.24   For each of the following functions, find $f'(x)$:

(a) $f(x) = (x^3 + 5x^2 + 7)^{9/2}$ (b) $f(x) = \dfrac{1}{\sqrt[3]{4x^3 + 6x + 7}}$

SOLUTION

(a)   Let

$$u(x) = x^3 + 5x^2 + 7 \quad \text{and} \quad g(u) = u^{9/2}$$

Then

$$g'\big(u(x)\big) = \tfrac{9}{2}u(x)^{9/2-1} = \tfrac{9}{2}u(x)^{7/2} = \tfrac{9}{2}(x^3 + 5x^2 + 7)^{7/2}$$

and

$$u'(x) = 3x^2 + 10x$$

Hence by the chain rule

$$f'(x) = g'\big(u(x)\big) \cdot u'(x) = \tfrac{9}{2}(x^3 + 5x^2 + 7)^{7/2}(3x^2 + 10x)$$

(b) Let

$$u(x) = 4x^3 + 6x + 7 \quad \text{and} \quad g(u) = u^{-1/3}$$

Then

$$g'\big(u(x)\big) = -\tfrac{1}{3}u(x)^{-1/3-1} = -\tfrac{1}{3}u^{-4/3} = -\tfrac{1}{3}(4x^3 + 6x + 7)^{-4/3}$$

and

$$u'(x) = 12x^2 + 6$$

Hence by the chain rule, since $f = g \circ u$,

$$f'(x) = g'\big(u(x)\big) \cdot u'(x) = -\tfrac{1}{3}(4x^3 + 6x + 7)^{-4/3}(12x^2 + 6)$$
$$= -2(4x^3 + 6x + 7)^{-4/3}(2x^2 + 1)$$

Example 4.25 If $y = x^2 \sqrt{x^2 - 1}$, find dy/dx.

SOLUTION Applying Theorem 4.5, we have

$$\frac{dy}{dx} = x^2 \frac{d}{dx}[(x^2 - 1)^{1/2}] + (x^2 - 1)^{1/2}\frac{d}{dx}(x^2)$$

and by the chain rule

$$\frac{d}{dx}[(x^2 - 1)^{1/2}] = \tfrac{1}{2}(x^2 - 1)^{(1/2)-1}\frac{d}{dx}(x^2 - 1)$$
$$= \tfrac{1}{2}(x^2 - 1)^{-1/2}(2x) = x(x^2 - 1)^{-1/2}$$

Hence

$$\frac{dy}{dx} = x^2[x(x^2 - 1)^{-1/2}] + (x^2 - 1)^{1/2}(2x)$$
$$= \frac{x^3}{\sqrt{x^2 - 1}} + 2x\sqrt{x^2 - 1} = \frac{x^3 + 2x(x^2 - 1)}{\sqrt{x^2 - 1}} = \frac{3x^3 - 2x}{\sqrt{x^2 - 1}}$$

Example 4.26 A snowball is rolling down a mountain side (see Figure 4.14). At a certain instant, the surface area is increasing at the rate of 50 sq ft/sec. If the radius of the snowball at that instant is 8 ft, how fast is the volume increasing?

SOLUTION In this model there are three functions of time: the volume $V(t)$, the surface area $S(t)$, and the radius $r(t)$. We assume that the snowball is a perfect sphere at all times and recall that the surface area of a sphere is

$$S(t) = 4\pi[r(t)]^2 \tag{4.26}$$

and that the volume of the sphere is

Figure 4.14

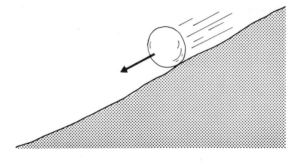

$$V(t) = \tfrac{4}{3}\pi[r(t)]^3 \tag{4.27}$$

Applying the chain rule, we find from Equation (4.26) that

$$S'(t) = 4\pi[2r(t)]\, r'(t) \tag{4.28}$$

Now, when $r = 8$, $S'(t) = 50$. Thus from Equation (4.28), we have

$$r'(t) = \frac{50}{8\pi(8)} = \frac{25}{32\pi} \text{ ft/sec}$$

From Equation (4.27)

$$V'(t) = \tfrac{4}{3}\pi \cdot 3[r(t)]^2\, r'(t) = \tfrac{4}{3}\pi(3)(8)^2\, \frac{25}{32\pi} = 200$$

Hence the volume of the snowball is increasing at the rate of 200 cu ft per sec at the instant in question.

Exercises 4.6

In Problems 1 through 4, use two methods to find $f'(x)$ and compare your results.

1. $f(x) = (3x + 1)^2$ 2. $f(x) = \left(x - \dfrac{1}{x}\right)^2$

3. $f(x) = (x + 1)^{-1}$ 4. $f(x) = (x + 1)^3$

In Problems 5 through 24, find $f'(x)$.

5. $f(x) = (x - 5)^4$ 6. $f(x) = (2 - 3x)^4$

7. $f(x) = (4x^2 + x)^6$ 8. $f(x) = (2x^2 + 1)^{1/2}$

9. $f(x) = (5x^3 + 6x^2 + 9)^{12}$ 10. $f(x) = (4x^2 + 5)^{3/2}$

11. $f(x) = 2x(4x - 1)^5$ 12. $f(x) = -3x(1 - 4x^2)^3$

13. $f(x) = 5x\sqrt{3x + 1}$ 14. $f(x) = -3x^2\sqrt{1 - 2x^2}$

15. $f(x) = (2x + 1)\sqrt{x^2 + 5}$ 16. $f(x) = \dfrac{3x}{(x^2 + 1)^5}$

17. $f(x) = \dfrac{-2x}{(2x^2 - x)^{3/2}}$ 18. $f(x) = \dfrac{1}{\sqrt{2x^2 + x - 1}}$

19. $f(x) = x^2 + 3 + \dfrac{1}{\sqrt{x^2 + 3}}$ 20. $f(x) = \sqrt{\dfrac{3x + 4}{5x + 1}}$

21. $f(x) = \sqrt[3]{\dfrac{x^2 + 2x + 4}{5x^2 + 7x + 3}}$ 22. $f(x) = \dfrac{2}{x}(x - x^{-1})^{1/2}$

23. $f(x) = \left(\dfrac{x+2}{1-2x}\right)^{1/2}$ 24. $f(x) = \left(\dfrac{3x^2 - x^{-1}}{4x - x^{-2}}\right)^{-3/2}$

25. An offshore oil well is leaking. The oil spreads on the surface of the water in a circle. If the radius r of the circle is increasing at the rate of 2 ft/sec, at what rate is the area A of the oil spill increasing when $r = 20$ ft? 50 ft? 200 ft?

26. A spherical weather balloon is being inflated. At a certain instant, the volume of the balloon is increasing at the rate of 500 cu in./min. If at that instant the radius of the balloon is 40 in., how fast is the surface area increasing?

27. A spherical snowball is melting. At a certain instant the volume is de-creasing $(V'(t) < 0)$ at the rate of 10 cu in./hr. At that instant, the radius of the snowball is 25 in.
    (a) How fast is the diameter decreasing?
    (b) How fast is the surface area decreasing?

28. Suppose in a model the pollution index $I(t)$ is related to a dose $A(t)$ of an antipollution agent by the equation

$$I \cdot A^{3/2} = k \qquad \text{(a constant)}$$

At a given instant, the index $I = 30\%$, and the dose $A = 9$ units. If at that moment the dose of the antipollution agent is decreasing at the rate of 3 units/min, how fast is the index I changing?

29. Smoke is being emitted into the atmosphere in the form of an inverted right circular cone (see Figure 4.15). At any instant, the diameter is

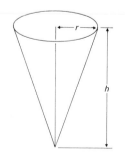

Figure 4.15

equal to the height h. If the rate of smoke emission is 200 cu ft/min, how fast is the height increasing when $h = 50$ ft? [*Hint*: Volume of cone = $\frac{1}{3}\pi r^2 h$.]

4.7 Implicit Differentiation

Suppose that f is a function given by the equation $y = f(x)$. Then we say that y is defined *explicitly* as a function of x. The functions we have en-countered so far have been of this form. However, a relationship be-tween x and y expressed in the form

Note It is possible that the relation $g(x, y) = 0$ does not define any function. For example,

$$2x^2 + y^2 + 1 = 0$$

does not define a function, since there are no values of x and y which satisfy this equation.

$$g(x, y) = 0$$

may also define one or more functions of x. In such a case the function or functions are said to be defined *implicitly*.

Thus in the equations

$$3x^3 + y^2 - 8y = 0 \tag{4.29}$$

$$x^2 - 4 - 3xy = 0 \tag{4.30}$$

and

$$x^2 + y^2 - 1 = 0 \tag{4.31}$$

y is defined implicitly as a function of x. Sometimes it is possible to solve an equation for y as a function of x. For example, from Equation (4.30), we have

$$y = \frac{x^2 - 4}{3x}$$

Equation (4.31) can be solved for y, yielding two functions

$$y = f_1(x) = \sqrt{1 - x^2} \quad \text{and} \quad y = f_2(x) = -\sqrt{1 - x^2}$$

Therefore Equation (4.31) implicitly defines two functions of x (see Figure 4.16).

Figure 4.16

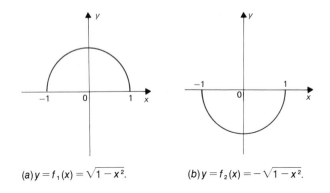

(a) $y = f_1(x) = \sqrt{1 - x^2}$.

(b) $y = f_2(x) = -\sqrt{1 - x^2}$.

We shall now indicate how to calculate the derivative of a function defined implicitly by an equation. The process involved is as follows.

Here we are assuming that y is a differentiable function of x defined implicitly by the equation $g(x, y) = 0$.

(a) Using the chain rule, differentiate both sides of the equation $g(x, y) = 0$.

(b) Solve the resulting equation for dy/dx.

Example 4.27 Find dy/dx if

$$x^3 + y^3 - 3xy = 0 \tag{4.32}$$

SOLUTION Assuming that y is a differentiable function of x, we differentiate both sides of Equation (4.32) to obtain

$$\frac{d}{dx}(x^3 + y^3 - 3xy) = \frac{d}{dx}(0)$$

or

$$\frac{d}{dx}(x^3) + \frac{d}{dx}(y^3) + \frac{d}{dx}(-3xy) = 0$$

This implies that

$$3x^2 + 3y^2 \frac{dy}{dx} - 3\left(x \frac{dy}{dx} + y\right) = 0$$

Collecting terms involving dy/dx and factoring, we obtain

$$(y^2 - x)\frac{dy}{dx} = y - x^2$$

Hence

$$\frac{dy}{dx} = \frac{y - x^2}{y^2 - x} \qquad \text{for} \quad y^2 \ne x$$

Observations:

(1) If we set $y = f(x)$, then

$$\frac{d}{dx}(y^3) = \frac{d}{dx}[f(x)]^3$$

$$= 3[f(x)]^2 f'(x)$$

$$= 3y^2 \frac{dy}{dx}$$

(2) $\dfrac{d}{dx}(xy) = \dfrac{d}{dx}[xf(x)]$

$$= xf'(x) + 1 \cdot f(x)$$

Here we have used the product rule for differentiation.

Example 4.28 Find dy/dx if

$$2x^2 + 3y^2 = 4 \qquad\qquad (4.33)$$

SOLUTION 1   Differentiating both sides of Equation (4.33) with respect to x, we obtain

$$\frac{d}{dx}(2x^2 + 3y^2) = \frac{d}{dx}(4)$$

$$\frac{d}{dx}(2x^2) + \frac{d}{dx}(3y^2) = \frac{d}{dx}(4)$$

$$4x + 6y \frac{dy}{dx} = 0$$

and so

$$\frac{dy}{dx} = -\frac{4x}{6y} = -\frac{2x}{3y}$$

Note again that

$$\frac{d}{dx}(y^2) = \frac{d}{dx}[f(x)]^2$$

$$= 2f(x) \cdot f'(x)$$

$$= 2y \frac{dy}{dx}.$$

SOLUTION 2   Suppose that we solve Equation (4.33) for y. We have

$$y = \frac{1}{\sqrt{3}}\sqrt{4 - 2x^2} \qquad (2 - x^2 \ge 0) \qquad\qquad (4.34)$$

or

$$y = -\frac{1}{\sqrt{3}}\sqrt{4 - 2x^2} \qquad (2 - x^2 \ge 0) \qquad\qquad (4.35)$$

From (4.34) we obtain

$$\frac{dy}{dx} = \frac{1}{\sqrt{3}} \cdot \frac{1}{2}(4 - 2x^2)^{-1/2} \frac{d}{dx}(4 - 2x^2) = \frac{1}{2\sqrt{3}}(4 - 2x^2)^{-1/2}(-4x)$$

$$= -\frac{2x}{\sqrt{3}(4 - 2x^2)^{1/2}} = -\frac{\sqrt{3}(2x)}{\sqrt{3}\sqrt{3}(4 - 2x^2)^{1/2}} = -\frac{2x}{3y}$$

which agrees with the preceding result we obtained. The reader is encouraged to repeat the solution process starting with Equation (4.35) and show that the same result is obtained again.

Example 4.29   Find the equation of the tangent line to the ellipse (see Figure 4.17)

$$9x^2 + 16y^2 = 144 \tag{4.36}$$

at the point $\left(\sqrt{7}, \frac{9}{4}\right)$.

Figure 4.17

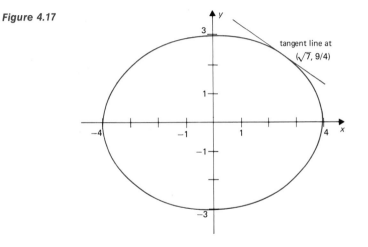

tangent line at
$(\sqrt{7}, 9/4)$

SOLUTION   Differentiating implicitly, we obtain

$$\frac{d}{dx}(9x^2 + 16y^2) = \frac{d}{dx}(144)$$

or

$$\frac{d}{dx}(9x^2) + \frac{d}{dx}(16y^2) = 0$$

$$18x + 32y \frac{dy}{dx} = 0$$

Hence

$$\frac{dy}{dx} = -\frac{9x}{16y}$$

and so the slope of the tangent line at the point $\left(\sqrt{7}, \frac{9}{4}\right)$ is

$$\frac{dy}{dx}\bigg|_{(\sqrt{7},9/4)} = -\frac{9(\sqrt{7})}{16\left(\frac{9}{4}\right)} = -\frac{\sqrt{7}}{4}$$

Therefore the equation of the tangent line at $\left(\sqrt{7}, \frac{9}{4}\right)$ is given by

$$y - \frac{9}{4} = -\frac{\sqrt{7}}{4}(x - \sqrt{7})$$

Using the point-slope form; we have $y - y_1 = m(x - x_1)$.

Simplifying, we have

$$4y + \sqrt{7}x - 16 = 0$$

Example 4.30   Two cars C_1 and C_2 are racing on two highways inter-secting at right angles (see Figure 4.18). Car C_1 is traveling south at the rate of 100 ft/sec and C_2 is traveling west at the rate of 120 ft/sec. At a certain instant, C_1 is 300 ft from point A and C_2 is 400 ft from A.

(a)   How fast is the distance between the two cars changing at that instant?

(b)   If the cars maintain the same speed and ignore the stop signs, will they collide?

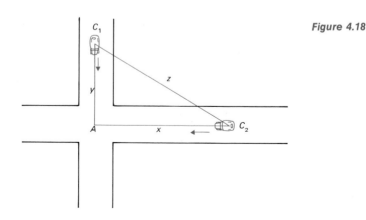

Figure 4.18

SOLUTION

(a)   Let x, y, and z be the distances from C_2 to A, C_1 to A, and C_1 to C_2, respectively. Here x, y, and z are functions of time t. Using the Pythagorean theorem, we have the relation

$$z^2 = x^2 + y^2 \qquad (4.37)$$

Differentiating both sides of Equation (4.37) with respect to t, we obtain

$$\frac{d}{dt}(z^2) = \frac{d}{dt}(x^2 + y^2) \quad \text{and} \quad 2z\frac{dz}{dt} = 2x\frac{dx}{dt} + 2y\frac{dy}{dt}$$

Dividing by 2, we have

$$z\frac{dz}{dt} = x\frac{dx}{dt} + y\frac{dy}{dt} \tag{4.38}$$

At the instant in question, $x = 400$ ft and $y = 300$ ft, and so

$$z = \sqrt{(400)^2 + (300)^2} = 500 \text{ ft}$$

Therefore from Equation (4.38)

Note that

$\frac{dx}{dt} = 120$ *ft/sec*

= *rate of change (velocity) of* C_2

$\frac{dy}{dt} = 100$ *ft/sec*

= *rate of change (velocity) of* C_1

$$\frac{dz}{dt} = \frac{1}{500}[(400)(120) + (300)(100)] = \frac{780}{5} = 156$$

Thus the distance between the two cars is changing at the rate of 156 ft/sec.

(b) Since car C_2 is 400 ft from point A and is traveling at 120 ft/sec, it will arrive at point A in $\frac{400}{120} = \frac{10}{3}$ sec later, while car C_1 will arrive at point A in $\frac{300}{100} = 3$ sec. In other words, the cars (we hope!) will miss one another by $\frac{1}{3}$ of a second.

Exercises 4.7

In Problems 1 through 4, find dy/dx using two methods (assume that y is a differentiable function of x).

1. $x^2 + y^2 = 25$
2. $y^2 = 3x^3$
3. $y^2 = \frac{x+2}{x-2}$
4. $x^3 + y^3 = 1$

In Problems 5 through 14, find dy/dx.

5. $y^2 = 3x^3 + 5x^2 + 7$
6. $y^3 + 3xy = x$
7. $x^2y^2 = a^2(x^2 - y^2)$ (a is a constant)
8. $\frac{x^2}{a^2} + \frac{y^2}{b^2} = 1$ (a and b are constants)
9. $x^5 + y^5 = 5(x^3y^2 + x^2y^3)$
10. $x^{2/3} + y^{2/3} = a^{2/3}$ (a is a constant)
11. $\sqrt{1-x^2} + \sqrt{1-y^2} = 8(x-y)$
12. $(x+4)^5 = 2y^3$
13. $2x^2 = \frac{x+y}{x-y}$
14. $y^2 = \frac{\sqrt[3]{x^2+7}}{3x}$

In Problems 15 through 18, find the equation of the tangent line to the given curve at the given point.

15. $2y^2 - 3x^2 = 6$ at (2, 3)
16. $x^2 + y^2 = 4$ at $(-\sqrt{3}, 1)$
17. $x - 1 - 2\sqrt{x} + \sqrt{y} = 0$ at (4, 1)

18. $\dfrac{y}{x} + \dfrac{x}{y} = 5xy$ at $\left(1, \frac{1}{2}\right)$

19. Show that the two circles

$$(x - 6)^2 + (y - 3)^2 = 20 \quad \text{and} \quad x^2 + y^2 + 2x + y = 10$$

have a common tangent line at $(2, 1)$.

20. A painter is on top of a ladder 25 ft long that is leaning against a vertical wall (Figure 4.19). If a prankster pulls the bottom of the ladder away from

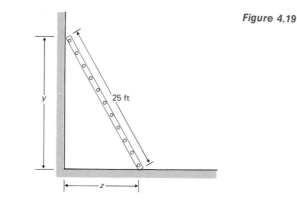

Figure 4.19

the wall at the rate of 6 in./sec, how fast is the painter sliding down when the bottom is 20 ft from the wall?

21. Ship A leaves port at 2:00 P.M. and travels due east at 15 knots. At 6:00 P.M. on the same day, ship B leaves the same port and travels south at 20 knots. How fast is the distance changing between A and B when B has traveled 80 nautical mi?

4.8   Higher Derivatives

Consider the function f, where $f(x) = x^3$. The derivative of f is a new function f', where $f'(x) = 3x^2$. We can again differentiate the function f'. If we do so, we get a new function f'', where $f''(x) = 6x$. If we further differentiate the function f'', we obtain $f'''(x) = 6$, and so on.

In general, if a function f has a derivative f' and if f' itself is differentiable, we denote the derived function of f' by f'' and call f'' the *second derivative* of f.

Continuing in this manner, we obtain the functions $f''', f^{(4)}, \ldots, f^{(n)}$, each of which is the derivative of the preceding function. Of course, here we are assuming that each of the preceding functions is differentiable. There are various notations used for the higher derivatives. If $y = f(x)$, the derivatives can be written as follows:

Note that if f is differentiable at c, f' is not necessarily differentiable at c. For example, if $f(x) = x^{4/3}$, than $f'(x) = \left(\frac{4}{3}\right) x^{1/3}$. Here we find that f is differentiable for all x. In particular, f is differentiable at $x = 0$. However, f' is not differentiable at $x = 0$, since $f''(x) = \left(\frac{4}{9}\right) x^{-2/3}$, which is not defined at $x = 0$.

To avoid confusing powers of f
with derivatives of f for n > 3,
we write $f^{(n)}$ to denote the nth
derivative of f and f^n to denote the
nth power of f.

first derivative $\qquad \dfrac{dy}{dx}, \dfrac{df}{dx}, y', f'$

second derivative $\qquad \dfrac{d^2y}{dx^2}, \dfrac{d^2f}{dx^2}, y'', f''$

third derivative $\qquad \dfrac{d^3y}{dx^3}, \dfrac{d^3f}{dx^3}, y''', f'''$

$$\vdots$$

nth derivative $\qquad \dfrac{d^ny}{dx^n}, \dfrac{d^nf}{dx^n}, y^{(n)}, f^{(n)}$

The reason for the peculiar location of the superscripts in the first two notations is that d/dx is a symbol of the operation of differentiation. Thus we have

$$\frac{dy}{dx} = \frac{d}{dx}(y)$$

$$\frac{d^2y}{dx^2} = \frac{d}{dx}\left(\frac{dy}{dx}\right) = \frac{d}{dx}\left(\frac{d}{dx}(y)\right)$$

$$\frac{d^3y}{dx^3} = \frac{d}{dx}\left(\frac{d^2y}{dx^2}\right) = \frac{d}{dx}\left\{\frac{d}{dx}\left[\frac{d}{dx}(y)\right]\right\}$$

and so on.

Example 4.31   Find the derivatives of all orders for the function defined by $y = x^5$.

SOLUTION   From Theorem 4.2, we have

$$\frac{dy}{dx} = 5x^4$$

and Theorem 4.3 then yields

$$\frac{d^2y}{dx^2} = \frac{d}{dx}\left(\frac{dy}{dx}\right) = \frac{d}{dx}(5x^4) = 20x^3$$

$$\frac{d^3y}{dx^3} = \frac{d}{dx}\left(\frac{d^2y}{dx^2}\right) = \frac{d}{dx}(20x^3) = 60x^2$$

$$\frac{d^4y}{dx^4} = \frac{d}{dx}\left(\frac{d^3y}{dx^3}\right) = \frac{d}{dx}(60x^2) = 120x$$

$$\frac{d^5y}{dx^5} = \frac{d}{dx}\left(\frac{d^4y}{dx^4}\right) = \frac{d}{dx}(120x) = 120$$

$$\frac{d^6y}{dx^6} = \frac{d}{dx}\left(\frac{d^5y}{dx^5}\right) = \frac{d}{dx}(120) = 0$$

and

$$\frac{d^n y}{dx^n} = 0 \qquad \text{for} \quad n > 6$$

In fact, it is not difficult to show (using mathematical induction) that if $y = x^n$, where n is a positive integer, then

$$\frac{d^n y}{dx^n} = n! \quad \text{and} \quad \frac{d^k y}{dx^k} = 0 \qquad \text{for } k > n$$

Example 4.32   If $\qquad\qquad\qquad x^2 + 4y^2 = 25, \qquad\qquad\qquad$ (4.39)

show that $\qquad\qquad\qquad \dfrac{d^2 y}{dx^2} = -\dfrac{25}{16y^3}, \quad y \neq 0$

SOLUTION   Differentiating both sides of the equation $x^2 + 4y^2 = 25$ with respect to x we obtain

Here we assume that y and dy/dx define differentiable functions.

$$\frac{d}{dx}(x^2 + 4y^2) = \frac{d}{dx}(25)$$

$$\frac{d}{dx}(x^2) + \frac{d}{dx}(4y^2) = 0$$

or

$$2x + 8y\frac{dy}{dx} = 0$$

Hence

$$\frac{dy}{dx} = -\frac{x}{4y}, \qquad y \neq 0 \qquad (4.40)$$

and so

$$\frac{d^2 y}{dx^2} = \frac{d}{dx}\left(-\frac{x}{4y}\right)$$

Theorem 4.7 then yields

$$\frac{d^2 y}{dx^2} = -\frac{1}{4}\left(\frac{y - x\dfrac{dy}{dx}}{y^2}\right) \qquad (4.41)$$

Substituting (4.40) into Equation (4.41), we have

$$\frac{d^2 y}{dx^2} = -\frac{1}{4}\left[\frac{y - x(-x/4y)}{y^2}\right] = -\frac{1}{4}\left[\frac{4y^2 + x^2}{4y^3}\right]$$

and from Equation (4.39), we obtain

$$\frac{d^2 y}{dx^2} = -\frac{25}{16y^3}, \qquad y \neq 0$$

If the motion of an object is described by distance as a function of time, say

$$s = f(t)$$

then, as we have indicated, the first derivative ds/dt gives the instantaneous velocity at time t. In this case the second derivative d^2s/dt^2 gives what is called the *instantaneous acceleration* of the object at time t. In other words, acceleration is the rate at which the velocity is changing with respect to time.

Example 4.33   A toy rocket is launched vertically upward with an initial velocity of 120 ft/sec. The distance s traveled t sec later is given by

$$s = 120t - 16t^2$$

Find the velocity and the acceleration. How high will the rocket rise?

Figure 4.20

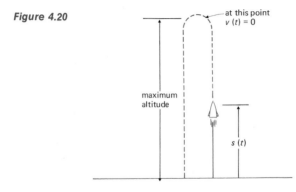

SOLUTION   Here we find that the velocity $v(t)$ is

$$v(t) = \frac{ds}{dt} = 120 - 32t$$

and the acceleration $a(t)$ is

$$a(t) = \frac{d^2s}{dt^2} = -32$$

This result tells us that the velocity of the rocket is changing at the rate of -32 ft/sec every second. The negative sign indicates that the velocity is decreasing; it will become zero ($v = 0$) when

$$120 - 32t = 0 \quad \text{or when} \quad t = \tfrac{120}{32} = \tfrac{15}{4} \text{ sec}$$

The distance traveled during this time interval is then

$$s\left(\tfrac{15}{4}\right) = 120\left(\tfrac{15}{4}\right) - 16\left(\tfrac{15}{4}\right)^2 = 450 - 225 = 225 \text{ ft}$$

In Problems 1 through 12, find the first and second derivatives.   Exercises 4.8

1. $f(x) = x^3 - 6$

2. $f(x) = 2x^5 - 3x^2 + 1$

3. $f(x) = x^3 + 2x^{3/2} - 4$

4. $f(x) = \sqrt{x} + \dfrac{1}{x}$

5. $f(x) = \sqrt{x^2 + 1}$

6. $f(x) = (2x^3 + 1)^{3/2}$

7. $f(x) = \dfrac{x+1}{x-1}$

8. $f(x) = x^2\sqrt{x} - 3x$

9. $f(x) = \sqrt{2x} + \dfrac{1}{\sqrt{2x}}$

10. $f(x) = \dfrac{1}{\sqrt{3x^2 + 1}}$

11. $f(x) = \dfrac{x^2 - 1}{x^2 + 1}$

12. $f(x) = \dfrac{1 - \sqrt{x}}{1 + \sqrt{x}}$

In Problems 13 through 16, find d^2y/dx^2.

13. $x + y = 6y^2$

14. $x^2 + y^2 = 1$

15. $x^{2/3} + y^{2/3} = 1$

16. $4x^2 + 9y^2 = 36$

In Problems 17 through 20, find $f'''(x)$.

17. $f(x) = x^4 - 3x^{-1} + 4$

18. $f(x) = \sqrt{4x + 1}$

19. $f(x) = \dfrac{1}{x-1}$

20. $f(x) = \dfrac{x}{(1-x)^2}$

* 21. If $y^3 + 3x^2 + 1 = 0$, show that

$$y^5 \frac{d^2y}{dx^2} + 2(x^2 - 1) = 0$$

22. If $x^{1/2} + y^{1/2} = 2$, show that

$$\frac{d^2y}{dx^2} = x^{-3/2}$$

For the equations of motion in Problems 23 through 26, find the velocity function $v(t)$ and the acceleration function $a(t)$.

23. $s(t) = 600 - 20t - 16t^2$

24. $s(t) = (4t + 7)^2$

25. $s(t) = 2t^3 - 12t^2 + 24t + 7$

26. $s(t) = t^3 + 6t^2 - 21t + 15$

27. For each of Problems 23 through 26, perform the following exercises.
     (a) Find the initial velocity $v(0)$.
     (b) Find the values of t^* for which $v(t^*) = 0$.
     (c) Compute the corresponding values of $a(t^*)$.

28. Let the reaction R of a body as a function of the dosage x of a drug be given by the equation,

$$R = 2x^2 \left(\frac{1}{4} - \frac{x}{6} \right)$$

     (a) Compute dR/dx and d^2R/dx^2.
     (b) Find dR/dx at $x = \frac{1}{4}, \frac{1}{3}, \frac{1}{2}, \frac{2}{3}, \frac{3}{4}$.

(c) Draw a graph of dR/dx vs x for $0 \le x \le 1$.

(d) Find d^2R/dx^2 at $x = \frac{1}{4}, \frac{1}{3}, \frac{1}{2}, \frac{2}{3}, \frac{3}{4}$.

Chapter 4 Review

Part 1   (Oral)

1.   Define the derivative of a function.

2.   What is meant by saying that f is differentiable at the point c?

3.   If a function f is differentiable at a point c, is it also continuous at c?

4.   If f is continuous at c, is it also differentiable at c?

5.   Define marginal cost and marginal revenue.

6.   Define instantaneous velocity and instantaneous acceleration.

7.   If f is a differentiable function and

   (a) $f(x) = u(x) + v(x)$, then what is $f'(x)$?

   (b) $f(x) = u(x)\, v(x)$, then what is $f'(x)$?

   (c) $f(x) = \dfrac{u(x)}{v(x)}$, then what is $f'(x)$?

8.   If f is a differentiable function at c, then what does $f'(c)$ represent geometrically?

9.   State Rolle's theorem.

10.   State the mean value theorem.

11.   What is the chain rule?

Part 2   (Written)

In Problems 1 through 4, use Definition 4.1 to find $f'(2)$.

1. $f(x) = 2x^2 + 1$

2. $f(x) = 10$

3. $f(x) = \sqrt{2x}$

4. $f(x) = \frac{1}{2}x^{-1}$

In Problems 5 through 14, find $f'(x)$ using the theorems in this chapter.

5. $f(x) = x^4 - 2x^{1/3} + 3$

6. $f(x) = 2\sqrt{x} + \dfrac{1}{2x}$

7. $f(x) = (3x^2 + x)(7x^2 + x^{-3} + 5)$

8. $f(x) = \dfrac{3x^2 + 1}{2x - 3}$

9. $f(x) = (3x^3 - 7x + 1)^{31}$

10. $f(x) = \sqrt{x}(4x^2 + 1)^{12}$

11. $f(x) = (x^2 + 1)\sqrt{2x + 1}$

12. $f(x) = \dfrac{x}{(x^2 + 2)^4}$

13. $f(x) = \dfrac{1}{x}(x^{-2} - x^2)^{-2}$

14. $f(x) = \left(\dfrac{x+1}{1-x}\right)^{3/2}$

15. Let $f(x) = \sqrt{x-1}$. Show that $f'(1)$ does not exist.

16. Suppose the total cost C of producing x units of a product is given by $C(x) = (2/x) + 3x + 150$. Find the following.

   (a) The marginal cost function

   (b) The marginal cost when 40 units are produced

   (c) The cost of manufacturing 40 units

17. An object thrown vertically upward moves according to the law $s(t) = -16t^2 + 64t$, where s is the distance measured in feet from the ground and t is the time measured in seconds. Find the following.

   (a) The initial velocity

   (b) The velocity at $t = 1$ sec

   (c) How many seconds it will take the object to reach its highest point

   (d) How high the object will rise

18. An object is thrown vertically upward from the roof of a building 336 ft high. If the equation of motion of the object is $s = -16t^2 + 96t$, find the following.

   (a) The instantaneous velocity of the object at 2 sec

   (b) How high the object will go

   (c) How long it will take for the object to reach the ground

19. Find the equation of the tangent line to the curve

$$y = 2x^2 - 3x + 1$$

at $x = 4$.

20. Find the point at which the slope of the tangent line to the curve $y = 2x^2 - 3x + 1$ is 5.

21. At what point on the curve $y = -2x^2 + 4x$ is the tangent line parallel to the x-axis?

22. Verify Rolle's theorem for the function $f(x) = x^2 - 3x + 2$, $[1, 2]$.

23. If $f(x) = -2x^2 - x + 3$, find a number c between -2 and $\frac{1}{2}$ which satisfies the equation

$$f'(c) = \frac{f\left(\frac{1}{2}\right) - f(-2)}{\frac{1}{2} - (-2)}$$

24. Find dy/dx if $x^3 + y^2 - 2xy = 0$.

25. Find the equation of the tangent line to the curve

$$2x^2 - y^2 + xy = 0$$

at $x = 1$.

26. Find d^2y/dx^2 if $2x + y = 3y^2$.

27. Find $f'''(x)$ given that $f(x) = 1 - 2x$.

28. A bomber is flying parallel to the ground at an altitude of 2 km and at a constant speed of 6 km/min. If the bomber flies directly over a tank, at what rate is the line-of-sight distance between the bomber and the tank changing 30 sec later?

29. A stone dropped into a still pond of water sends out concentric circular ripples. If the radius of the outermost ripple increases at the rate of 1 m/sec, find how quickly the area of the disturbed surface is increasing at the end of 10 sec.

30. Two roads cross each other at right angles. One person starts from the junction point along one road with a constant speed of 1.5 m/sec and another person walks toward the junction point from a distance of 12.5 m on the other road with a uniform speed of 1.2 m/sec. Find the rate of increase of the distance between the two people at the end of 5 sec.

31. A spherical iron ball is heated and its radius is found to increase at the rate of 0.25 cm per minute. Find the rates at which the surface area of the sphere and its volume increase when the radius of the sphere has become 12 cm.

32. Water is flowing into a vertical circular cylinder of diameter 4 ft at the rate of 10 cu ft/min. Find how quickly the surface of water rises in the tank.

Numerical Applications Using Calculators

We shall use a pocket calculator to find the derivative of a function from its definition. Consider the following example.

Example 4.34 $f(x) = x^2 + 1$. By definition,

$$f'(x) = \lim_{h \to 0} \frac{f(x+h) - f(x)}{h} = \lim_{h \to 0} \frac{(x+h)^2 + 1 - (x^2 + 1)}{h}$$

Suppose we want to find $f'(2)$,

$$f'(2) = \lim_{h \to 0} \frac{(2+h)^2 + 1 - (2^2 + 1)}{h}$$

We obtain Table 4.3 by using a calculator.

Table 4.3

h	$(2+h)^2+1$	$(2+h)^2+1-(2^2+1)$	$\dfrac{(2+h)^2+1-(2^2+1)}{h}$
0.1	5.41	0.41	4.1
0.01	5.0401	0.0401	4.01
0.001	5.004001	0.004001	4.001
0.0001	5.00040001	0.00040001	4.0001
0.00001	5.00004	0.00004	4
−0.1	4.61	−0.39	3.9
−0.01	4.9601	−0.0399	3.99
−0.001	4.996001	−0.003999	3.999
−0.0001	4.99960001	−0.00039999	3.9999
−0.00001	4.99996	−0.00004	4

Therefore it is reasonable to conclude that $f'(2) = 4$.

Example 4.35 Find $f'(5)$ if $f(x) = \sqrt{x}$. Again, by definition,

$$f'(5) = \lim_{h \to 0} \frac{f(5+h) - f(5)}{h} = \lim_{h \to 0} \frac{\sqrt{5+h} - \sqrt{5}}{h}$$

We use the calculator to obtain Table 4.4.

Table 4.4

h	$5+h$	$\sqrt{5+h}$	$\sqrt{5+h} - \sqrt{5}$	$\dfrac{(\sqrt{5+h} - \sqrt{5})}{h}$
0.1	5.1	2.258317958	0.022249981	0.22249981
0.01	5.01	2.238302929	0.002234952	0.2234952
0.001	5.001	2.236291573	0.000223596	0.223596
0.0001	5.0001	2.236090338	0.000022361	0.22361
0.00001	5.00001	2.23609070214	0.000002237	0.2237
−0.1	4.9	2.213594362	−0.022473615	0.22473615
−0.01	4.99	2.23383079	−0.002237187	0.2237187
−0.001	4.999	2.23584436	−0.000223617	0.223617
−0.0001	4.9999	2.236045617	−0.00002236	0.2236
−0.00001	4.99999	2.236065741	−0.000002236	0.2236

From Table 4.4 we see that $f'(5) \approx 0.2236$.

Check   If $f(x) = \sqrt{x}$, then $f'(x) = 1/(2\sqrt{x})$. Thus

$$f'(5) = \frac{1}{2\sqrt{5}}$$

$$\sqrt{5} \approx 2\ 236067977, \qquad 2\sqrt{5} \approx 4.472135954$$

and

$$\frac{1}{2\sqrt{5}} \approx 0.2236067978$$

Exercises Use a pocket calculator to compute the derivative of each of the following functions at the indicated value.

1. $f(x) = 2x^2 + x - 1$ at $x = 3$

2. $f(x) = \dfrac{1}{x}$ at $x = -1$

3. $f(x) = \sqrt{x - 1}$ at $x = 3$

4. $f(x) = 2x^3 - \dfrac{1}{\sqrt{x}}$ at $x = 5$

5. $f(x) = \dfrac{x + 1}{x - 1}$ at $x = 10$

6. $f(x) = \dfrac{1}{x^2 + 2x}$ at $x = 1$

7. $f(x) = x + \sqrt{x}$ at $x = 2$

8. $f(x) = (x - 1)(x^2 + 2)$ at $x = -2$

9. $f(x) = (x + 1)^8$ at $x = 0.1$

10. $f(x) = (2x + 1)^{2/3}$ at $x = 1$

More on the Derivative

CHAPTER 5

5.1 Increasing and Decreasing Functions

In Chapter 3 we indicated that a continuous function can usually be visualized as a function whose graph can be drawn over an interval without removing pencil from paper. Thus if we know that the given function is continuous, we can sketch its graph by plotting enough points and then connecting the successive points by a curve. The derivative, by telling us about the slope of the tangent lines to the graph, provides much information about the behavior of the function and thus helps us in graphing it. To illustrate other relations between the derivative and the graph, let us consider the function f whose graph is given in Figure 5.1.

We note from the graph that the function is *increasing* (the graph is

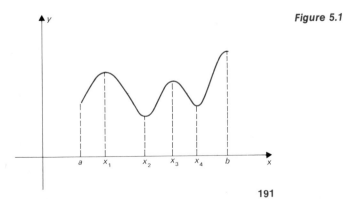

Figure 5.1

rising) in the intervals (a, x_1), (x_2, x_3), and (x_4, b), and that the function is *decreasing* (the graph is falling) in the intervals (x_1, x_2) and (x_3, x_4). Moreover, the slope of the tangent line (the derivative) is positive over the intervals where the function is increasing, and is negative over the intervals where the function is decreasing. Thus it seems safe to guess that an increasing function has a positive derivative and a decreasing function has a negative derivative.

Increasing and decreasing functions: A function f is increasing (decreasing) at a point c, if it is increasing (decreasing) over an interval containing c.

▶ **Definition 5.1** *A function f is said to be* increasing *over an interval if, for any two points* x_1, x_2 *of the interval such that* $x_1 < x_2$, *we have* $f(x_1) < f(x_2)$. *Similarly, f is* decreasing *over the interval if* $x_1 < x_2$ *implies that* $f(x_1) > f(x_2)$.

The following theorem shows how the derivative helps in determining whether a function is increasing or decreasing.

▶ **Theorem 5.1** *Let f be a differentiable function. Then f is increasing over any interval on which* $f'(x)$ *is positive, and f is decreasing over any interval on which* $f'(x)$ *is negative.*

PROOF   Suppose $f'(x) > 0$ for each x in the open interval (a, b). Let x_1, x_2 be any two points in this interval with $x_1 < x_2$. Then by the mean value theorem (see Chapter 4), there exists a point $c \in (a, b)$, such that

$$f(x_2) - f(x_1) = f'(c)(x_2 - x_1) \tag{5.1}$$

Since $x_1 < x_2$, we have $x_2 - x_1 > 0$ and by hypothesis, $f'(c) > 0$. Consequently, the right-hand side of (5.1) is positive. Hence

$$f(x_2) - f(x_1) > 0 \quad \text{or} \quad f(x_2) > f(x_1) \tag{5.2}$$

This proves that for any two points x_1, x_2 in (a, b) with $x_1 < x_2$, we have $f(x_1) < f(x_2)$. Therefore by definition, f is increasing over (a, b).

A similar argument proves that if $f'(x) < 0$ for each x in (a, b), then f is decreasing over (a, b).

We point out that the converse of Theorem 5.1, "If a function f is differentiable and increasing over (a, b), then $f'(x) > 0$ for any $x \in (a, b)$," is not true.

For example, let us consider the function f given by

$$f(x) = x^3 \qquad (-1 < x < 1)$$

Then we have

$$f'(x) = 3x^2$$

Since $f'(x) > 0$ for all $x \neq 0$, the function is increasing over $(-1, 0) \cup (0, 1)$. In fact, f is increasing over $(-1, 1)$: If $x < 0$, then

$x^3 < 0$, and if $x > 0$, then $x^3 > 0$ (see Figure 5.2). However, $f'(0) = 0$

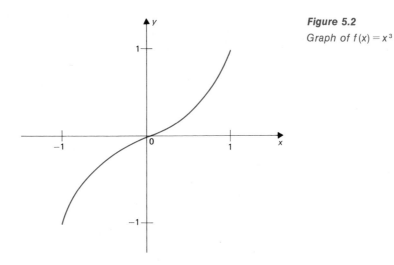

Figure 5.2
Graph of $f(x) = x^3$

(not positive). From the preceding discussion, we conclude that the points where the derivative is zero or where the derivative is undefined require separate consideration.

In conclusion, the sign of $f'(x)$ gives us the following information about the graph of the original function f.

Sign of $f'(x)$	*Graph of f*
(i) $f'(x) > 0$ on an interval (a, b)	f is increasing on (a, b); graph of f is rising
(ii) $f'(x) < 0$ on an interval (a, b)	f is decreasing on (a, b); graph of f is falling

Table 5.1

Example 5.1 Find the intervals on which the graph of the function f, defined by

$$f(x) = x^2 - 2x + 3$$

is increasing and decreasing. Sketch the graph of f.

SOLUTION We have

$$f'(x) = 2x - 2 = 2(x - 1)$$

It is easy to see that for all $x \in (-\infty, 1)$, $f'(x) < 0$. Thus the function f is decreasing on the interval $(-\infty, 1)$. Also, for all $x \in (1, \infty)$, $f'(x) > 0$. Hence the function f is increasing on the interval $(1, \infty)$. By using this information and plotting a few points $(x, f(x))$, we obtain the graph of f shown in Figure 5.3.

Figure 5.3

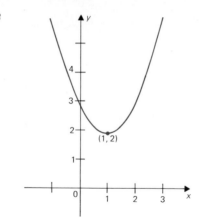

Example 5.2 Find the intervals in which the function f defined by $f(x) = x^3 - 6x^2 + 9x - 7$ is increasing and decreasing. Sketch the graph of f.

SOLUTION We have

$$f(x) = x^3 - 6x^2 + 9x - 7$$

Differentiating, we obtain

$$f'(x) = 3x^2 - 12x + 9 = 3(x^2 - 4x + 3)$$

or

$$f'(x) = 3(x - 1)(x - 3) \qquad (5.3)$$

Now, $f'(x)$ is zero only for the values $x = 1$ and $x = 3$. These two points where the derivative is zero divide the x-axis into three sets:

$$I_1 = \{x \mid -\infty < x < 1\}$$
$$I_2 = \{x \mid 1 < x < 3\}$$
$$I_3 = \{x \mid 3 < x < \infty\}$$

If $x < 1$, then $x < 3$, and $(x - 1)(x - 3) > 0$. Hence $f'(x) > 0$ and f is increasing. Similarly, if $x > 3$, then $x > 1$, and $(x - 1)(x - 3) > 0$. Therefore $f'(x) > 0$ and f is increasing. If $1 < x < 3$, then $(x - 1) > 0$ and $(x - 3) < 0$. As a result, $(x - 1)(x - 3) < 0$. Hence $f'(x) < 0$ and f is decreasing.

Observe that for $x = 1$, $f(1) = -3$, and for $x = 3$, $f(3) = -7$.

With this information, we can sketch the graph of f roughly as in Figure 5.4. As we draw the graph from left to right, it rises to the point $(1, -3)$, then falls to the point $(3, -7)$, and rises thereafter.

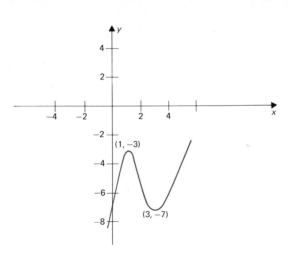

Example 5.3   A manufacturer can sell x units of a product per day if the unit price is $p = 8 - 0.01x$. Determine whether the total revenue is increasing or decreasing when x is 200, 300, 500, and 600. Sketch a graph.

SOLUTION   The total revenue is

$$R(x) = xp(x)$$

or
$$R(x) = 8x - 0.01x^2 \tag{5.4}$$

Taking the derivative on both sides of Equation (5.4), we obtain

$$\frac{dR}{dx} = 8 - 0.02x \tag{5.5}$$

At $x = 200$,
$$\left.\frac{dR}{dx}\right|_{x=200} = 8 - (0.02)(200) = 4$$

Since this result is a positive number, the total revenue is increasing at $x = 200$. Similarly, for $x = 300$, we have

$$\left.\frac{dR}{dx}\right|_{x=300} = 8 - (0.02)(300) = 2$$

Hence $R(x)$ is increasing at $x = 300$. At $x = 500$,

$$\left.\frac{dR}{dx}\right|_{x=500} = 8 - (0.02)(500) = -2$$

a negative number, and so the total revenue is decreasing. Finally, at $x = 600$, the total revenue is decreasing, since

$$\left.\frac{dR}{dx}\right|_{x=600} = 8 - (0.02)(600) = -4$$

In fact, we find that

$$\frac{dR}{dx} > 0 \qquad \text{whenever} \quad x < 400$$

and

$$\frac{dR}{dx} < 0 \qquad \text{whenever} \quad x > 400$$

Thus the total revenue is increasing on any interval where $x < 400$ and is decreasing when $x > 400$. A rough sketch of the graph of R is shown in Figure 5.5.

Figure 5.5

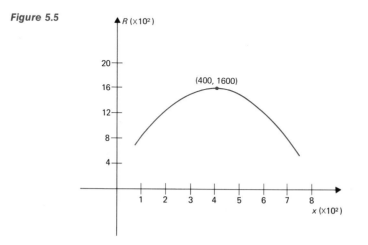

Example 5.4 Find the intervals on which the function f defined by

$$f(x) = \frac{x+1}{x-1}$$

is increasing and decreasing. Sketch the graph of f.

SOLUTION We have

$$f(x) = \frac{x+1}{x-1}$$

so that

$$f'(x) = \frac{(x-1) - (x+1)}{(x-1)^2} = -\frac{2}{(x-1)^2}$$

Now for all $x \neq 1$, $f'(x) < 0$. Consequently, the function f is decreasing over the interval $(-\infty, 1) \cup (1, \infty)$. Note that the function f is discontinuous at $x = 1$. From the previous information about f, the plotting of a few additional points, and the observation that

$$\lim_{x \to -\infty} f(x) = 1 \qquad \text{(Why?)} \quad \text{and} \quad \lim_{x \to \infty} f(x) = 1 \qquad \text{(Why?)}$$

we can sketch the graph of f as shown in Figure 5.6.

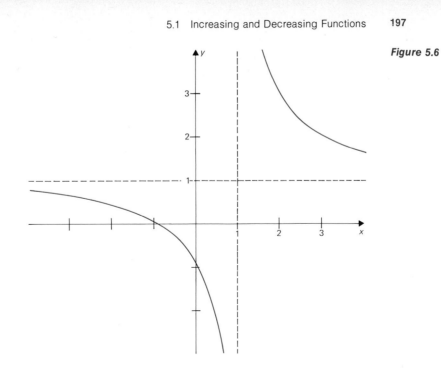

Figure 5.6

In Problems 1 through 22, find the intervals where f is increasing and decreasing, and then sketch the graph of f.

Exercises 5.1

1. $f(x) = 2x - 1$

2. $f(x) = -3x + 4$

3. $f(x) = -\frac{1}{2}x - 1$

4. $f(x) = 4x^2$

5. $f(x) = -2x^2$

6. $f(x) = x^2 - 2x + 1$

7. $f(x) = x^2 + 6x + 3$

8. $f(x) = -3x^2 + 2x$

9. $f(x) = x^2 - 3x + \frac{1}{4}$

10. $f(x) = -\frac{1}{2}x^2 + 4x - 1$

11. $f(x) = x^3 - 3x^2 - 9x + 15$

12. $f(x) = x^3 - 12x$

13. $f(x) = 3x^5 - 5x^3$

14. $f(x) = 4x^3 + 3x^2 - 6x - 1$

15. $f(x) = 2x^4 - x^2$

16. $f(x) = 4x^3 - 15x^2 - 18x + 2$

17. $f(x) = x^4 - 8x^3 + 22x^2 - 24x + 1$

18. $f(x) = \dfrac{1}{x}$

19. $f(x) = \dfrac{1}{x - 1}$

20. $f(x) = \dfrac{x - 1}{x + 1}$

21. $f(x) = x - \dfrac{1}{x}$

22. $f(x) = -2x + \dfrac{2}{x}$

23. Complete the proof of Theorem 5.1.

24. Suppose that the rate R of an autocatalytic reaction is given by

$$R = kx(4 - x) \qquad 0 \le x \le 4$$

where k is a positive constant and x is the amount of substance produced. Find the intervals where the rate R is increasing and where it is decreasing. Sketch the graph.

25. A dress manufacturer has found that the number x of dresses sold per month is related to the price p dollars per dress by the equation

$$p = 2 - \left(\frac{x}{300}\right)^2$$

Determine whether the marginal revenue is increasing or decreasing when x is 200, 300, 500, and 600. Sketch a graph of the marginal revenue.

5.2   Extreme Values of a Function

In many practical problems, we are interested in determining the maximum or minimum values of a function. Geometrically, of course, these values correspond to high and low points on a function's graph. Presently we shall see how the derivative of a function is helpful in locating these values. We begin our discussion with the following definition.

Critical points on the graph ▶
of a function f

Definition 5.2   *Let f be a function. A point c for which*

$$f'(c) = 0$$

is called a critical number in the domain of f, and the point $(c, f(c))$ is called a critical point of the graph of f.

By our definition, the slope of the tangent line to the graph of $y = f(x)$ at a critical point $(c, f(c))$ is zero. Consequently, the tangent line at the critical point is parallel to the x-axis. The function whose graph is given in Figure 5.7 has three critical points: $(x_1, f(x_1))$, $(x_2, f(x_2))$ and $(x_3, f(x_3))$.

Figure 5.7

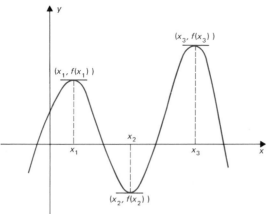

To compute the critical numbers of a function f, we need to solve the equation

$$f'(x) = 0$$

for x.

Example 5.5   Find the critical numbers of

(a) $f(x) = 2x^3 - 3x^2 - 36x + 5$ (b) $g(x) = 5 + \dfrac{1}{x}$

SOLUTION

(a) Differentiating, we get

$$f'(x) = 6x^2 - 6x - 36$$

so the critical numbers must satisfy the equation

$$6x^2 - 6x - 36 = 0 \quad \text{or} \quad x^2 - x - 6 = 0 \quad \text{or} \quad (x + 2)(x - 3) = 0$$

Thus $x = -2$ and $x = 3$ are the critical numbers of the function f.

(b) Differentiating, we get

$$g'(x) = -\frac{1}{x^2}$$

Since $-1/x^2$ is never zero, the function g has no critical numbers.

Definition 5.3   *A function f has a* relative maximum *value at c if there is an open interval I in the domain of f containing c, such that*   ◀ *Relative extremum*

$$f(c) \ge f(x) \qquad \text{for all} \quad x \in I$$

Similarly, f has a relative minimum *value at c if there is an open interval I in the domain of f containing c, such that*

$$f(c) \le f(x) \qquad \text{for all} \quad x \in I$$

The term relative extremum *covers both situations.*

In Figure 5.7 the function f has relative maxima at the points x_1 and x_3, and a relative minimum at x_2.

Definition 5.4   *A function f is said to have an* absolute maximum *at a if $f(a) \ge f(x)$ for all x in the domain of f. Similarly, f has an* absolute minimum *at b if $f(b) \le f(x)$ for all x in the domain of f.*   ◀ *Absolute extremum*

Let us discuss some of the preceding concepts in the following example.

Example 5.6   Find the extreme values of the following functions.

(a) $f(x) = x^2$ $(-\infty < x < \infty)$

(b) $f(x) = x^2$ $(-2 \le x \le 2)$

(c) $f(x) = x^2$ $(-1 \le x < 3)$

SOLUTION   The graphs of the functions are given in Figure 5.8.

The relative and, in fact, the absolute minimum value for (a), (b), and (c) is at $x = 0$, since $f(0) = 0 \leq x^2 = f(x)$ for all x in the domain of each

Figure 5.8

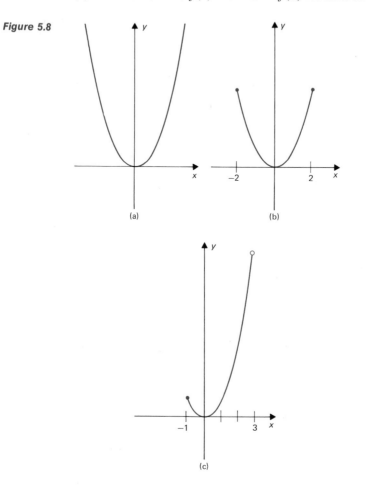

(a)

(b)

(c)

of the functions. Obviously, the function $f(x) = x^2$ ($-\infty < x < \infty$) has no relative or absolute maxima. The function

$$f(x) = x^2 \qquad (-2 \leq x \leq 2)$$

has an absolute maximum value at $x = -2$ and at $x = 2$. The function

$$f(x) = x^2 \qquad (-1 \leq x < 3)$$

has no relative or absolute maximum value. (Why?)

We shall now prove a theorem which gives a connection between the critical numbers and the extreme values of a function.

▶   **Theorem 5.2**   *Let f be differentiable at c, and suppose f has a relative extremum at c. Then*

$$f'(c) = 0$$

In other words, c is a critical number of f.

PROOF   We recall that

$$f'(c) = \lim_{h \to 0} \frac{f(c+h) - f(c)}{h} \qquad (5.6)$$

The limit in (5.6) is the same if h approaches zero through positive or negative values.

Suppose that f has a relative maximum at c. Since f is differentiable at c, it is defined in some open interval I_1 in the domain of f containing c. Also, because f has a relative maximum at c, there must be an open interval I_2 containing c such that

$$f(x) \le f(c) \qquad \text{for all} \quad x \in I_2$$

Thus for all h, such that $c + h \in I = I_1 \cap I_2$, we have

$$f(c+h) - f(c) \le 0 \qquad (5.7)$$

Now if $h < 0$, we have

$$\frac{f(c+h) - f(c)}{h} \ge 0$$

If we let h approach zero through negative values, we have

$$f'(c) = \lim_{h \to 0^-} \frac{f(c+h) - f(c)}{h} \ge 0 \qquad (5.8)$$

Similarly, if $h > 0$, we have, from (5.7),

$$\frac{f(c+h) - f(c)}{h} \le 0$$

So if we let h approach zero through positive values, we have

$$f'(c) = \lim_{h \to 0^+} \frac{f(c+h) - f(c)}{h} \le 0 \qquad (5.9)$$

From (5.8) and (5.9), we conclude that $f'(c) \le 0 \le f'(c)$, and hence that $f'(c) = 0$.

The proof for the case when f has a relative minimum at c is similar and is left as an exercise.

We advise the reader not to jump to the conclusion that a local extremum always exists when the derivative is zero. The theorem does not say that. The reader should carefully reread the theorem. Let us see why such care is necessary by considering some examples.

The function

$$\frac{f(c+h) - f(c)}{h}$$

is nonnegative; it cannot have a negative limit.

Example 5.7   Find the critical numbers and the extreme values for the function

$$f(x) = x^3 \qquad (-\infty < x < \infty)$$

SOLUTION   The graph of this function is sketched in Figure 5.9. Solving the equation

Figure 5.9

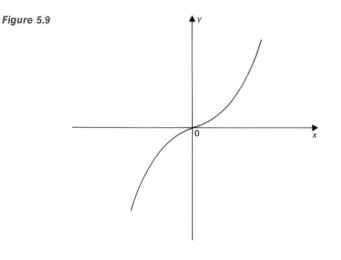

$$f'(x) = 3x^2 = 0$$

to find all the critical numbers of f, we find that $x = 0$ is the only critical number.

We determine that $f(0) = 0$ is *not* a relative extreme value. Since for $h > 0$, $f(-h) = -h^3 < 0$ and $f(h) = h^3 > 0$, $f(0)$ is neither greater nor less than both $f(-h)$ and $f(h)$.

As the graph shows, this function f is an increasing function. It has no extreme values.

The next example shows that a function may have an extremum, either relative or absolute, without the derivative being zero at that point.

Example 5.8   Find the extreme values of the function

$$f(x) = |x| (-\infty < x < \infty)$$

Figure 5.10

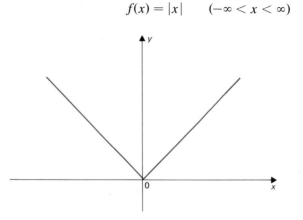

SOLUTION   The graph of this function is sketched in Figure 5.10. Clearly, $f(0) = 0$ is a relative minimum value (in fact, an absolute minimum value). However, $f'(0)$ is not equal to zero, because $f'(0)$ does not exist! (See Example 4.5, Section 4.1.)

The above examples illustrate that the derivative does not always tell us where the extreme values are. Rather, it tells us where the extreme values are not. If the derivative f' is positive (or negative) at a point c, then f is increasing (or decreasing) on an interval containing the point c. Thus there cannot be an extreme value of f at c.

Consequently, if we wish to locate the extreme values of a function, only three kinds of points need be examined.

(i)   Points where the derivative is zero

(ii) Points where the derivative does not exist

(iii) Endpoints of the intervals over which the function is defined

Remark   Example 5.7 illustrates that a function need not have an extreme value. The following theorem, whose proof is beyond the scope of this book, shows that an important class of functions always has maximum and minimum values.

Theorem 5.3   *If a function f is continuous on a closed interval, then f has a maximum value and a minimum value on that interval.*

◀ For a function continuous on a closed interval, see Problem 28, Exercises 3.4.

The following theorem is used in many problems to test for relative maxima and minima.

Theorem 5.4   *Let a function f have a continuous second derivative at a critical number c.*

◀ Second-derivative test

(i)   *If $f''(c) > 0$, then f has a relative minimum at c.*

(ii) *If $f''(c) < 0$, then f has a relative maximum at c.*

The proof of this theorem will be sketched in Section 5.5.

Example 5.9   Find the relative maxima and minima for the function

$$f(x) = x^3 - 6x^2 + 9x - 7 \qquad (0 \le x \le 5) \qquad (5.10)$$

Find the absolute extrema.

SOLUTION   We have

$$f'(x) = 3x^2 - 12x + 9 = 3(x - 1)(x - 3)$$

and the critical numbers are obtained from the equation

$$f'(x) = 0 \quad \text{or} \quad 3(x - 1)(x - 3) = 0$$

namely, $x = 1$ and $x = 3$. Now

$$f''(x) = 6x - 12$$

Thus we have

$$f''(1) = 6(1) - 12 = -6 < 0$$

Hence $f(1)$ is a relative maximum. Similarly,

$$f''(3) = 6(3) - 12 = 6 > 0$$

Hence $f(3)$ is a relative minimum.

Next, we check the endpoints. We find that $f(0) = -7$ and $f(5) = 13$. To compute the absolute extrema, we compare

$$f(0) = -7 \qquad f(1) = -3 \qquad f(3) = -7 \qquad f(5) = 13$$

The function has an absolute maximum at $x = 5$ and an absolute minimum at $x = 0$ and $x = 3$.

Example 5.10   Find the relative maxima and minima for the function

$$f(x) = x + \frac{9}{x} \tag{5.11}$$

SOLUTION   We have

$$f'(x) = 1 - \frac{9}{x^2}$$

Since $f(x)$ is undefined for $x = 0$, the function cannot have a relative maximum or a relative minimum value at 0.

Now from the equation

$$f'(x) = 1 - \frac{9}{x^2} = 0$$

we have

$$x^2 - 9 = 0$$

and the critical numbers are $x = 3$ and $x = -3$. Since

$$f''(x) = \frac{18}{x^3}$$

we have

$$f''(3) = \frac{18}{(3)^3} = \frac{2}{3} > 0$$

Thus by Theorem 5.4, $x = 3$ yields a relative minimum value for f. This minimum value is given by

$$f(3) = 3 + \frac{9}{3} = 6$$

Also, we find that

$$f''(-3) = \frac{18}{(-3)^3} = -\frac{2}{3} < 0$$

Thus $x = -3$ yields a relative maximum value for f. This maximum value is given by

$$f(-3) = (-3) + \frac{9}{(-3)} = -6 \qquad \text{(See Figure 5.11.)}$$

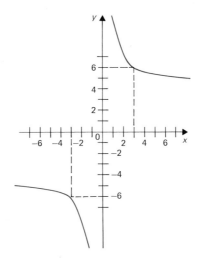

Figure 5.11

1. Consider the function whose graph is shown in Figure 5.12. State whether the function has a relative maximum, relative minimum, absolute maximum, or absolute minimum at each of the points $x_1, x_2, x_3, x_4, x_5, x_6,$ $x_7,$ and x_8.

Exercises 5.2

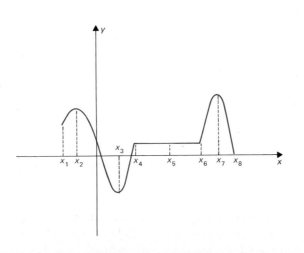

Figure 5.12

In Problems 2 through 25, find the critical numbers and determine the relative maxima and the relative minima. Find the absolute extrema if any exist.

2. $f(x) = x^2 + 4x - 5$ $(0 \le x \le 2)$ 3. $f(x) = 5 - 4x - x^2$ $(-4 \le x \le 4)$

4. $f(x) = 2x^3 - 6x^2 + 9$ 5. $f(x) = x^3 + 3x^2 + 3x$

6. $f(x) = x^3 - 3x + 2$ 7. $f(x) = x^3 + 3x + 3$

8. $f(x) = 2x^4 + x^2$ 9. $f(x) = x^3 + 3x^2 - 12x + 7$

10. $f(x) = 2x^4 - x^2$ $(-1 \le x \le 1)$ 11. $f(x) = x^3 - 3$ $(-1 \le x \le 1)$

12. $f(x) = x^4 + 3x^2 - 6$ 13. $f(x) = \dfrac{x}{x+1}$

14. $f(x) = \dfrac{1}{x^2 + 1}$ 15. $f(x) = \dfrac{1 + x^2}{x^2}$

16. $f(x) = \dfrac{1 - x^2}{x^3}$ 17. $f(x) = (x^2 - 4)^2$ $(-2 \le x \le 4)$

18. $f(x) = x^3(x + 1)$ 19. $f(x) = (x - 2)^2(x + 4)$

20. $f(x) = x^{2/3}$ 21. $f(x) = x^3 + 3x$

22. $f(x) = (x + 1)^{2/3}$ 23. $f(x) = (x + 2)^{2/3}(x - 1)^{1/3}$

24. $f(x) = x^{1/3}(x - 1)^{4/3}$

25. $f(x) = \dfrac{x^{1/3}}{(x - 1)^{4/3}}$

26. Complete the proof of Theorem 5.2.

27. Consider the cubic polynomial

$$f(x) = x^3 + ax^2 + bx + c$$

(a) For what values of a and b will f have critical numbers at $x = 4$ and $x = -2$?
(b) Determine the nature of the critical points.

28. Consider the function $f(x) = 3x^2 - x + 1$. Show that $f(x) > 0$ for all x. [*Hint:* Determine the minimum value of $f(x)$.]

29. If $f(x) = -x^2 + x - 4$, show that $f(x) < 0$ for all x.

30. Given that the selling price x is related to the profit P by the equation

$$P = 5000x - 125x^2$$

(a) For what range of values of x is the profit increasing?
(b) For what range of values of x is the profit decreasing?
(c) Determine the value of x that would yield maximum profit.

5.3 Applications of Maxima and Minima I

Although many students may understand the concepts involved in the theory of maxima and minima for a curve, they often find it difficult to solve word problems. In the next two sections we shall illustrate how the theory developed in Section 5.2 can be applied to practical situations. We shall give examples of how to translate words into abstract symbols.

Example 5.11   Using a 120-ft length of fencing, a farmer wishes to contain a cow in a rectangular plot of land along the bank of a river. What should the dimensions of the rectangle be to provide the cow with maximum grazing ground? (Assume that no fence is needed along the river, which is flowing along a straight edge.)

SOLUTION   We illustrate in steps a method for solving problems such as this one.

Step 1   Draw a figure showing the relevant features (see Figure 5.13).

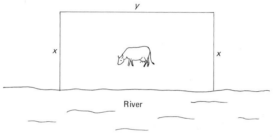

Figure 5.13

Step 2   Using Figure 5.13, find a relationship between the variables involved, and express the quantity to be maximized or minimized (in this case, the area) as a function of one of the variables. Since the length of fencing is 120 ft, we have

$$120 = 2x + y \tag{5.12}$$

and the area of the rectangle is

$$A = xy \tag{5.13}$$

Solving for y in Equation (5.12) we obtain

$$y = 120 - 2x \tag{5.14}$$

Substitution of (5.14) into (5.13) yields

$$A = x(120 - 2x)$$

or

$$A = 120x - 2x^2 \tag{5.15}$$

Thus the area is expressed in terms of a single variable x.

Step 3   Compute the derivative of $A(x)$:

$$\frac{dA}{dx} = 120 - 4x \tag{5.16}$$

Step 4   Locate the critical numbers by solving the equation

$$\frac{dA}{dx} = 0$$

Thus $120 - 4x = 0$ or $x = 30$

Step 5 Apply the second-derivative test to determine which of the critical points yields the maximum or minimum value of the function. In this case there exists only one critical number, namely, $x = 30$. From Equation (5.16), we find that

$$\frac{d^2A}{dx^2} = -4 < 0 \qquad (5.17)$$

and by Theorem 5.4, the function A has a maximum value at $x = 30$. (Note again that if you obtain more than one critical point, select those that render the problem meaningful.) To solve for y, we use Equation (5.14):

$$y = 120 - 2(30) = 60$$

In conclusion, the farmer encloses maximum area if $x = 30$ and $y = 60$. The area enclosed is

$$A = (30)(60) = 1800 \text{ sq ft}$$

Example 5.12 Find the two positive numbers having maximum product whose sum is 30.

SOLUTION   In this case we do not have the aid of a figure. So we proceed with Step 2 and find the product as a function of a single variable.
    Let the two numbers be x and y. We have

$$x + y = 30 \qquad (5.18)$$

and the product P is

$$P = xy \qquad (5.19)$$

Using equation (5.18), we express one variable in terms of the other, say

$$y = 30 - x \qquad (5.20)$$

Substitution of (5.20) into (5.19) yields the function P, where

$$P(x) = x(30 - x) = 30x - x^2 \qquad (5.21)$$

The derivative of P with respect to x is

$$\frac{dP}{dx} = 30 - 2x \qquad (5.22)$$

Therefore

$$\frac{dP}{dx} = 0 \qquad \text{when} \quad x = 15$$

Now from Equation (5.22), we have

$$\frac{d^2P}{dx^2} = -2 < 0 \qquad (5.23)$$

and by Theorem 5.4, the function P has a maximum value when $x = 15$. From Equation (5.18), we obtain $y = 15$.

Example 5.13 A soup manufacturing company wishes to pack 25 cu in. of mushroom soup in a can having the form of a right circular cylinder. Find the dimensions of the can if the surface area is to be a minimum.

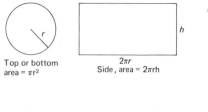

Top or bottom
area = πr²

2πr
Side, area = 2πrh

Figure 5.14

SOLUTION Again, we start by drawing a figure (see Figure 5.14). Clearly, the surface area of the metal used is

$$A = 2\pi r^2 + 2\pi rh \qquad (5.24)$$

where $2\pi r^2$ is the area of the top and the bottom. Here the area is a function of r and h. To obtain A as a function of a single variable, we use the formula for the volume of the can:

$$V = \pi r^2 h = 25 \qquad (5.25)$$

Solving for h, we have

$$h = \frac{25}{\pi r^2} \qquad (r \neq 0) \qquad (5.26)$$

Substitution of (5.26) into (5.24) yields the function A, where

$$A(r) = 2\pi r^2 + 2\pi r \frac{25}{\pi r^2} \quad \text{or} \quad A(r) = 2\pi r^2 + \frac{50}{r}$$

Therefore we have expressed A as a function of a single variable r. Of course, we could have expressed A as a function of h only. The student should do this and show that the end result is the same.

Next we compute the derivative of A

$$\frac{dA}{dr} = 4\pi r - \frac{50}{r^2} \quad \text{(Why?)}$$

Therefore from the Equation $dA/dr = 0$, we have

$$4\pi r - \frac{50}{r^2} = 0 \quad \text{or} \quad 4\pi r^3 - 50 = 0$$

and

$$r^3 = \frac{50}{4\pi} = \frac{25}{2\pi} \quad \text{or} \quad r = \sqrt[3]{\frac{25}{2\pi}}$$

Applying the second-derivative test, we find that

$$\frac{d^2A}{dr^2} = 4\pi + \frac{100}{r^3}$$

and at $r = \sqrt[3]{25/2\pi}$,

$$\frac{d^2A}{dr^2} = 4\pi + \frac{100}{(\sqrt[3]{25/2\pi})^3} = 4\pi + (100)\left(\frac{2\pi}{25}\right) = 12\pi \quad (>0)$$

By Theorem 5.4, therefore, A has a minimum value at $r = \sqrt[3]{25/2\pi}$. The corresponding value of h is obtained from Equation (5.26):

$$h = \frac{25}{\pi(25/2\pi)^{2/3}} = \frac{25}{\pi}\left(\frac{2\pi}{25}\right)^{2/3}$$

We now give an example of an apparently similar problem for which an extremum does not exist.

Example 5.14   A cylindrical package having 18 cu in. volume is to be wrapped in a rectangular sheet of paper. What should the dimensions of the cylinder be to minimize the surface area of the paper?

SOLUTION   Let the dimensions of the rectangular paper be x by y (see Figure 5.15). Now the volume of the cylinder is

$$V = \pi r^2 x = 18 \tag{5.27}$$

Figure 5.15

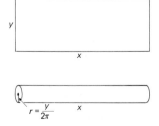

Since $r = y/2\pi$, we have

$$\pi\left(\frac{y}{2\pi}\right)^2 x = 18$$

or

$$x = \frac{18(4\pi)}{y^2} \tag{5.28}$$

The area A of the paper is

$$A = xy \tag{5.29}$$

Substituting (5.28) into (5.29), we obtain A in terms of a single variable.

$$A(y) = y\,\frac{72\pi}{y^2} = \frac{72\pi}{y} \tag{5.30}$$

The derivative of A with respect to y is

$$\frac{dA}{dy} = -\frac{72\pi}{y^2}$$

Since dA/dy is never zero, A has no critical number and hence no minimum value.

1. A rectangle is to be constructed having a given perimeter P. Find the dimensions of the rectangle of maximum area.

2. A box with square base is to be constructed to hold 64 cu in. Find the dimensions of the box of minimum surface area.

3. A rectangular piece of aluminum sheet is 48 in. $\times$ 90 in. An open rectangular container is to be made by cutting a square from each corner

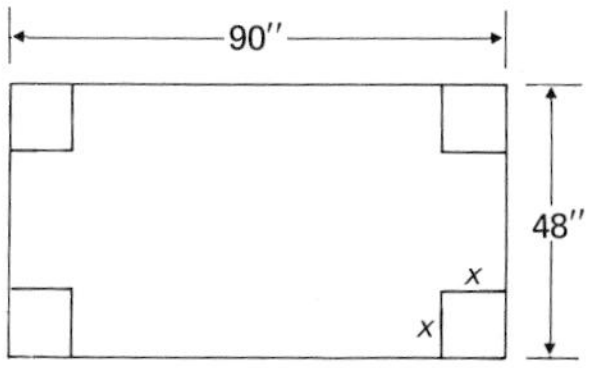

Figure 5.16

and bending up the sides. Find the dimensions of the container of maximum volume.

4. A sheet of paper for a poster is 18 sq ft in area. The margins at the top and bottom are 9 in. each and the margin on each side is 6 in. What are the dimensions of the paper if the printed area is maximum?

5. Find the dimensions of the isosceles triangle of maximum area if the perimeter is to be 24 in.

6. Show that the rectangle of largest area that can be inscribed in a circle is a square.

7. Find the dimensions of the rectangle of largest area that can be inscribed in a semicircle having diameter $2r$.

8. Find the dimensions of the rectangle of largest area that can be inscribed in the ellipse

$$\frac{x^2}{a^2} + \frac{y^2}{b^2} = 1$$

9. Consider an isosceles triangle with sides 5, 5, and 6. Find the dimensions of the rectangle of largest area that can be inscribed in the triangle so that one side is along the base.

10. A piece of wire 20 in. long is to be cut in two pieces, one to form a circle and the other a square. How should the wire be cut in order that the sum of the two areas enclosed by the wire be minimal?

11. An open cylindrical tank with circular base is to be constructed of sheet metal so as to contain a volume πa^3 of water. Find the height and the radius of the base so that the quantity of sheet metal required may be minimal.

12. A 288-cu ft pool is to have a square top. The sides are to be built of glass and the bottom of mosaic. The cost per unit area of glass is three times the cost of mosaic. Find the dimensions of the pool having minimum cost.

13. Two cars are traveling along two roads which cross each other at right angles at A. Both cars are traveling toward A at 30 ft/sec. Initially, their distances from A are 1500 ft and 2100 ft, respectively. At what time is the distance between the two cars a minimum? Find this distance.

14. A rectangular box with square bottom and top is to contain 1000 cu ft. The cost of material per square foot for the bottom is 25¢; for the top, 15¢; and for the sides, 20¢. The labor charge for making the box is $3. Find the dimensions of the box when the cost is minimal.

5.4 Applications of Maxima and Minima II

The theory of maxima and minima has many applications in a variety of disciplines. In this section we shall consider examples from the sciences, business, and economics.

Example 5.15 An experiment shows that the rate of an autocatalytic reaction is proportional to the amount of substance produced times the amount of the original substance. For what value of the substance produced will the rate of the reaction be maximal?

SOLUTION   Let c denote the amount of the original substance at the beginning of the reaction. At a later time, if the amount of the new substance formed is x, then the amount of original substance left is $c - x$. Therefore if R is the rate of the reaction, then

$$R = kx(c - x) = ckx - kx^2 \qquad (0 \leq x \leq c) \qquad (5.31)$$

where $k > 0$ is a constant of proportionality. For the rate of the reaction to be a maximum, we must have for some value of x,

$$\frac{dR}{dx} = ck - 2kx = 0 \qquad (5.32)$$

Equation (5.32) is satisfied if $x = c/2$. Thus when the concentration of the new substance formed is one-half the amount of the original substance at the beginning of the reaction, the rate of the reaction is maximal.

We now show by application of the second-derivative test that the rate of the reaction is a maximum when $x = c/2$. From Equation (5.32), we have

$$\frac{d^2R}{dx^2} = -2k$$

Since $k > 0$, $d^2R/dx^2 < 0$, and by Theorem 5.4, R attains a maximum at $x = c/2$.

Example 5.16 We return to the problem of the response of the body to a dose of a drug (see Example 4.14, Section 4.4). Suppose the response R of the body is found to be related to the dosage x of the drug administered according to the equation

$$R = kx^2 - \frac{x^3}{3} \qquad (k > 0) \qquad (5.33)$$

The rate of change of R is

$$\frac{dR}{dx} = 2kx - x^2 \qquad (5.34)$$

Equation (5.34) tells us that the body is responding as long as $0 < x < 2k$. For what dosage of the drug is the rate of change dR/dx a maximum?

Why is the domain $0 < x < 2k$?

SOLUTION The function we wish to maximize is dR/dx. Set

$$y = \frac{dR}{dx}$$

Then y has critical points for values of x for which $dy/dx = 0$:

$$\frac{dy}{dx} = \frac{d}{dx}\left(\frac{dR}{dx}\right) = \frac{d^2R}{dx^2} = 2k - 2x = 0 \qquad (5.35)$$

Equation (5.35) is satisfied if $x = k$. Application of the second-derivative test yields

$$\frac{d^2y}{dx^2} = \frac{d^3R}{dx^3} = -2 < 0 \qquad (5.36)$$

The second-derivative test indicates that for $x = k$, the function dR/dx attains a maximum. In other words, when the concentration of the drug in the body is 50% of the amount administered, the greatest changes in the reaction will occur. This type of analysis is used to determine the optimum amount of the drug to be administered to give the greatest changes in the reaction for a small change in dosage.

Example 5.17   A Florida orange grower finds that an orange tree produces (on the average) 400 oranges per year if no more than 16 trees are planted in a unit area. For each additional tree planted per unit area, the grower finds that the yield decreases by 20 oranges per tree. How many trees should the grower plant per unit area to maximize the yield?

   SOLUTION   If x is the number of trees planted per unit area, then the number N of oranges per tree depends on x, that is, $N = N(x)$. Since each tree produces 20 fewer oranges for every additional tree over 16, we have for $x \geq 16$

$$N = 400 - 20(x - 16) \qquad (5.37)$$

and the total number T of oranges produced per unit area is

$$T = xN = x[400 - 20(x - 16)]$$

or

$$T = 720x - 20x^2 \qquad (5.38)$$

What assumptions are we making about the function T in order to take its derivative?

The first-derivative condition for maximum yields

$$\frac{dT}{dx} = 720 - 40x = 0 \qquad (5.39)$$

Solving Equation (5.39), we obtain $x = 18$; that is, we find that T has a critical point at $x = 18$. Applying the second-derivative test, we determine that

$$\frac{d^2T}{dx^2} = -40$$

which is a negative number, and so, by Theorem 5.4 the yield is maximum if 18 trees are planted per unit area.

Example 5.18   A firm manufactures swim suits and sells its product at $24 per unit. The total cost C (dollars) of producing x swim suits is given by

$$C = 150 + \frac{39}{10}x + \frac{3}{1000}x^2 \qquad (5.40)$$

Write the profit P as a function of x and determine the number of swim suits that the firm should produce and sell to achieve maximum profit.

SOLUTION The firm's profit is its total revenue minus its total cost; that is,

$$P = R(x) - C(x) \qquad (5.41)$$

The total revenue from the sale of x swim suits at \$24 per swim suit is

$$R(x) = 24x \qquad (5.42)$$

Substituting (5.40) and (5.42) into (5.41), we find that

$$P = 24x - \left(150 + \frac{39}{10}x + \frac{3}{1000}x^2\right) \qquad (5.43)$$

To achieve maximum profit, we must have

$$\frac{dP}{dx} = 24 - \frac{39}{10} - \frac{3}{500}x = 0 \qquad (5.44)$$

Solving Equation (14) for x, we obtain

$$\frac{201}{10} - \frac{3}{500}x = 0$$

Thus the critical number of P is $x = 3350$. Applying the second-derivative test, we find that

$$\frac{d^2P}{dx^2} = -\frac{3}{500}$$

which is a negative number, and so, by Theorem 5.4, a production level of 3350 swim suits yields a maximum profit. The profit in dollars at this level of production is

$$P = 24(3350) - \left[150 + \frac{39}{10}(3350) + \frac{3}{1000}(3350)^2\right] = \$33,517.50$$

Note that if the firm should increase (or decrease) its output by one or more swim suits, the firm's profit will decrease.

It is of interest to observe that the marginal revenue is $dR/dx = 24$ and that the marginal cost is also

$$\left.\frac{dC}{dx}\right|_{x=3350} = \left.\frac{39}{10} + \frac{3}{500}x\right|_{x=3350} = 24$$

at a production level of 3350 swim suits. (See Problem 9 in Exercises 5.4).

Exercises 5.4

1. If the rate of a certain autocatalytic reaction is given by

$$r(x) = 12x(200 - x)$$

where x is the amount of the new substance formed and 200 is the amount of the original substance, for what value of x is the rate of the reaction a maximum? What is the maximum rate of the reaction?

2. If the number of bacteria (in thousands) present in a culture is given by

$$N(t) = 250t - 50t^2 + 20$$

where t is the time, at what time will the population of the bacteria be a maximum? What is the maximum population?

3. The concentration of hydrogen ion in a solution is given by

$$X = H + \frac{10^{-5}}{H}$$

For what value of H is the concentration a minimum?

4. Experiments show that the velocity of the flow of air through the respiratory system during coughing is given by

$$v = \frac{k}{\pi} r^2 (r_0 - r)$$

where k is a constant, r_0 is the radius of the windpipe when there is no pressure, and r is the radius of the windpipe after pressure builds up. Find the value of r for which the velocity of the flow of air is maximum.

5. For each of the following cost functions, find the production level x for which the cost is a minimum.

    (a) $C = 2x^3 - 8x^2 + 10x$

    (b) $C = 800 - \dfrac{x}{2} + \dfrac{1}{1000} x^2$

    (c) $C = \dfrac{x}{2} + \dfrac{5 \cdot 10^5}{x} + 10^3$

6. A firm produces x units of a product. If the profit P in dollars earned in the manufacture and sale of x units is

$$P = 20x - \frac{1}{500} x^2$$

find the number x that yields maximum profit. What is the maximum profit?

7. A firm produces mink coats and sells each coat for $1500. The cost of producing x coats per year is

$$C = 9 + 3x + 0.015x^2$$

    (a) Write an expression for the profit in terms of x.
    (b) Find a number x that yields maximum profit.
    (c) What is the maximum annual profit?

8. A manufacturer finds that it is possible to sell x units per month of a certain item if the price is $p = 18 - 0.001x$ dollars. Find the production level

that will maximize profits if the cost of producing x items is $C = 4x + 1500$ dollars. What is the price of each item at this production level?

9. A firm's production level is such that the profit is maximum. Find a relationship between the firm's marginal revenue and marginal cost.

10. If 25 apple trees are planted per acre in a certain area, the average yield per tree is 450 apples. For each tree in excess of 25 per acre, the average yield is reduced by 15 apples per tree.
    (a) Show that the total number of apples produced per acre is

$$T = 825x - 15x^2 \qquad (x \geq 25)$$

    (b) Show that if 28 trees are planted per acre, the yield is maximum. (Remember that you cannot plant a fraction of a tree.)

11. The owner of a day nursery finds that if a monthly fee of $75 per child is charged, 20 children will enroll. To increase the enrollment, the owner finds that the fee will have to be decreased by $3 for each additional child.
    (a) If x ($x \geq 20$) is the number of children enrolled, show that the total revenue is $r = 135x - 3x^2$.
    (b) Show that the maximum revenue is obtained by enrolling either 22 or 23 children.

12. A group of students arrange a chartered flight from New York to London, Paris, Heidelberg, Barcelona, and back to New York with a three-day stay at each European city. The charge per person is $499 if 100 students go on the flight. If more than 100 participate, the charge per student is reduced by an amount equal to $4 times the number of students above 100.
    (a) Find the total revenue.
    (b) Find the number of students that will furnish maximum revenue.
    (c) What is the cost per student if the maximum revenue is obtained?

5.5 The Second Derivative

In Section 5.2 we stated (without proof) a theorem giving a test for relative maxima and relative minima using the second derivative. In this section we shall sketch a proof of that theorem and study the relationship between the second derivative of a function and its graph.

Consider the graph of the function f in Figure 5.17. As x increases from a to b, the graph of the function bends in a counterclockwise direction, and from b to c, it bends in a clockwise direction. We say that the graph of the function is *concave upward* between a and b and that it is *concave downward* between b and c. At the point $x = b$, the direction of bending changes. The point $(b, f(b))$ is called the *point of inflection* of the curve. It can be shown that if the graph of a function f is concave upward in an interval I, then the slopes of the tangent lines increase as the curve bends in the counterclockwise direction. Similarly, if the graph of a function f is concave downward in an interval J, then the slopes of the tangent lines decrease as the curve bends in the clockwise direction.

Figure 5.17

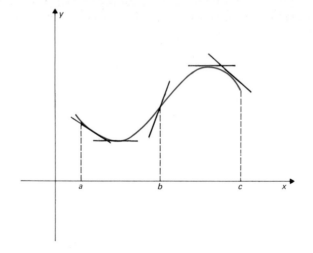

We showed in Section 5.1 that if the derivative of a function is positive on an interval, then the function is increasing on that interval. Since f'' is found from f' in the same way that f' is found from f, we note that if f'' is positive over an interval, then f' is increasing on that interval. Consequently the graph of the function f is concave upward on that interval. Similar remarks apply if f'' is negative over an interval. We have the following property.

▶   *Property 1   If $f''(x) > 0$ on the interval I, the graph of f is concave upward on I. If $f''(x) < 0$ on the interval I, the graph of f is concave downward on I.*

Example 5.19   Discuss the concavity of the curve

$$y = f(x) = x^4 - 6x^3 + 12x^2 + 5x + 7$$

SOLUTION   Differentiating, we have

$$f'(x) = 4x^3 - 18x^2 + 24x + 5$$

and

$$f''(x) = 12x^2 - 36x + 24 = 12(x - 1)(x - 2)$$

For

$$x < 1, \qquad f''(x) > 0$$

Explain why.

$$1 < x < 2, \qquad f''(x) < 0$$

$$x > 2, \qquad f''(x) > 0$$

Hence the curve is concave upward in the intervals $(-\infty, 1)$ and $(2, \infty)$, and the curve is concave downward in the interval $(1, 2)$.

Formally, we say that a curve is concave upward on an open interval I if and only if the curve lies above its tangent lines for all $x \in I$. Similarly, a curve is said to be concave downward on I, if and only if the curve lies below its tangent lines for all $x \in I$. Note that the tangent line at a point of inflection crosses this curve.

In Example 5.19 it is easy to see that $f''(1) = 0$. Furthermore, the curve is concave upward on one side of $(1, f(1))$ and concave downward on the other side of $(1, f(1))$. In other words, the concavity changes at $(1, f(1))$. Thus the point $(1, f(1))$ is a point of inflection. Similarly, the point $(2, f(2))$ is also a point of inflection. We can state the definition formally.

Definition 5.5 *A point* $(a, f(a))$ *is called a* point of inflection *of the graph of a function f if there exists an interval* $(a - c, a + c)$ *in the domain of f such that the graph of f is concave upward in* $(a - c, a)$ *and concave downward in* $(a, a + c)$, *or vice versa.*   ◀ Point of inflection

The following theorem enables us to find the possible points of inflection.

Theorem 5.5 If $(a, f(a))$ *is a point of inflection of the graph of the*   ◀
function f and if f'' is continuous at a, then $f''(a) = 0$.

The proof follows directly from Property 1.

Theorem 5.5 asserts that under the assumption that f'' is continuous, the points of inflection are among those points $(x, f(x))$ for which $f''(x) = 0$. However, if $f''(c) = 0$, $(c, f(c))$ is not then necessarily a point of inflection, as the following example illustrates.

Example 5.20 Show that $(0, 0)$ is not a point of inflection for the curve

$$y = f(x) = x^4$$

SOLUTION We have

$$f'(x) = 4x^3 \quad \text{and} \quad f''(x) = 12x^2$$

Thus $f''(x) = 0$ if $x = 0$. However, $(0, 0)$ is not a point of inflection, since $f''(x) > 0$ for every $x \neq 0$. Therefore the concavity of the graph of f does not change at $x = 0$. Hence $(0, 0)$ is not an inflection point.

From Theorem 5.5, we conclude that the point $(a, f(a))$ is a candidate for being the point of inflection if either $f''(a) = 0$ or f'' is discontinuous at a. Thus we follow the following procedure to find the points of inflection. For the function f defined by $y = f(x)$:   *Procedure for finding points of inflection.*

(i) Find $f''(x)$.

(ii)   Determine those values of x for which $f''(x) = 0$ or f'' is discontinuous. Let us call these values of x the *hypercritical numbers*.

That is, if $f''(x) > 0$ for
$x \in (c - a, a)$ and $f''(x) < 0$ for
$x \in (a, a + c)$, or vice versa.

(iii)   Check to see if $f''(x)$ changes sign at a hypercritical number. If $f''(x)$ changes sign at the hypercritical number $x = a$, then $(a, f(a))$ is a point of inflection of f provided that a is in the domain of f.

Example 5.21   For the function f defined by

$$y = f(x) = x^3 - 3x^2 + 1$$

(a)   Find the points of inflection.

(b)   Find the equations of the tangent lines to the graph of f at the points of inflection.

SOLUTION

(a)   (i)   Differentiating, we have

$$f'(x) = 3x^2 - 6x \quad \text{and} \quad f''(x) = 6x - 6 = 6(x - 1)$$

(ii) f'' is continuous everywhere and $f''(x) = 0$, when $x = 1$. Thus $x = 1$ is the only hypercritical number.

(iii)   We now check to see if f'' changes sign at the hypercritical number $x = 1$. In an interval around $x = 1$, if $x < 1$, then $f''(x) < 0$, and if $x > 1$, then $f''(x) > 0$. Therefore when $x < 1$, the curve is concave downward and when $x > 1$, the curve is concave upward. Since $x = 1$ is in the domain of f, the only point of inflection for f is $(1, f(1)) = (1, -1)$.

(b)   To find the equation of the tangent line to the graph of f at $(1, -1)$, we first find the slope of this line. The slope m is given by

$$f'(1) = 3(1)^2 - 6(1) = -3 \qquad \text{(Why?)}$$

Using the point-slope form of the equation of a line, we have

$$y - (-1) = -3(x - 1) \quad \text{or} \quad y + 1 = -3x + 3 \quad \text{or} \quad y = -3x + 2$$

which is the required equation of the tangent line to the graph of f at the point of inflection $(1, -1)$.

We now restate and prove Theorem 5.4 which gives us the sufficient conditions for the extremum of a function.

▶   **Theorem 5.4**   *Let f be a function and a be a number such that $f'(a) = 0$. Suppose that f'' is continuous at a. Then*

(i) $f(a)$ *is a relative maximum value of f if f''(a) < 0.*

(ii) $f(a)$ *is a relative minimum value of f if f''(a) > 0.*

PROOF

(i) Since f'' is continuous at a and $f''(a) < 0$, it follows (see Problem 29, Exercises 3.4) that there exists an interval $(a - c, a + c)$ such that $f''(x) < 0$ for every x in $(a - c, a + c)$. Therefore, by Property 1, the graph of f is concave downward in the interval $(a - c, a + c)$. Since $f'(a) = 0$, the tangent line at $(a, f(a))$ is horizontal, and the graph of the function f is below the tangent line between $x = a - c$ and $x = a + c$. Thus $f(a)$ is a relative maximum value of f.

(ii) Similar to (i).

In the next example we shall illustrate the use of our knowledge of increasing and decreasing functions, critical points, relative extrema, concavity and points of inflection to sketch the graph of a given function f.

Example 5.22   Sketch the graph of

$$f(x) = x^4 - 6x^2 + 1$$

SOLUTION

1. We first find $f'(x)$ to determine

    (i)   The critical numbers that may yield the relative extrema

    (ii) The intervals where f is increasing and decreasing

    $$f'(x) = 4x^3 - 12x = 4(x^3 - 3x)$$
    $$= 4x(x^2 - 3) = 4x(x - \sqrt{3})(x + \sqrt{3})$$

    Then

    $$f'(x) = 0 \quad \text{when} \quad x = 0, \quad x = \sqrt{3}, \quad \text{and} \quad x = -\sqrt{3}$$

    Therefore 0, $\sqrt{3}$, and $-\sqrt{3}$ are critical numbers. When

    $$x < -\sqrt{3}, \quad f'(x) < 0 \quad \text{and} \quad f \text{ is decreasing}$$
    $$-\sqrt{3} < x < 0, \quad f'(x) > 0 \quad \text{and} \quad f \text{ is increasing}$$
    $$0 < x < \sqrt{3}, \quad f'(x) < 0 \quad \text{and} \quad f \text{ is decreasing}$$
    $$\sqrt{3} < x, \quad f'(x) > 0 \quad \text{and} \quad f \text{ is increasing}$$

2. We next find $f''(x)$ to

    (i)   Test the critical numbers for relative extrema

    (ii) Find the intervals where f is concave upward and concave downward

(iii)   Find the points of inflection

Since

$$f''(x) = 12x^2 - 12 = 12(x^2 - 1) = 12(x - 1)(x + 1)$$

$$f''(0) = -12 < 0$$

f has a relative maximum at $x = 0$. Since

$$f''(-\sqrt{3}) = 24 > 0$$

f has a relative minimum at $x = -\sqrt{3}$. Since

$$f''(\sqrt{3}) = 24 > 0$$

f has a relative minimum at $x = \sqrt{3}$. When

$$x < -1, \qquad f''(x) > 0 \quad \text{and} \quad f \text{ is concave upward,}$$

$$-1 < x < 1, \qquad f''(x) < 0 \quad \text{and} \quad f \text{ is concave downward,}$$

$$1 < x, \qquad f''(x) > 0 \quad \text{and} \quad f \text{ is concave upward.}$$

The concavity of the curve changes at 1 and -1. Consequently, $(1, f(1)) = (1, -4)$ and $(-1, f(-1)) = (-1, -4)$ are the points of inflection

3. A rough sketch of the graph of f is shown in Figure 5.18. We do not need to plot additional points.

Figure 5.18

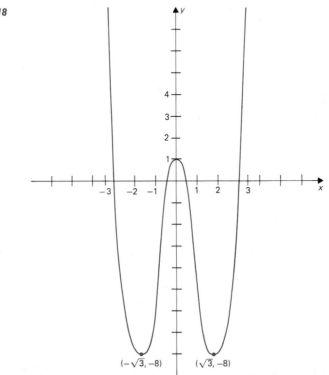

$(-\sqrt{3}, -8)$ $(\sqrt{3}, -8)$

In Problems 1 through 20, perform the following.

(a) Find the intervals over which the given function is increasing or decreasing.
(b) Find the relative extrema.
(c) Find the intervals over which the function is concave upward or concave downward.
(d) Find the points of inflection.
(e) Sketch the graph of f.

1. $f(x) = x^2 - 3x + 4$ 2. $f(x) = -2x^2 + 3x - 2$

3. $f(x) = x^3 - 1$ 4. $f(x) = -x^3 + 4$

5. $f(x) = 4x^{2/3}$ 6. $f(x) = -3x^{5/3}$

7. $f(x) = \frac{1}{3}x^3 + 3x^2$ 8. $f(x) = x^3 - 3x^2$

9. $f(x) = 2x^3 + 4x^2 + 2x + 1$ 10. $f(x) = x^3 - 5x^2 + 8x - 4$

11. $f(x) = x^2 - 3x^3 + 3x^4$ 12. $f(x) = -x^3 + 3x^2 - 3x + 7$

13. $f(x) = \frac{x^3}{6} - \frac{x^5}{20}$ 14. $f(x) = \frac{1}{x}$

15. $f(x) = \frac{1}{1-x}$ 16. $f(x) = x + \frac{1}{x}$

17. $f(x) = -3x + \frac{3}{x}$ 18. $f(x) = \frac{x}{x^2 + 3}$

19. $f(x) = \frac{x^2}{1+x^2}$ 20. $f(x) = \frac{x}{(x+3)^2}$

21. Find the equation of the tangent line to the graph of f at each point of inflection in Problems 8 through 11 above.

22. Determine a value of k so that the function

$$f(x) = x^2 + k/x$$

has a point of inflection at $x = 2$.

23. If k is a positive integer, then show that
(a) $f(x) = x^{2k+1}$ has $(0, 0)$ as a point of inflection,
(b) $f(x) = x^{2k}$ does not have $(0, 0)$ as a point of inflection.

24. Consider the fourth-degree polynomial function,

$$f(x) = x^4 + ax^3 + bx^2 + cx + d$$

and suppose that $3a^2 = 4b$. Show that f has no points of inflection.

25. If the cost function $C(x)$ is given by

$$C(x) = 6x - 0.03x^2 + 0.00005x^3$$

determine whether the curve is concave upward, concave downward, or has a point of inflection at $x = 100$, $x = 200$, $x = 300$, and $x = 400$.

5.6 The Differential

Let $y = f(x)$ define a differentiable function of x. In the previous chapter we introduced the notation dy/dx to represent the derivative of f with respect to x without clarifying the meaning of the symbols dy and dx. During the nineteenth century much controversy shrouded these symbols. Historically, the symbols date back to Leibniz, who attempted to define dy and dx as infinitesimals (very small quantities) "greater than zero but less than any positive real number." Leibniz was aware that there were no such real numbers. Nevertheless, he was very successful in solving many problems using these symbols. Some years later, Cauchy offered an alternative definition of the symbols. In this section we shall define these symbols and show how they are used in mathematics and the sciences.

Let us return to the definition of the derivative of a function. Suppose that f, given by $y = f(x)$, is a differentiable function of x. Let Δx be the *increment*, or change, in the variable x. Then the *increment* Δy in the functional value $f(x)$ is given by

Here we are using the customary Δx (read "delta x") in place of h as used in Chapter 4.

$$\Delta y = f(x + \Delta x) - f(x)$$

For a particular function f, Δy depends on both the numbers x and Δx. We may write the derivative f' as

$$f'(x) = \lim_{\Delta x \to 0} \frac{f(x + \Delta x) - f(x)}{\Delta x} = \lim_{\Delta x \to 0} \frac{\Delta y}{\Delta x}$$

In many practical problems (see Example 5.25), we are interested in computing Δy. However, the computation of Δy may be quite difficult or tedious. Therefore we use a close approximation to Δy, called the *differential*.

Augustin Louis Cauchy

Augustin Louis Cauchy (1789–1857) was born in Paris at the outset of the French Revolution. His father was a lawyer and later became Secretary of the Senate under Napoleon. The first eleven years of Cauchy's life were spent in the village of Arcueil where the family had gone to escape the Reign of Terror.

At the age of 18, Cauchy entered civil engineering school, and two years later Napoleon commissioned him to participate in the construction of the harbor at Cherbourg. In 1815 he was appointed professor at the École Polytechnique and at the Sorbonne. With the revolution of 1830, he left France in voluntary exile. When Cauchy returned to Paris eight years later, he resumed his posts without giving allegiance to the government.

Cauchy introduced rigorous methods in mathematical analysis. He created the theory of functions of a complex variable and produced over 500 papers on all branches of mathematics.

Definition 5.6　Let $y = f(x)$ *define a differentiable function of* x. *Then we have the following.*　◀

(*i*)　The differential dx *of the independent variable* x *is defined by* $dx = \Delta x$, *where* Δx *is any real number.*

(*ii*)　The differential dy *of the dependent variable* y *is defined by* $dy = f'(x)\, dx$.

A few remarks are in order.

Remark 1　The definition $dx = \Delta x$ is completely arbitrary and is, therefore, an independent variable whose range is the set of real numbers.

Remark 2　As we have indicated earlier, dy is used as an approximation for Δy (see the following examples), and in many cases it will be a relatively simpler matter to calculate dy rather than Δy.

Remark 3　An important result of this definition is the fact that the ratio of dy to dx is $f'(x)$.

Example 5.23　Let $y = f(x) = x^2 - 3x + 2$. Compare Δy and dy for

(a)　$x = 3$,　$\Delta x = -\frac{1}{2}$　　　　(b)　$x = 3$,　$\Delta x = \frac{1}{8}$

SOLUTION　Since

$$f(x) = x^2 - 3x + 2 \quad \text{and} \quad f'(x) = 2x - 3$$

then by definition,

$$dy = f'(x)\, dx = (2x - 3)\, dx \qquad\qquad (5.45)$$

Also

$$\Delta y = f(x + \Delta x) - f(x)$$
$$= [(x + \Delta x)^2 - 3(x + \Delta x) + 2] - (x^2 - 3x + 2)$$

or

$$\Delta y = 2x\,\Delta x + (\Delta x)^2 - 3\,\Delta x \qquad\qquad (5.46)$$

(a)　Substituting $x = 3$ and $dx = \Delta x = -\frac{1}{2}$ into Equations (5.45) and (5.46), respectively, we obtain

$$dy = (2 \cdot 3 - 3)\left(-\tfrac{1}{2}\right) = -\tfrac{3}{2}$$
$$\Delta y = 2 \cdot 3\left(-\tfrac{1}{2}\right) + \left(-\tfrac{1}{2}\right)^2 - 3\left(-\tfrac{1}{2}\right)$$
$$= -3 + \tfrac{1}{4} + \tfrac{3}{2} = -\tfrac{3}{2} + \tfrac{1}{4}$$

Thus

$$|\Delta y - dy| = \tfrac{1}{4}$$

(b)   Setting $x = 3$ and $\Delta x = \frac{1}{8}$, we get

$$dy = (2 \cdot 3 - 3)\,\tfrac{1}{8} = \tfrac{3}{8}$$

and

Note that $|\Delta y - dy| = (\Delta x)^2$ in this example.

$$\Delta y = 2 \cdot 3 \cdot \tfrac{1}{8} + \left(\tfrac{1}{8}\right)^2 - 3 \cdot \tfrac{1}{8} = \tfrac{3}{8} + \tfrac{1}{64}$$

or

$$|\Delta y - dy| = \tfrac{1}{64}$$

We observe that even in such simple cases it is easier to calculate dy than Δy. We also note in this example that dy approximates Δy better when Δx is changed from $-\frac{1}{2}$ to $\frac{1}{8}$ while $x = 3$ is kept fixed.

Let us now consider the geometrical meaning of the differential. Let $P(x, y)$ and $Q(x + \Delta x, y + \Delta y)$ be two points on the graph of $y = f(x)$ (see Figure 5.19). Then the slope of the tangent line at P is given by

Figure 5.19

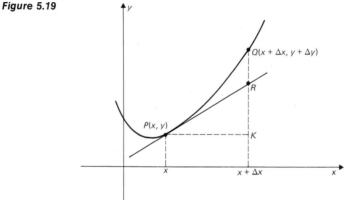

$f'(x) = dy/dx$. But the slope of the tangent line is also given by the ratio RK/PK, and since $PK = \Delta x = dx$, it follows that $RK = f'(x)\,dx = dy$.

Now $\Delta y = KQ$. Thus the difference between Δy and dy is represented geometrically by the segment RQ.

Applications   In many problems we come across a quantity x which is measured by a faulty instrument. Due to this error in the increment of x, there will be a corresponding error in the computed value of y, where $y = f(x)$. The error in y (that is, Δy) is approximated by dy. If the error in y is dy, then dy/y is called the *relative error* in y and $100\,dy/y$ is called the *percentage error* in y.

Note   *The nearest perfect fourth root to 626 is 625, and so we take $x = 625$ and $\Delta x = 1$.*

Example 5.24   Calculate approximately the fourth root of 626.

SOLUTION   Since we know that the fourth root of 625 is 5, we shall approximate the amount by which $x^{1/4}$ increases when x is increased by 1. Let

$$y = x^{1/4} \quad \text{and} \quad y + \Delta y = \sqrt[4]{x + \Delta x}$$

Then

$$dy = \tfrac{1}{4}x^{-3/4}\, dx \quad \text{or} \quad dy = \frac{dx}{4x^{3/4}}$$

If $x = 625$ and $dx = \Delta x = 1$, then

$$y = (625)^{1/4} = 5 \quad \text{and} \quad dy = \frac{1}{4(625)^{3/4}} = \frac{1}{4(125)} = \frac{1}{500} = 0.002$$

Now

$$(626)^{1/4} = y + \Delta y \approx y + dy \approx 5 + 0.002 \approx 5.002$$

Example 5.25 A student in a physics laboratory measures the side of a cube as 2 ft (see Figure 5.20) with a possible error of 0.002 ft. What is the greatest possible error in the volume computed by the student?

Figure 5.20

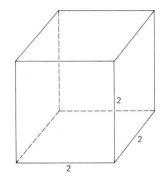

SOLUTION   The possible error of 0.002 ft in the side x of the cube signifies that the value 2 ft is subject to a correction which may amount to as much as 0.002 ft in either direction. In other words, the error in x may be as much as $dx = \Delta x = \pm 0.002$. Then, since Δx is small, the computed value of the volume

$$V = f(x) = x^3$$

is subject to an approximate error given by

$$dV = f'(x)\, dx = 3x^2\, dx$$

Here $x = 2$ and $\Delta x = \pm 0.002$. Therefore

$$dV = \pm 3(2)^2(0.002) = \pm 0.024$$

Thus the computed volume of 8 cu ft may be too large or too small by 0.024 cu ft, approximately.

The relative error in this case is

$$\frac{dV}{V} = \frac{0.024}{8} = 0.003$$

The relative error in x is

$$\frac{dx}{x} = \frac{0.002}{2} = 0.001$$

The percentage error in the side of the cube and its volume will be 0.1 and 0.3, respectively.

Exercises 5.6

In Problems 1 through 6, find dy, Δy, and $|\Delta y - dy|$, when $x = 4$ and $\Delta x = 0.5$.

1. $y = x^3$
2. $y = 2x^3 - 7x$
3. $y = x^2 + 5x + 3$
4. $y = \sqrt{x}$
5. $y = \dfrac{2x}{3 + x}$
6. $y = \dfrac{1}{\sqrt{x}}$

In Problems 7 through 15, use differentials to find an approximate value for the given quantity.

7. $\sqrt{10}$
8. $\sqrt{122}$
9. $\sqrt[5]{33}$
10. $\sqrt{0.0143}$
11. $\sqrt{255}$
12. $\sqrt[3]{7.5}$
13. $\sqrt[3]{25}$
14. $(1.98)^5$
15. $\dfrac{1}{\sqrt[5]{31}}$

16. The diameter of a circle is 4 ft. By approximately how much does the area decrease if the radius decreases by 1/30 in.?

17. The diameter of a sphere is 6 in. with a possible error of 0.05 in. What is the greatest possible error in the computed area and the relative possible error?

18. A cubic box is to hold 1000 cu ft. What is the allowable error in the edge of the cube if the error in the volume is not to exceed 2 cu ft?

19. If the area of a circle increases at a constant rate, show that the rate of increase of the perimeter varies inversely as the radius.

20. A spherical shell has an inner radius of 10 cm and a thickness of $\frac{1}{4}$ cm. Find the approximate volume of the shell.

21. The total cost of producing x units is C dollars, where

$$C = \tfrac{1}{100}x^2 + 5x + 200$$

  (a) Find the differential dC.
  (b) When the level of production changes from 100 to 101, what is the value of dC?
  (c) Determine the error involved when dC is used as an approximation for ΔC.
  (d) Determine the relative and percentage errors.

22. The total cost C (dollars) of producing x shirts is

$$C = \frac{x^3}{1500} - \frac{2}{5}x^2 + 45x$$

and each shirt is sold at \$10.
(a) Find the total profit P.
(b) Find the differential dP.
(c) When the production level changes from $x = 350$ to $x = 355$, what will be an approximate change in P?

Part 1 (*Oral*)   Chapter 5 Review

Define or discuss the following:

1. Increasing function
2. Decreasing function
3. Critical number
4. Relative maximum, absolute maximum
5. Relative minimum, absolute minimum
6. Concave upward, concave downward
7. Point of inflection
8. Hypercritical number
9. Differential

Part 2 (*Written*)

In Problems 1 through 20, find $f'(x)$ and $f''(x)$. Find the intervals over which f is increasing, decreasing, concave upward, and concave downward. Locate relative extrema. Find the points of inflection. Graph each function.

1. $f(x) = 3x^2 - 7x + 2$ 2. $f(x) = 2x^2 + x + 1$

3. $f(x) = x(3 - x)$ 4. $f(x) = x^3 - 27x$

5. $f(x) = x^2(x + 9)$ 6. $f(x) = x^3 - 3x^2 + 3x - 9$

7. $f(x) = \sqrt{x + 3}$ 8. $f(x) = x^4 - x^2$

9. $f(x) = 3x^{1/3}$ 10. $f(x) = \frac{1}{x + 1}$

11. $f(x) = 2x + \frac{1}{x}$ 12. $f(x) = \sqrt{4 - x^2}$

13. $f(x) = \frac{-2}{x + 3}$ 14. $f(x) = -3x + \frac{3}{x}$

15. $f(x) = \frac{4}{x^2}$ 16. $f(x) = \frac{1}{5}x^5 - 1$

17. $f(x) = 4x^3 + 3x^2 - 6x - 1$ 18. $f(x) = x^4 - 14x^2 + 24x + 5$

19. $f(x) = 2x^4 + x^2 + 2x$ 20. $f(x) = \dfrac{2x - 1}{x + 3}$

21. Find the relative and absolute extrema (if any) for the function $f(x) = 2x + 3$, when

     (a) $x \in (0, 1)$ (b) $x \in [1, 3)$ (c) $x \in (-\infty, \infty)$

22. Repeat Problem 21 for $f(x) = 1/x$.

23. Find two positive numbers whose sum is 50 and whose product is maximum.

24. Find two numbers x and y that have a sum of 80 and such that $x^3 + y^3$ is minimum.

25. Of all the rectangles with an area of 400 sq ft, find the following.
     (a)   The one with the smallest perimeter
     (b)   The one with the shortest diagonal

26. Find the dimensions of the most economical can (with closed top) if it is to hold 40 cu in. of a liquid.

27. A rectangular field along the bank of a river is to have an area of 1800 sq m. What must the dimensions of the field be to use the least amount of fencing material in enclosing the field? (Assume that no fence is needed along the river, which is flowing along a straight line.)

28. A figure consists of a rectangle surmounted by a semicircle. If the perimeter of the figure is $(8 + 2\pi)$ in. and the area is maximum, find the dimensions of the rectangle and the radius of the semicircle.

29. Two sources of light are 10 ft apart. Assuming that the intensity of illumination varies inversely as the square of the distance from the source, find the point of maximum illumination on the straight line joining the two sources if one source has six times the candle power of the other.

30. P and Q are two points on a straight coast at a distance of 10 mi. L is a point in the sea, where PL is perpendicular to PQ and $PL = 3$ mi. A person wishes to go from L to Q as quickly as possible partly by rowing on the sea and partly by running along the coast. The person is able to row at 4 mph and run at 5 mph. How far from P should the person land?

31. Let the reaction R of the body to a dose d of a drug be given by

$$R(d) = d^2 \left(\frac{c}{2} - \frac{d}{3} \right)$$

where $c > 0$ is the maximum amount of the drug that can be administered. Find the dose for which the rate of increase of R is maximum. In other words, find d when $R'(d)$ is maximum.

32. An *average cost function* $f(x)$ is given by

$$f(x) = \frac{C(x)}{x}$$

where $C(x)$ is the cost of producing x number of items. Suppose that

$$C(x) = x^2 + 2x + 9 \qquad (0 < x \le 20)$$

For what value of x is the average cost a minimum?

33. If $y = 8 - \left(\frac{1}{3}\right)x^3$, find dy and Δy when $x = 3$ and $\Delta x = 0.5$.

34. Compute $\sqrt[5]{31}$ approximately.

35. A cubical box is to have a capacity of 1000 cu in. What is the allowable error in measuring the edge of the cube if the error in the volume is not to exceed 3 cu in.

Numerical Applications Using Calculators

Differentials

One of the applications of differentials is to approximate the value of certain algebraic expressions which are otherwise difficult to evaluate. This can be better accomplished by the aid of a pocket calculator. Let us consider the example of approximating the fourth root of 626. We were able to obtain

$$(626)^{1/4} = 5.002$$

by using the differential (see Example 5.24 for details). If we take the fourth power of 5.002, we have

$$(5.002)^4 \approx 626.0006 \quad \text{and} \quad |626 - 626.0006| = 6.0 \times 10^{-4}$$

This gives us some idea how good the approximation is. Suppose that we are not satisfied with the value obtained. How do we proceed to find a better approximation for the fourth root of 626? The procedure can be carried out by repeating the same process as explained in this chapter with different values of x, y, and Δy. Let us conduct the process using a calculator.

We have $y = x^{1/4}$ and $y - \Delta y = \sqrt[4]{x - \Delta x}$. Again,

$$dy = \frac{dx}{4x^{3/4}}$$

Take $x = 626.0006$ and $dx = \Delta x = 0.0006$; then $y = (626.006)^{1/4} = 5.002$ and

$$dy = \frac{0.0006}{4(626.0006)^{3/4}} = 1.198561151 \times 10^{-6}$$

Therefore

$$(626)^{1/4} \approx y - \Delta y = 5.002 - 1.198561151 \times 10^{-6} = 5.001998801$$

Let us find out how good the new approximation is:

$$(5.001998801)^4 \approx 626.0000002$$

and

$$|626 - 626.0000002| = 2.0 \times 10^{-7}$$

Indeed, the newly approximated value is better than the one previously obtained. The same procedure can be applied to obtain still better approximations. As the number of digits after the decimal point get larger, the process becomes more cumbersome for a calculator and it is best to seek the help of a computer.

Exercises

Using differentials, approximate each of the following:

1. $\sqrt[5]{35}$ 2. $\sqrt[4]{80}$ 3. $\sqrt{0.0143}$

4. $\sqrt{24}$ 5. $\sqrt[6]{63}$ 6. $\sqrt[3]{25}$

Exponential and Logarithmic Functions

CHAPTER 6

6.1 The Exponential Function

In Section 1.5 we introduced the notation

$$b^n = \underbrace{b \cdot b \cdot b \cdots b}_{n \text{ factors}}$$

where b is a real number and n is a positive integer. The reader is encouraged to review the properties of rational exponents discussed in Section 1.5.

Now consider the expression $3^{\sqrt{2}}$. We can approximate $3^{\sqrt{2}}$ as follows:

$$3^1 < 3^{1.4} < 3^2$$

$$3^{1.4} < 3^{1.41} < 3^{1.5}$$

Note $\sqrt{2} = 1.4142\ldots$

and

$$3^{1.41} < 3^{1.414} < 3^{1.42}$$

and so on. Following this procedure indefinitely, we can make the difference between the right member and the left member of the inequality as small as we wish and we can find a value of $3^{\sqrt{2}}$ to satisfy each successive inequality. In fact, this case will hold for any number b^x, where b is a positive number and x is any real number. We shall use this idea to define an *exponential function*.

Exponential function ▶

Definition 6.1 *If $b > 0$ and $b \neq 1$, then the function f defined by $f(x) = b^x$ is called an* **exponential function**. *The number b is called the* **base** *of the function. The domain of f is the set of all real numbers and the range of f is the set of all positive numbers.*

Note *We require that $b > 0$ because if $b < 0$, then for some x, such as $x = \frac{1}{2}$, b^x is not a real number.*

Note that we require that $b \neq 1$ because in this case $f(x) = 1^x = 1$, which represents a constant function.

We illustrate the behavior of the function $f(x) = b^x$ in the following example.

Example 6.1 Sketch the graph of each of the following functions:

(a) $f(x) = 2^x$ (b) $f(x) = \left(\frac{1}{2}\right)^x$

SOLUTION

(a) Assigning certain values for x and finding the corresponding values $f(x)$, we obtain the following table.

x	-2	-1	0	1	2
$f(x)$	$\frac{1}{4}$	$\frac{1}{2}$	1	2	4

It can be shown that the exponential function is continuous. Thus locating the points in the preceding table, we construct a graph of f as shown in Figure 6.1.

Figure 6.1

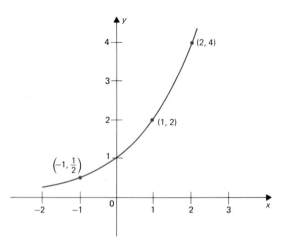

(b) First we construct the following table:

x	-2	-1	0	1	2
$f(x)$	4	2	1	$\frac{1}{2}$	$\frac{1}{4}$

A graph of the function $f(x) = \left(\frac{1}{2}\right)^x$ is shown in Figure 6.2.

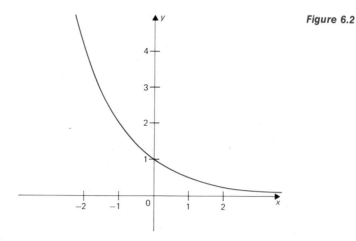

Figure 6.2

In general, the exponential function $f(x) = b^x$ has the following properties:

Property 1   The exponential function is continuous over the real line.   ◄

Property 2   The graph of f intersects the y-axis at $y = 1$.   ◄

Property 3   If $b > 1$, then f is an increasing function, and   ◄
$$f(x) \to +\infty \quad as \quad x \to +\infty$$
and $$f(x) \to 0 \quad as \quad x \to -\infty$$

Property 4   If $0 < b < 1$, then f is a decreasing function, and   ◄
$$f(x) \to +\infty \quad as \quad x \to -\infty$$
and $$f(x) \to 0 \quad as \quad x \to +\infty$$

The general shape of the graphs of $f(x) = b^x$ for $b > 1$ and $0 < b < 1$ are shown in Figures 6.3 and 6.4, respectively.

Figure 6.3
Graph of $f(x) = b^x \, (b > 1)$

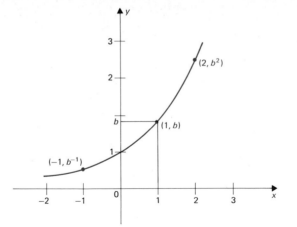

Figure 6.4
Graph of $f(x) = b^x \, (0 < b < 1)$

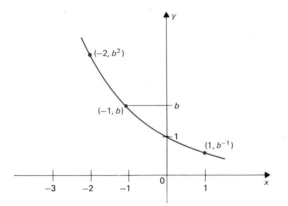

Example 6.2 If $f(x) = b^x$, show that

$$f(x_1 + x_2) = f(x_1) \, f(x_2)$$

SOLUTION From Definition 6.1, we have

By the properties of exponents

$$f(x_1 + x_2) = b^{x_1 + x_2} = b^{x_1} b^{x_2} = f(x_1) \, f(x_2)$$

In many fields a certain value of b is important in the applications of mathematics. This value is the irrational number e, where

Note *The number e was introduced by Euler, who showed that it is an irrational number. The student is encouraged to read about the interesting history of e.*

$$e = 2.71828 \ldots$$

The exponential function with base $b = e$ is often denoted by exp, and we write

$$\exp(x) = e^x$$

In many real-life situations involving processes of growth or decay, exponential functions appear frequently. For example, exponential functions play a role in banking problems, population growth, radioactive decay, and many others. Some of these applications will be illustrated in the next two chapters.

In Problems 1 through 16, sketch a graph of the given exponential function.   Exercises 6.1

1. $f(x) = 3^x$ 2. $f(x) = \left(\frac{1}{3}\right)^x$

3. $f(x) = 4^x$ 4. $f(x) = \left(\frac{1}{4}\right)^x$

5. $f(x) = 3^{x+1}$ 6. $f(x) = \left(\frac{1}{4}\right)^{-x}$

7. $f(x) = 2 \cdot 3^x$ 8. $f(x) = 5^x$

9. $f(x) = 2^{x-1}$ 10. $f(x) = 3 \cdot 2^x$

11. $f(x) = 3^{x-1}$ 12. $f(x) = 2^{-x^2}$

13. $f(x) = 3^{-x^2}$ 14. $f(x) = e^x$ [*Hint*: Use tables to compute e^x.]

15. $f(x) = e^{-x}$ 16. $f(x) = e^{2x}$

17. If $f(x) = b^x$, show that $f(x_1 - x_2) = \dfrac{f(x_1)}{f(x_2)}$

In Problems 18 through 25, the graph of the exponential function $f(x) = b^x$ contains the given point. Determine the base b.

18. $\left(2, \frac{1}{9}\right)$ 19. $\left(-2, \frac{1}{9}\right)$

20. $\left(3, \frac{1}{8}\right)$ 21. $(-4, 16)$

22. $(-2, 16)$ 23. $(-1, 5)$

24. $\left(\frac{1}{2}, 2\right)$ 25. $\left(-\frac{1}{2}, 2\right)$

6.2 The Logarithmic Function

The invention of logarithms by John Napier was a milestone in the advancement of mathematics. Whereas most mathematical concepts

---

John Napier

John Napier (1550–1617) was born at Merchiston, near Edinburgh, Scotland. His early education was received at St. Andrews, and he completed his studies in Europe. In 1572 he married and had two children. His wife, Elizabeth, died seven years later. He remarried and fathered ten more children. Theology and mathematics occupied Napier. He took an active part in church politics and was the author of a theological treatise.

His work with equations involved imaginary roots, which he referred to as "a great algebraic secret." Napier devised several mechanical aids for calculations including the Rabologia (familiarly known as "Napier's bones"). He spent many years preparing the Table of Logarithms, which was published in 1614. This invention was welcomed by the mathematical world. Napier also wrote the *Constructio* and *Descripto*, which explained logarithms.

evolve from previous ideas, the logarithm is an exception. Napier invented logarithms to simplify the computation of cumbersome algebraic operations, though his original work was unwieldy. However, together with Henry Briggs (1561-1631), he constructed tables for $\log_{10} x$. These tables were completed and published by Briggs after Napier's death. The invention of logarithms paved the way for the famous astronomer Johann Kepler (1571-1630) to complete the calculations that led to the third law of planetary motion. Having received Napier's tables, Kepler refused to use them until he had derived the formula responsible for their construction. Kepler invented other formulas and constructed his own tables, which he published in 1624. Napier's invention led to the logarithmic and exponential functions which today play an important role in the sciences.

In this section we shall define the logarithmic function and indicate the relationship between logarithmic and exponential functions.

The exponential function $f(x) = b^x$ has domain $\{x | x \in R\}$ and range $\{y | y > 0\}$. We note that f is one-to-one. Therefore f has an inverse f^{-1} with domain $\{x | x > 0\}$ and range $\{y | y \in R\}$. The inverse of an exponential function is called a *logarithmic function*.

Logarithmic function ▶ **Definition 6.2** *Let $b > 0$ and $b \neq 1$. Then, the* logarithmic function *with base b, denoted by*

$$f^{-1}(x) = \log_b x \tag{6.1}$$

(read "logarithm to the base b of x") is defined to be the inverse of the exponential function $f(x) = b^x$.

The graphs of f and f^{-1} are shown in Figure 6.5 for $b > 1$, in Figure 6.6 for $0 < b < 1$. Observe that the graph of f^{-1} is the reflection of the

Figure 6.5

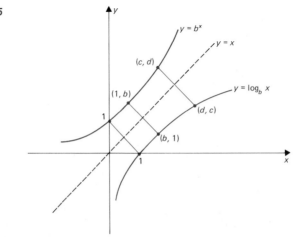

Figure 6.6

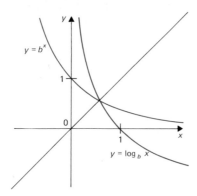

graph of f through the line $y = x$. Now, if f^{-1} is the inverse of the function f, we have

$$x = f(f^{-1}(x))$$

Therefore for the exponential function we have

$$x = b^{f^{-1}(x)} \qquad\qquad (6.2)$$

From (6.1) and (6.2), we obtain

$$x = b^{\log_b x} \qquad\qquad (6.3)$$

Equation (6.3) implies that if b is a positive real number and $b \neq 1$, then

$$y = \log_b x \text{ is equivalent to } x = b^y.$$

What happens if we allow $b = 1$? Check the definition of one-to-one.

Example 6.3 Find the unknown x, y, or b in each of the following:

(a) $y = \log_2 8$ (b) $4 = \log_3 x$

(c) $\frac{1}{2} = \log_b 3$ (d) $y = \log_2 \sqrt{2}$

SOLUTION

(a) The equation $y = \log_2 8$ is equivalent to

$$2^y = 8 \quad \text{or} \quad 2^y = 2^3$$

Thus $y = 3$

(b) The equation $4 = \log_3 x$ is equivalent to $3^4 = x$
     Thus

$$x = 81$$

(c) The equation $\frac{1}{2} = \log_b 3$ is equivalent to $b^{1/2} = 3$
     Thus

$$b = 9$$

(d)   The equation $y = \log_2 \sqrt{2}$ is equivalent to $2^y = \sqrt{2}$
Thus

$$y = \tfrac{1}{2}$$

With the aid of Equation (6.3), we can establish the following theorem easily.

▶   *Theorem 6.1   If a, b, and c are positive numbers and $b \neq 1$, then*

$$\log_b (ac) = \log_b a + \log_b c \tag{6.4}$$

$$\log_b \frac{a}{c} = \log_b a - \log_b c \tag{6.5}$$

$$\log_b 1 = 0 \tag{6.6}$$

$$\log_b b = 1 \tag{6.7}$$

$$\log_b a^r = r \log_b a \quad \text{if r is a real number} \tag{6.8}$$

We prove (6.4), (6.6), and (6.8) and leave (6.5) and (6.7) for the reader as an exercise.

PROOF OF (6.4)   Let $u = \log_b a$ and $v = \log_b c$. We have

$$a = b^u \quad \text{and} \quad c = b^v$$

and

$$a \cdot c = b^u \cdot b^v = b^{u+v} \tag{6.9}$$

Equation (6.9) is equivalent to

$$\log_b (a \cdot c) = u + v = \log_b a + \log_b c$$

Recall that $b^0 = 1$.   **PROOF OF (6.6)**   Let $u = \log_b 1$. Then $b^u = 1$ and we have $u = 0$. Therefore $\log_b 1 = 0$.

PROOF OF (6.8)   Let $u = \log_b a$. Then $b^u = a$ and

$$(b^u)^r = a^r \quad \text{or} \quad b^{ur} = a^r \tag{6.10}$$

From Equations (6.3) and (6.10), we have

$$a^r = b^{\log_b a^r} \quad \text{and} \quad b^{ur} = b^{\log_b a^r}$$

Since the exponential function is one-to-one,

$b^x = b^y \Rightarrow x = y$ $(b > 1)$.
Substitution of $u = \log_b a$

$$ur = \log_b a^r \quad \text{and} \quad r \log_b a = \log_b a^r$$

Example 6.4   Use logarithms of x, y, and z (assume that x, y, and z are positive real numbers) to express the following term.

$$\log_b \left(\frac{x^2 z^{2/3}}{y^4} \right)$$

SOLUTION Using Theorem 6.1, we have

$$\log_b \left(\frac{x^2 z^{2/3}}{y^4} \right) = \log_b (x^2 z^{2/3}) - \log_b y^4$$

$$= \log_b x^2 + \log_b z^{2/3} - \log_b y^4$$

$$= 2 \log_b x + \tfrac{2}{3} \log_b z - 4 \log_b y$$

Example 6.5 Simplify each of the following expressions.

(a) $\log_2 4^2$ (b) $\log_3 (9^{1/2} \cdot 3^{1/4})$

SOLUTION

(a) $\log_2 4^2 = \log_2 (2^2)^2 = \log_2 2^4 = 4 \log_2 2 = 4$

(b) $\log_3 (9^{1/2} \cdot 3^{1/4}) = \log_3 9^{1/2} + \log_3 3^{1/4} = \log_3 (3^2)^{1/2} + \log_3 3^{1/4}$

$$= \log_3 3 + \tfrac{1}{4} \log_3 3 = 1 + \tfrac{1}{4} = \tfrac{5}{4}$$

Example 6.6 Given that $\log_{10} 2 = 0.301$ and $\log_{10} 3 = 0.477$, compute the value of each of the following expressions.

(a) $\log_{10} 6$ (b) $\log_{10} 200$ (c) $\log_{10} \sqrt{1.5}$

SOLUTION

(a) $\log_{10} 6 = \log_{10} (2 \cdot 3) = \log_{10} 2 + \log_{10} 3 = 0.301 + 0.477 = 0.778$

(b) $\log_{10} 200 = \log_{10} (2 \cdot 10^2) = \log_{10} 2 + \log_{10} 10^2$

$$= 0.301 + 2 \log_{10} 10 = 0.301 + 2 = 2.301$$

(c) $\log_{10} \sqrt{1.5} = \log_{10} \left(\tfrac{3}{2} \right)^{1/2} = \tfrac{1}{2} \log_{10} \left(\tfrac{3}{2} \right)$

$$= \tfrac{1}{2} [\log_{10} 3 - \log_{10} 2] = \tfrac{1}{2} [0.477 - 0.301] = 0.088$$

Example 6.7 Solve each of the following equations for x:

(a) $\log_2 (2x + 1) = 4$ (b) $\log_2 x + \log_2 (3x - 2) = 3$

SOLUTION

(a) The equation $\log_2 (2x + 1) = 4$ is equivalent to

$$2x + 1 = 2^4 \quad \text{or} \quad x = \tfrac{15}{2}$$

(b) We have from Theorem 6.1,

$$\log_2 x + \log_2 (3x - 2) = \log_2 x(3x - 2)$$

Thus

$$\log_2 (3x^2 - 2x) = 3 \quad \text{or} \quad 3x^2 - 2x = 2^3$$

or

$$3x^2 - 2x - 8 = 0 \quad \text{or} \quad (3x + 4)(x - 2) = 0$$

Hence

$$x = 2 \quad \text{or} \quad x = -\tfrac{4}{3}$$

Since $x = -\tfrac{4}{3}$ does not satisfy the original equation, the only solution is $x = 2$.

Exercises 6.2   In Problems 1 through 6, express the given statement by an equivalent logarithmic statement.

1. $3^4 = 81$
2. $2^5 = 32$
3. $5^{-3} = \frac{1}{125}$
4. $\left(\frac{1}{3}\right)^{-2} = 9$
5. $10^{-3} = 0.001$
6. $27^{2/3} = 9$

In Problems 7 through 10, express the given statement by an equivalent exponential statement.

7. $\log_2 32 = 5$
8. $\log_{1/2} 8 = -3$
9. $\log_{10} \frac{1}{1000} = -3$
10. $\log_{16} \frac{1}{8} = -\frac{3}{4}$

In Problems 11 through 20, simplify the given expressions.

11. $\log_2 2^3$
12. $\log_3 9^2$
13. $\log_3 \dfrac{9^{2/3}}{3^{1/2}}$
14. $\log_4 (16^{1/5} \cdot 64^{1/4})$
15. $\log_2 \dfrac{2^3}{64^{1/2}}$
16. $3^{\log_3 3}$
17. $4^{\log_4 2}$
18. $27^{\log_8 2}$
19. $b^{\log_b 10 - \log_b 5}$
20. $\log_b (\log_b b^b)$

In Problems 21 through 28, find the unknown x, y, or b.

21. $y = \log_2 \frac{1}{8}$
22. $4 = \log_2 x$
23. $\log_b 625 = 4$
24. $\log_2 (4x + 3) = 3$
25. $\log_{10} (3x + 1) = 1$
26. $\log_3 x + \log_3 (2x + 1) = 2$
27. $\log_2 (x - 1) - \log_2 (x + 1) = 3$
28. $\log_{10} 10^{-\log_{10} x^2} = -4$

29.   Prove (6.5) of Theorem 6.1.

30.   Prove (6.7) of Theorem 6.1.

In Problems 31 through 38, use $\log_{10} 3 = 0.4771$, $\log_{10} 5 = 0.699$, and $\log_{10} 7 = 0.8451$ to compute the value of each of the given expressions.

31. $\log_{10} 35$ 32. $\log_{10} 15$

33. $\log_{10} 1.5$ 34. $\log_{10} 500$

35. $\log_{10} \sqrt{7}$ 36. $\log_{10} 0.003$

37. $\dfrac{\log_{10} 5}{\log_{10} 7}$ 38. $\log_{10} \frac{5}{7}$

39.   Show that $x^x = b^{x \, \log_b x}$.

40.   Show that $\log_b x = \dfrac{\log_a x}{\log_a b}$

41.   Show that $\log_b a = \dfrac{1}{\log_a b}$

6.3   Differentiation of the Logarithmic and Exponential Functions

Consider the expression

$$(1 + p)^{1/p}$$

We can find its value for various values of p:

$p = 1$	$(1 + 1)^1 = 2.0000$
$p = \frac{1}{2}$	$\left(1 + \frac{1}{2}\right)^2 = 2.2500$
$p = \frac{1}{4}$	$\left(1 + \frac{1}{4}\right)^4 = 2.4414$
$p = \frac{1}{100}$	$\left(1 + \frac{1}{100}\right)^{100} = 2.7048$
$p = \frac{1}{1000}$	$\left(1 + \frac{1}{1000}\right)^{1000} = 2.7181$

We can show that $\lim_{p \to 0} (1 + p)^{1/p}$ exists and is the irrational number e, that is,

$$\lim_{p \to 0} (1 + p)^{1/p} = e \approx 2.71828 \qquad (6.11)$$

To obtain a formula for the derivative of $\log_b x$, Equation (6.11) will be needed. Now we prove the following theorem.

Theorem 6.2   *If $f(x) = \log_b x$, then $f'(x) = (1/x) \log_b e$ $(x > 0)$.*   ◀

PROOF   We write the difference quotient:

$$\frac{f(x+h)-f(x)}{h} = \frac{1}{h}[\log_b (x+h) - \log_b x]$$

Using Theorem 6.1.

$$= \frac{1}{h}\left[\log_b\left(\frac{x+h}{x}\right)\right] = \frac{1}{h}\log_b\left(1+\frac{h}{x}\right)$$

$$= \frac{1}{x}\frac{x}{h}\log_b\left(1+\frac{h}{x}\right) = \frac{1}{x}\log_b\left(1+\frac{h}{x}\right)^{x/h}$$

Letting $p = h/x$, we have

$$\frac{f(x+h)-f(x)}{h} = \frac{1}{x}\log_b (1+p)^{1/p}$$

As $h \to 0$, $p \to 0$, and

$$\lim_{h\to 0}\frac{f(x+h)-f(x)}{h} = \lim_{p\to 0}\left[\frac{1}{x}\log_b (1+p)^{1/p}\right]$$

Hence

$$f'(x) = \frac{1}{x}\lim_{p\to 0}\log_b (1+p)^{1/p}$$

This is permissible because of the continuity of the logarithmic function.

$$= \frac{1}{x}\log_b \lim_{p\to 0}(1+p)^{1/p}$$

Using Equation (6.11)

$$= \frac{1}{x}\log_b e$$

If u is a differentiable function of x, we can use the chain rule to extend the result of Theorem 6.2 and obtain

For $u(x) > 0$
$$\frac{d}{dx}\left(\log_b u\right) = \frac{1}{u}(\log_b e)\frac{du}{dx} \qquad (6.12)$$

Example 6.8   If $f(x) = \log_b (2x^2 - 3x)$, find $f'(x)$.

SOLUTION   Let $u = 2x^2 - 3x$, then $du/dx = 4x - 3$. Using Equation (6.12) we obtain

For $2x^2 - 3x > 0$
$$f'(x) = \frac{1}{2x^2 - 3x}(\log_b e)(4x - 3) = \frac{4x - 3}{2x^2 - 3x}\log_b e$$

We see that the expression in (6.12) can be simplified if we select $b = e$:

$\log_e e = 1$
$$\frac{d}{dx}(\log_e u) = \frac{1}{u}(\log_e e)\frac{du}{dx} = \frac{1}{u}\frac{du}{dx} \qquad (6.13)$$

The expression $\log_e x$ is called the *natural logarithm* (or *Napierian logarithm*) of x and the following notation is commonly used

$$\log_e x = \ln x$$

(read "*ell-en* of x").

Example 6.9 Find $f'(x)$.

(a) $f(x) = \ln \sqrt{x^2 + 1}$

(b) $f(x) = \ln \dfrac{(x^2 + 1)^{20} (3x - 1)^3}{(2x^2 + 3)^4}$

SOLUTION

(a)   Letting $u = \sqrt{x^2 + 1}$, then we have $\dfrac{du}{dx} = \dfrac{x}{\sqrt{x^2 + 1}}$. Therefore

$$f'(x) = \frac{1}{\sqrt{x^2 + 1}} \cdot \frac{x}{\sqrt{x^2 + 1}} = \frac{x}{x^2 + 1}$$

We could simplify the problem by writing

$\ln \sqrt{x^2 + 1} = \ln(x^2 + 1)^{1/2}$

$= \frac{1}{2} \ln(x^2 + 1)$

(b)   First we simplify the function by using the rules of logarithms. Thus

$$\ln \frac{(x^2 + 1)^{20} (3x - 1)^3}{(2x^2 + 3)^4}$$

Note $\ln \dfrac{AB}{C} = \ln A + \ln B - \ln C$

$$= 20 \ln (x^2 + 1) + 3 \ln (3x - 1) - 4 \ln (2x^2 + 3)$$

Therefore

$$f'(x) = \frac{d}{dx} [20 \ln (x^2 + 1) + 3 \ln (3x - 1) - 4 \ln (2x^2 + 3)]$$

$$= 20 \frac{d}{dx} [\ln (x^2 + 1)] + 3 \frac{d}{dx} [\ln (3x - 1)] - 4 \frac{d}{dx} [\ln (2x^2 + 3)]$$

$$= 20 \cdot \frac{2x}{x^2 + 1} + 3 \cdot \frac{3}{3x - 1} - 4 \cdot \frac{4x}{2x^2 + 3}$$

Note $\dfrac{d}{dx} [\ln(x^2 + 1)]$

$$= \frac{40x}{x^2 + 1} + \frac{9}{3x - 1} - \frac{16x}{2x^2 + 3}$$

$= \dfrac{1}{x^2 + 1} \cdot \dfrac{d}{dx}(x^2 + 1) = \dfrac{1}{x^2 + 1} \cdot 2x$

Now we develop a formula for the derivative of the function $f(x) = b^x$.

Theorem 6.3   If $f(x) = e^x$, then $f'(x) = e^x$.

◀

PROOF   Let $y = e^x$. Then $\ln y = x$ and by differentiating, we find that

$$\frac{d}{dx} [\ln y] = \frac{d}{dx} [x], \qquad \frac{1}{y} \frac{dy}{dx} = 1 \quad \text{or} \quad \frac{dy}{dx} = y$$

We are using implicit differentiation.

that is,

$$f'(x) = e^x$$

If u is a differentiable function of x, then by the chain rule

$$f(x) = e^u \Longrightarrow f'(x) = e^u \frac{du}{dx} \qquad\qquad (6.14)$$

Example 6.10 Find $f'(x)$.

(a) $f(x) = e^{x^2 + 4}$ (b) $f(x) = 3e^{1/x}$

SOLUTION

(a) Let $u = x^2 + 4$, then $du/dx = 2x$. Using (6.14), we find that

$$f'(x) = 2xe^{x^2 + 4}$$

(b) Let $u = 1/x$, then $du/dx = -(1/x^2)$. Therefore

$$f'(x) = -\frac{3}{x^2} e^{1/x}$$

▶ **Theorem 6.4** *If b is a positive number and* $f(x) = b^x$, *then* $f'(x) = b^x \ln b$.

Recall that $b = e^{\ln b}$

PROOF Let $y = b^x$. Then

$$y = e^{(\ln b)x}$$

Set $u = (\ln b)x$ and $\dfrac{du}{dx} = \ln b$. Then

$$y = e^u$$

and by differentiating, we find that

$$\frac{dy}{dx} = e^u \frac{du}{dx} = e^{(\ln b)x} \ln b = b^x \ln b$$

that is,

$$f'(x) = b^x \ln b$$

In general, if u is a differentiable function of x, then

$$f(x) = b^u \Rightarrow f'(x) = b^u (\ln b) \frac{du}{dx} \qquad (6.15)$$

Example 6.11 Find $f'(x)$.

(a) $f(x) = 3^x$ (b) $f(x) = 5^{3x^2 - 2x}$

(c) $f(x) = 3^{x-1} \cdot 2^x$

SOLUTION

Application of Theorem 6.4

(a) $f'(x) = 3^x \ln 3$

Application of (6.15)

(b) $f'(x) = 5^{3x^2 - 2x} (\ln 5)(6x - 2)$

Using the formula for the derivative of the product of two functions

(c) $f'(x) = (3^{x-1}) \dfrac{d}{dx} (2^x) + 2^x \dfrac{d}{dx} (3^{x-1})$

$$= 3^{x-1} \cdot 2^x \ln 2 + 2^x \cdot 3^{x-1} \ln 3$$

$$= 3^{x-1} \cdot 2^x (\ln 2 + \ln 3) \qquad \textit{Factoring}$$

$$= 3^{x-1} \cdot 2^x \ln 6 \qquad \textit{Using Theorem 6.1}$$

We can make use of logarithms to differentiate complicated functions, as in the following example.

Example 6.12 If

$$y = \frac{(x^2 + 3)^{10}(x - 1)^{1/2}}{(x^2 + 3x + 5)^{80}} \text{ find } \frac{dy}{dx}.$$

SOLUTION   We have

$$\ln y = \ln \frac{(x^2 + 3)^{10}(x - 1)^{1/2}}{(x^2 + 3x + 5)^{80}}$$

$$= 10 \ln (x^2 + 3) + \tfrac{1}{2}\ln (x - 1) - 80 \ln (x^2 + 3x + 5)$$

Differentiation yields

$$\frac{1}{y} \frac{dy}{dx} = 10 \; \frac{1}{x^2 + 3} \frac{d}{dx} (x^2 + 3) + \frac{1}{2} \frac{1}{x - 1} \frac{d}{dx} (x - 1)$$

$$- 80 \; \frac{1}{x^2 + 3x + 5} \frac{d}{dx} (x^2 + 3x + 5)$$

$$= \frac{20x}{x^2 + 3} + \frac{1}{2(x - 1)} - \frac{80(2x + 3)}{x^2 + 3x + 5}$$

Substituting for y and solving for dy/dx, we obtain

$$\frac{dy}{dx} = \frac{(x^2 + 3)^{10}(x - 1)^{1/2}}{(x^2 + 3x + 5)^{80}} \left[\frac{20x}{x^2 + 3} + \frac{1}{2(x - 1)} - \frac{80(2x + 3)}{x^2 + 3x + 5} \right]$$

The process used in Example 6.12 is called "logarithmic differentiation."

In Problems 1 through 24, find $f'(x)$.

Exercises 6.3

1. $f(x) = e^{4x}$

2. $f(x) = 4e^{4x^2 + 5}$

3. $f(x) = e^{2/x^2}$

4. $f(x) = e^{3x^2 - 4x + 1}$

5. $f(x) = x^3 e^{2x}$

6. $f(x) = (2x^2 + 3) e^x$

7. $f(x) = e^{-2x^2}$

8. $f(x) = (e^{-2x})^2$

9. $f(x) = \ln (x^2 + 4)$

10. $f(x) = \ln (e^x + 1)$

11. $f(x) = e^x \ln x$

12. $f(x) = \dfrac{e^x - 1}{e^x + 1}$

13. $f(x) = \ln (e^{2x^2 + 4} + 1)$

14. $f(x) = (e^{x^2} + 4)e^{5x^2 - 1}$

15. $f(x) = 5^x$

16. $f(x) = 5^{-x}$

17. $f(x) = 2^{x-1} 5^x$

18. $f(x) = \dfrac{2^{x-1}}{5^x}$

19. $f(x) = e^{2x} + (2x)^e$

20. $f(x) = \ln (x + \sqrt{x^2 + 1})^2$

21. $f(x) = \log_{10} x$

22. $f(x) = \log_{10} (\log_{10} x)$

23. $f(x) = \log_3 (x^2 + e^x)$

24. $f(x) = \log_2 [\log_5 (x^3 + x^{1/2})]$

In Problems 25 through 27, use the rules of logarithms to simplify the expression and find $f'(x)$.

25. $f(x) = \ln \dfrac{(x + 1)^{16}(2x^2 + x)^8}{\sqrt{x^2 + 4}}$

26. $f(x) = \ln \dfrac{(x^2 + 5)^7 (x^3 + 1)^{1/3}}{(x^2 + 1)^{5/4}}$

27. $f(x) = \ln \dfrac{(e^{2x} + 6)^7 \sqrt{x + 4}}{(e^{-x} + e^x)^5}$

In Problems 28 through 30, use the procedure of Example 6.12 to find dy/dx.

28. $y = \dfrac{(3x^2 + 1)^5 (x^2 + 1)^{3/2}}{(x^2 + 4)^{10}}$

29. $y = \dfrac{(x^2 + x + 1)^{60}(x^2 - 5x)^4}{\sqrt{x^2 - 4x + 10}}$

30. $y = \dfrac{(10x^2 + 5x + 1)^{5/2}(x^2 - 3x + 1)^{9/7} \, e^{-x^3 + 5 + 7}}{(x^3 + 2x^2 + 5)^8}$

Chapter 6 Review

Part 1   (Oral)

Define or discuss each of the following:

1. Exponential function
2. Properties of exponential functions
3. Logarithmic function
4. Properties of logarithmic functions
5. The number e
6. Natural logarithm

Part 2   (Written)

1. Sketch the graphs of each of the following functions:

(a) $f(x) = 2^{-x}$ (b) $f(x) = \left(\frac{1}{2}\right)^{2-x}$ (c) $f(x) = 3^x$

(d) $f(x) = \log_3 x$ (e) $f(x) = \log_{1/3} x$

2. Let $f(x) = 2^x - 2^{-x}$ and $g(x) = 2^x + 2^{-x}$. Find

(a) $f(x) + g(x)$ (b) $f(x) - g(x)$ (c) $f(x) \cdot g(x)$

(d) $[f(x)]^2 - [g(x)]^2$ (e) $f(x^2) \cdot g(x^2)$

3. Find the number:

(a) $\log_5 125$ (b) $\log_{\sqrt{8}} 64$ (c) $\log_3 \frac{1}{27}$

(d) $3^{\log_3 7}$

4. Express $\log \sqrt{x^{-3}y^4/z^5}$ in terms of logarithms of x, y, and z.

5. Express $3 \log x - \frac{3}{2} \log xy^2 + \frac{4}{3} \log z$ as one logarithm.

In Problems 6 through 22, find $f'(x)$.

6. $f(x) = e^{3x}$ 7. $f(x) = e^{-5x}$

8. $f(x) = 2e^{x^2-x}$ 9. $f(x) = xe^{-x}$

10. $f(x) = xe^{1/x}$ 11. $f(x) = x \ln x$

12. $f(x) = \left(x^2 - \frac{1}{x}\right)e^{x+1}$ 13. $f(x) = \left(x + \frac{1}{x}\right)\ln(x+3)$

14. $f(x) = 3^x$ 15. $f(x) = \dfrac{e^x - e^{-x}}{e^x + e^{-x}}$

16. $f(x) = 4^{-x}$ 17. $f(x) = e^x + x^e + e^e$

18. $f(x) = \log_5 x$ 19. $f(x) = \log_2 (x^3 - x^2)$

20. $f(x) = \log_{10} \dfrac{(x-1)^5(x+2)^7}{(x-2)^4}$ 21. $f(x) = \log_{10} \dfrac{(x^2 - 5x + 3)^{1/2}}{x(x-5)}$

22. $f(x) = \log_{10} \dfrac{(e^x - 5x)\sqrt{x^2-3}}{(x^2-3)^4}$

In Problems 23 through 28, use logarithms to find dy/dx.

23. $y = (x-1)^4(x^2 + 5x)(x+3)^5$

24. $y = \dfrac{(x^2 - 2x + 3)^{40}(x-1)^{1/2}}{(x+5)^{3/4}}$

25. $y = \dfrac{(x^2 - x - 1)^{1/2}(x^2 - 4x + 1)^{-3/5}}{(5x^2 - 2x + 1)^6}$ 26. $y = x^x$

27. $y = 3^x + x^3 + x^x$ 28. $y = x^{xe^x}$

Numerical Applications Using Calculators

Most calculators are equipped to perform certain operations dealing with exponential and logarithmic functions. In particular, the two operations $\ln x$ (find the natural logarithm of x) and e^x (raising the real number e to the xth power) are usually available in a pocket calculator. Some models also have the feature of finding x^y, where x and y are real numbers with certain restrictions. It is very handy to use a pocket calculator equipped with such features to solve problems involving exponential and logarithmic functions. Let us first use a calculator to find $\lim_{p \to 0} (1 + p)^{1/p}$.

Table 6.1

p	$1+p$	$(1+p)^{1/p}$
0.1	1.1	2.59374246
0.01	1.01	2.704813815
0.001	1.001	2.716923842
0.0001	1.0001	2.718145918
0.00001	1.00001	2.718254646
0.000001	1.000001	2.718281828
0.0000001	1.0000001	2.718281828
−0.1	0.9	2.867971989
−0.01	0.99	2.731999012
−0.001	0.999	2.719642125
−0.0001	0.9999	2.718417746
−0.00001	0.99999	2.718281828
−0.000001	0.999999	2.718281828
−0.0000001	0.9999999	2.718281828

Therefore $\lim_{p \to 0} (1 + p)^{1/p} \approx 2.718281828$, which is approximately the value we obtain for e from a calculator.

We shall now construct the following table with the help of a calculator:

Table 6.2

x	e^x	2.71828^x	$\ln x$	$e^{\ln x}$	$\ln e^x$
1	2.718281828	2.71828	0	1	0.9999999999
2	7.389056099	7.389046161	0.6931471805	1.99999999	2.
3	20.08553692	20.08549639	1.098612289	3.000001	3.
4	54.59815003	54.59800311	1.386294361	3.99999999	4.
5	148.4131591	148.41266	1.609437912	4.999999998	5.

Suppose that we are to evaluate the expression

$$y = \frac{\ln 5}{0.3} - 2e^{0.21} + \sqrt{0.5} - 7^{0.37}$$

We use a calculator and obtain

$$\ln 5 = 1.609437912 \qquad \frac{\ln 5}{0.3} = 5.36479304$$

$$e^{0.21} = 1.23367806 \qquad 2e^{0.21} = 2.46735612$$

$$\sqrt{0.5} = 0.7071067812 \qquad 7^{0.37} = 2.054406$$

Finally, we have

$$y = \frac{\ln 5}{0.3} - 2e^{0.21} + \sqrt{0.5} - 7^{0.37} = 1.550137701$$

Exercises   Use a calculator to evaluate each of the following.

1. (a) $\lim_{p \to \infty} \left(1 + \frac{1}{p}\right)^p$ (b) $\lim_{p \to \infty} \left(1 + \frac{3}{p}\right)^p$

2. $y = \ln (2^x + xe^3)$ for $x = 2, 2.1, \ldots, 3$

3. $y = e^{x^2 \ln 5}$ for $x = 1, 1.001, 1.002, \ldots, 1.005$

4. $y = \dfrac{e^x - e^{-x}}{e^x + e^{-x}}$ for $x = 0.1, 0.2, \ldots, 1$

5. $y = \ln \left(\dfrac{e^x + e^{-x}}{2}\right)$ for $x = 10, 10.1, \ldots, 10.5$

6. Use the definition of derivative to compute $f'(x)$ with $f(x) = \ln x$ for $x = 5, 5.1, \ldots, 6$. Compare your result with $1/x$.

If a, b, and c are three positive real numbers, then it can be shown that

$$\log_a b \cdot \log_b c = \log_a c$$

In particular,

$$\log_a b = \frac{1}{\log_b a} \quad \text{if} \quad \log_b a \neq 0$$

Hence

$$\log_a e \cdot \log_e c = \log_a c \quad \text{or} \quad \log_a c = \frac{\ln c}{\ln a}$$

Use a calculator and the preceding formula to compute the following:

1. $\log_3 5$ 2. $\log_7 \frac{1}{2}$

3. $\log_{99} 45$ 4. $\log_{71} 35^3$

5. $\log_{31} 502$

CHAPTER 7 The Integral

The problem of finding the area of a region bounded by curves is an old one. Archimedes (287–212 B.C.) successfully solved one case see Figure 7.1); however, his method could not be applied generally. The problem was not solved until Newton and Leibniz created calculus.

Elementary calculus is often divided into two major branches: differential calculus and integral calculus. Differential calculus deals with the concept of the derivative. Integral calculus deals with another basic concept—that of the *integral*. We have seen that the derivative, from a geometrical point of view, is related to the slope of a tangent line to a curve, and we have used this notion to study problems in the sciences and in economic theory. From a geometrical point of view, the integral is related to the area of a region. In this chapter we shall

Figure 7.1

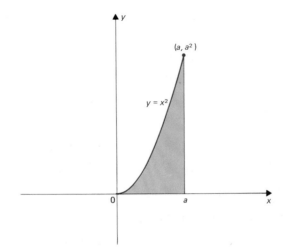

252

study some elementary aspects of the integral and develop techniques for calculating areas of regions bounded by curves. We shall also show how the integral is related to certain problems arising in the sciences and economic theory. First we introduce the notion of antiderivatives and then we provide a definition of the definite integral. We shall relate the problem of tangents to the problem of areas.

7.1   The Antiderivative

Consider the function f defined by $f(x) = 2x$. Then the function F defined by $F(x) = x^2$ has the property that

$$F'(x) = f(x)$$

for all x.

Similarly, if $f(x) = 3x^2$, then $F(x) = x^3$ has the property that

$$F'(x) = f(x)$$

Definition 7.1   Suppose f is a continuous function defined on an open interval I. Then a function F is called an antiderivative *of f on I if*

$$F'(x) = f(x)$$

for all $x \in I$.

◄ Antiderivative

The process involved in calculating F for a given f is called antidifferentiation. If the student thinks of differentiation as a forward process, then antidifferentiation is a reverse process

Thus in each of our previous examples the function F is an antiderivative of f.

We observe that for $f(x) = 2x$, the function $G(x) = x^2 + C$, where C is a constant, is also an antiderivative of f. We find that

$$G'(x) = (x^2 + C)' = 2x = f(x)$$

Theorem 7.1 states the general case.

Theorem 7.1   Suppose that F and G are differentiable functions on the open interval (a, b) and continuous at a and b. If $F'(x) = G'(x)$ for all $x \in (a, b)$, then

$$F(x) = G(x) + C, \qquad x \in (a, b)$$

when C is a constant.

◄

PROOF   Let $h(x) = F(x) - G(x)$. Then

$$h'(x) = F'(x) - G'(x) = 0$$

By the mean value theorem (see Theorem 4.9), there exists an $x_0 \in (a, x)$ such that

$$h'(x_0) = \frac{h(x) - h(a)}{x - a}$$

Since $h'(x_0) = 0$, we have $h(x) = h(a)$ for every $x \in (a, b)$. Hence $h(x) = C$ (a constant). Consequently,

$$h(x) = F(x) - G(x) = C$$

or

$$F(x) = G(x) + C$$

We write

$$\int f = F + C$$

Historically, the symbol ∫ has evolved from the letter S, which in turn evolved from the summation symbol Σ. The notation Σ for the integral sign was first used by Euler and in the following sections we shall indicate the relationship between ∫ and Σ.

and call

$\int$ the integral sign

f the integrand

$\int f$ the antiderivative or the indefinite integral of f

∫ f is also called the general antiderivative of f, or simply the antiderivative of f.

The constant C is called the constant of integration. Another notation we will use for the indefinite integral of f is

$$\int f(x)\, dx$$

Example 7.1   Find the antiderivative of f, where $f(x) = x^2$.

SOLUTION   We have

$$\int f(x)\, dx = \int x^2\, dx = \tfrac{1}{3}x^3 + C$$

It is not difficult to check the answer. Remember that the derivative of the indefinite integral should be equal to the integrand. Thus in Example 7.1 we have

$$\frac{d}{dx}\left(\frac{x^3}{3} + C\right) = \frac{d}{dx}\left(\frac{x^3}{3}\right) + \frac{d}{dx}(C) = x^2 + 0 = x^2$$

We observe that

$$\int 4x^3\, dx = x^4 + C \qquad \int x^4\, dx = \frac{x^5}{5} + C$$

$$\int x^5\, dx = \frac{x^6}{6} + C \qquad \int x^{-1/2}\, dx = \frac{x^{1/2}}{\frac{1}{2}} + C$$

Note $\dfrac{d}{dx}\left(\dfrac{x^{r+1}}{r+1} + c\right)$

$= x^r$ for $r \neq -1$.

In general, we have

$$\int x^r\, dx = \frac{x^{r+1}}{r+1} + C \qquad (r \neq -1) \qquad (7.1)$$

The case r = −1 will be discussed in Section 7.6

Some important properties of the indefinite integral are stated in

Theorem 7.2 *If f and g have antiderivatives and k is a constant, then* ◀

(i) $\displaystyle\int kf(x)\,dx = k\int f(x)\,dx \qquad (k \neq 0)$

(ii) $\displaystyle\int [f(x) + g(x)]\,dx = \int f(x)\,dx + \int g(x)\,dx$

PROOF OF **(i)** We have

$$\frac{d}{dx}\left[k\int f(x)\,dx\right] = k\frac{d}{dx}\left[\int f(x)\,dx\right] = kf(x)$$

Therefore by Definition 7.1, $k\int f$ is the antiderivative of kf.

The proof of (ii) is left for the student as an exercise (see Problem 22).

Example 7.2 Calculate each of the following indefinite integrals:

(a) $\displaystyle\int (2 + x)\,dx$

(b) $\displaystyle\int \sqrt{x}\,dx$

(c) $\displaystyle\int \frac{x + 3}{x^3}\,dx, \quad x \neq 0$

(d) $\displaystyle\int \frac{x^2 - 4}{x - 2}\,dx, \qquad x \neq 2$

SOLUTION

(a) $\displaystyle\int (2 + x)\,dx = \int 2\,dx + \int x\,dx$

$\qquad = 2\int dx + \int x\,dx = 2x + C_1 + \frac{x^2}{2} + C_2$

$\qquad = 2x + \frac{x^2}{2} + C$

Application of Theorem 7.2

Application of Equation (7.1)

yields $\int dx = \int x^0\,dx = \dfrac{x^{0+1}}{0+1} = x$

$C = C_1 + C_2$

(b) $\displaystyle\int \sqrt{x}\,dx = \int x^{1/2}\,dx = \frac{x^{1/2+1}}{\frac{1}{2}+1} + C = \frac{2}{3}x^{3/2} + C$

(c) $\displaystyle\int \frac{x+3}{x^3}\,dx = \int \left(\frac{1}{x^2} + \frac{3}{x^3}\right)dx, \qquad x \neq 0$

$\qquad = \int (x^{-2} + 3x^{-3})\,dx = \int x^{-2}\,dx + 3\int x^{-3}\,dx$

$\qquad = \frac{x^{-1}}{-1} + 3\left(\frac{x^{-2}}{-2}\right) + C = -\frac{2x+3}{2x^2} + C$

The two constants of integration are combined.

(d) $\displaystyle\int \frac{x^2-4}{x-2}\,dx = \int \frac{(x-2)(x+2)}{x-2}\,dx, \qquad x \neq 2$

$\qquad = \int (x+2)\,dx = \int x\,dx + 2\int dx = \frac{x^2}{2} + 2x + C$

Example 7.3 Suppose that the American Gadget Company has estimated that the marginal cost at a level of production of x gadgets is given by

$$C'(x) = 3 + \frac{8}{\sqrt{x}}$$

and the fixed cost is $20,000. Find the cost function $C(x)$.

SOLUTION   Since

$$C'(x) = 3 + \frac{8}{\sqrt{x}} = 3 + 8x^{-1/2}$$

the antiderivative of $C'(x)$ is $C(x)$ given by

$$C(x) = \int (3 + 8x^{-1/2}) \, dx$$

$$= 3x + 8(2x^{1/2}) + K \qquad (K \text{ is a constant})$$

$$= 3x + 16x^{1/2} + K$$

We find K by using the fact that the fixed cost is $20,000, that is, if $x = 0$, then $C(x) = 20,000$. Consequently,

$$20,000 = 3(0) + 16(0) + K \quad \text{or} \quad K = 20,000$$

Hence the cost function is

$$C(x) = 3x + 16x^{1/2} + 20,000$$

Exercises 7.1   In Problems 1 through 20, calculate the given indefinite integrals.

1. $\int \frac{1}{2} dx$

2. $\int \frac{1}{2} x \, dx$

3. $\int 3x^3 \, dx$

4. $\int (x + x^3) \, dx$

5. $\int (x^2 + 2x + 1) \, dx$

6. $\int x(x - 2) \, dx$

7. $\int \frac{1}{x^2} dx, \quad x \neq 0$

8. $\int \frac{3}{x^3} dx, \quad x \neq 0$

9. $\int (x - 1)^2 \, dx$

10. $\int \frac{x + 4}{x^4} dx, \quad x \neq 0$

11. $\int 2\sqrt{x} \, dx$

12. $\int \frac{1}{\sqrt{x}} dx, \quad x > 0$

13. $\int x^{-1/3} \, dx, \quad x \neq 0$

14. $\int (3x^{1/3} + 2x^{-1/3}) \, dx, \quad x \neq 0$

15. $\int \frac{x^2 - 1}{x + 1} dx, \quad x \neq -1$

16. $\int x^{1/3}(2 + x^{1/3}) \, dx$

17. $\int \frac{x^3 - 2}{\sqrt{x}} dx, \quad x > 0$

18. $\int \frac{x^3 - 27}{x - 3} dx, \quad x \neq 3$

19. $\int \frac{x - 4}{\sqrt{x} + 2} dx, \quad x \geq 0$

20. $\int \frac{x + 1}{x^{1/3} + 1} dx, \quad x \neq -1$

21. Give examples of functions f and g having antiderivatives such that

(a) $\int f(x)g(x) \, dx \neq \left(\int f(x) \, dx \right) \left(\int g(x) \, dx \right)$

(b) $\displaystyle\int xf(x)\,dx \ne x \int f(x)\,dx$

22. Prove Theorem 7.2, part (ii).

23. The marginal cost of the ABC Shirt Company for producing x shirts is given by

$$C'(x) = \frac{12}{\sqrt[3]{x}}$$

and the fixed cost is $10,000. Find the cost function $C(x)$. Find also the total cost of producing 2000 shirts.

24. Find the cost function $C(x)$, if the marginal cost of producing x items is given by

$$C'(x) = 2x^2 + 5x - 3$$

It is given that the fixed cost is $5000.

25. The marginal profit for a Pizza Parlor is given by

$$P'(x) = 3x + 5000$$

where x (in thousands) is the number of pizzas sold. If there is a loss of $200 if no pizzas are sold, find the profit function $P(x)$.

26. Find the equation of the curve whose tangent line at x has a slope given by

$$f'(x) = 3x^2 - 5x + 8$$

if the point $(1, -6)$ is on the curve.

7.2 The Definite Integral

In this section we shall introduce the concept of the definite integral and in the next section we shall relate this concept to antidifferentiation.

What is meant by the area of a region in the plane? If the region is bounded by straight line segments, it is easy to calculate its area by dividing it into rectangles and triangles. Thus the area $A(R)$ of the region R in Figure 7.2 is given by

$$A(R) = A(R_1) + A(R_2) + A(R_3) + A(R_4)$$

Figure 7.2

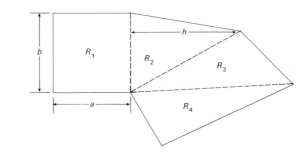

where the area of the rectangular region R_1 with length a and height b is

$$A(R_1) = ab$$

the area of the triangular region R_2 with base b and height h is

$$A(R_2) = \tfrac{1}{2}bh$$

and so on.

The problem of finding areas is more complicated if the region has curvilinear boundaries (see Figure 7.3). In this section we shall find

Figure 7.3

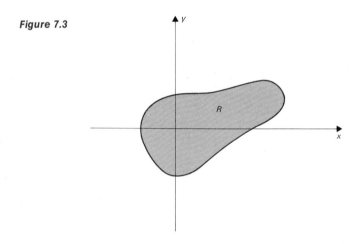

that the area of a region in the plane with curvilinear boundaries can be represented as the limit of a sequence. Our exposition will be intuitive in nature and will give the general idea of the precise definition which follows.

Consider the region R bounded by the graph of $f(x) = x^2$ and the lines $x = 0$, $x = 2$ and $y = 0$ (see Figure 7.4). We see that R is contained

Figure 7.4

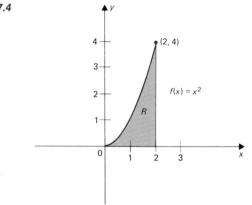

in a rectangular region whose area is 8 square units. Then, assuming that R has an area,

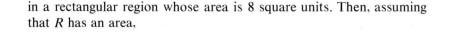

$$0 < A(R) < 8$$

and we have an approximation of the area of the region R.

Suppose that we divide the interval $[0, 2]$ into two parts and consider the point $(1, 1)$ on the graph of f. We draw two sets of rectangles as shown in Figure 7.5. Obviously,

$$A(R_1) < A(R) < A(R_2) + A(R_3)$$

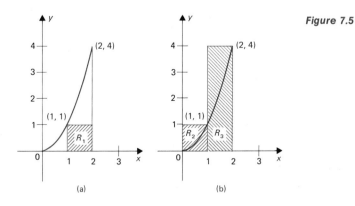

Figure 7.5

(a)   (b)

where

$$A(R_1) = (1)(1) = 1$$

$$A(R_2) = (1)(1) = 1$$

$$A(R_3) = (1)(4) = 4$$

Thus

$$1 < A(R) < 1 + 4 \quad \text{or} \quad 1 < A(R) < 5$$

Next we divide the interval $[0, 2]$ into four parts and consider the points $\left(\frac{1}{2}, \frac{1}{4}\right)$, $(1, 1)$, and $\left(\frac{3}{2}, \frac{9}{4}\right)$ on the graph of f. We draw two sets of rectangles as shown in Figure 7.6. Here we find that

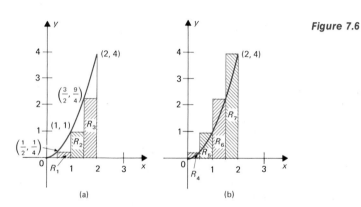

Figure 7.6

(a)   (b)

$$A(R_1) + A(R_2) + A(R_3) < A(R) < A(R_4) + A(R_5) + A(R_6) + A(R_7) .$$

where

$$A(R_1) = \left(\tfrac{1}{2}\right)\left(\tfrac{1}{4}\right) = \tfrac{1}{8}$$

$$A(R_2) = \left(\tfrac{1}{2}\right)(1) = \tfrac{1}{2}$$

$$A(R_3) = \left(\tfrac{1}{2}\right)\left(\tfrac{9}{4}\right) = \tfrac{9}{8}$$

$$A(R_4) = \left(\tfrac{1}{2}\right)\left(\tfrac{1}{4}\right) = \tfrac{1}{8}$$

$$A(R_5) = \left(\tfrac{1}{2}\right)(1) = \tfrac{1}{2}$$

$$A(R_6) = \left(\tfrac{1}{2}\right)\left(\tfrac{9}{4}\right) = \tfrac{9}{8}$$

$$A(R_7) = \left(\tfrac{1}{2}\right)(4) = 2$$

Hence

$$\tfrac{1}{8} + \tfrac{1}{2} + \tfrac{9}{8} < A(R) < \tfrac{1}{8} + \tfrac{1}{2} + \tfrac{9}{8} + 2$$

or

$$\tfrac{14}{8} < A(R) < \tfrac{30}{8}$$

It appears that by increasing the number of subintervals of [0, 2], the areas of the corresponding rectangles yield a better approximation for $A(R)$.

We divide the interval [0, 2] into n subintervals by choosing numbers

$$0 = x_0 < x_1 < x_2 < \cdots < x_{n-1} < x_n = 2$$

This is called a *partition* of the interval [0, 2]. To simplify our problem, we choose a partition of n equal subintervals. Then each subinterval is of length $2/n$. (Why?) Now the set of rectangles in Figure 7.7(a)

Figure 7.7

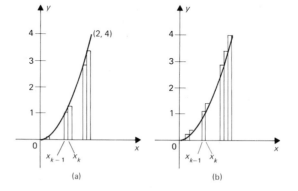

(a)   (b)

all have length $2/n$ and height $f(x_{k-1})$, $k = 1, 2, \ldots, n$. Thus starting at the left endpoint, we have

Area of first rectangle $= \dfrac{2}{n} f(x_0) = \dfrac{2}{n} f(0) = \dfrac{2}{n} \cdot 0 = 0$

Area of second rectangle $= \dfrac{2}{n} f(x_1) = \dfrac{2}{n} f\left(\dfrac{2}{n}\right) = \dfrac{2}{n} \left(\dfrac{2}{n}\right)^2$

Area of third rectangle $\quad = \dfrac{2}{n} f(x_2) = \dfrac{2}{n} f\left(\dfrac{4}{n}\right) = \dfrac{2}{n} \left(\dfrac{4}{n}\right)^2$

$\cdots$ $\qquad\qquad\qquad\qquad \cdots$

Area of kth rectangle $\quad = \dfrac{2}{n} f(x_{k-1}) = \dfrac{2}{n} f\left(\dfrac{2k-2}{n}\right) = \dfrac{2}{n} \left[\dfrac{2(k-1)}{n}\right]^2$

Therefore the sum of the areas of the rectangles in Figure 7.7 (a) is given by

$$s_n = 0 + \frac{2}{n}\left(\frac{2}{n}\right)^2 + \frac{2}{n}\left(\frac{4}{n}\right)^2 + \cdots + \frac{2}{n}\left[\frac{2(k-1)}{n}\right]^2 + \cdots + \frac{2}{n}\left[\frac{2(n-1)}{n}\right]^2$$

$$= \sum_{k=1}^{n} \frac{2}{n}\left[\frac{2(k-1)}{n}\right]^2 = \frac{8}{n^3} \sum_{k=1}^{n} (k-1)^2$$

The rectangles in Figure 7.7 (b) all have length $2/n$ and height $f(x_k)$, $k = 1, 2, \ldots, n$. Hence

Area of kth rectangle $\quad = \dfrac{2}{n} f(x_k) = \dfrac{2}{n} f\left(\dfrac{2k}{n}\right) = \dfrac{2}{n} \left(\dfrac{2k}{n}\right)^2$

Thus the sum of the areas of the rectangles in Figure 7.7 (b) is given by

$$S_n = \frac{2}{n}\left(\frac{2}{n}\right)^2 + \frac{2}{n}\left(\frac{4}{n}\right)^2 + \cdots + \frac{2}{n}\left(\frac{2n}{n}\right)^2 = \frac{2}{n}\sum_{k=1}^{n}\left(\frac{2k}{n}\right)^2 = \frac{8}{n^3}\sum_{k=1}^{n} k^2$$

Using the formula

$$\sum_{k=1}^{n} k^2 = 1^2 + 2^2 + 3^2 + \cdots + n^2 = \frac{n(n+1)(2n+1)}{6} \qquad (7.2)$$

we can write

$$S_n = \frac{8}{n^3}\left[\frac{n(n+1)(2n+1)}{6}\right] = \frac{8}{3} + \frac{4}{n} + \frac{4}{3n^2}$$

and

$$s_n = \frac{8}{n^3}\left\{\frac{(n-1)\,n[2(n-1)+1]}{6}\right\} = \frac{8}{3} - \frac{4}{n} + \frac{4}{3n^2}$$

Obviously,

$$s_n < A(R) < S_n \quad \text{or} \quad \frac{8}{3} - \frac{4}{n} + \frac{4}{3n^2} < A(R) < \frac{8}{3} + \frac{4}{n} + \frac{4}{3n^2}$$

As n increases indefinitely, it is clear that

$$\lim_{n\to\infty} s_n = \lim_{n\to\infty} S_n = \lim_{n\to\infty} \left(\frac{8}{3} + \frac{4}{n} + \frac{4}{3n^2}\right) = \frac{8}{3}$$

Here we say that

$$A(R) = \frac{8}{3}$$

Use mathematical induction to verify this result. The following formulas will also be needed in this section.

$$\sum_{k=1}^{n} k = 1 + 2 + 3 + \cdots + n$$

$$= \frac{n(n+1)}{2} \qquad (7.2a)$$

$$\sum_{k=1}^{n} c = c + c + c + \cdots + c$$

$$= cn \qquad (7.2b)$$

It is possible to use the same type of procedure for a function f defined on the closed interval $[a, b]$ without any reference to the concept of area. We do this to define the *definite integral*.

Note   *f is not necessarily nonnegative.*

Consider a continuous function f defined on a closed interval $[a, b]$. Let

$$a = x_0 < x_1 < x_2 < \cdots < x_{n-1} < x_n = b$$

In practice we choose $c_k = x_k$.

be a partition of $[a, b]$ and choose numbers $c_1, c_2, \ldots, c_n$, such that

$$x_{k-1} \le c_k \le x_k, \qquad k = 1, 2, \ldots, n$$

Corresponding to each point $c_1, c_2, \ldots, c_n$, we find $f(c_1), f(c_2), \ldots, f(c_n)$ and form the sum (see Figure 7.8)

Figure 7.8

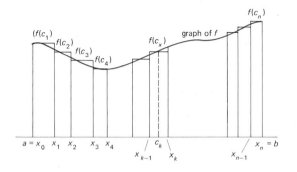

$$f(c_1)(x_1 - x_0) + f(c_2)(x_2 - x_1) + \cdots + f(c_k)(x_k - x_{k-1})$$

$$+ \cdots + f(c_n)(x_n - x_{n-1}) = \sum_{k=1}^{n} f(c_k)(x_k - x_{k-1})$$

This sum is called a Riemann sum. Let

$$S_n = \sum_{k=1}^{n} f(c_k) \, dx_k \qquad (7.3)$$

where $dx_k = x_k - x_{k-1}$ for $k = 1, 2, \ldots, n$.

Question   What happens to S_n when n increases?

It can be shown that S_n approaches a fixed number A as $n \to \infty$ and each $dx_k \to 0$. Furthermore, the number A is independent of the partition used and choice of the numbers c_k. Note the restriction on the partition: As $n \to \infty$, we require that

$$dx_k \to 0 \qquad \text{for} \quad k = 1, 2, \ldots, n$$

The limit A is called the definite integral of f over $[a, b]$ and is denoted by the symbol

$$\int_a^b f(x) \, dx$$

that is,

$$\int_a^b f(x)\, dx = \lim_{\substack{n \to 0 \\ dx_k \to 0}} \sum_{k=1}^{n} f(c_k)\, dx_k \tag{7.4}$$

The numbers a and b are called the *lower* and *upper limits of integration*, respectively.

Remark 1 In practice, to simplify the computations, we divide the interval $[a, b]$ into n subintervals of equal length. Then

$$dx_k = \frac{b - a}{n}, \qquad k = 1, 2, \ldots, n$$

and in this case $dx_k \to 0$ as $n \to \infty$.

Remark 2 If $f(x) \geq 0$ on $[a, b]$, then the definite integral $\int_a^b f(x)\, dx$ represents the area $A(R)$ of the region R bounded by the curve $y = f(x)$, the x-axis, and the lines $x = a$ and $x = b$.

Remark 3 The similarity between $\int_a^b f(x)\, dx$ and the antiderivative symbol is intentional. A remarkable relationship exists between the definite integral and the antiderivative, and this relationship is stated precisely in the fundamental theorem of calculus (see Section 7.3).

Example 7.4 Calculate $\int_0^1 f(x)\, dx$, where $f(x) = x + 1$.

SOLUTION (See Figure 7.9.)

Step 1 Divide the interval $[0, 1]$ into n equal subintervals. Each subinterval is of length

$$dx_k = \frac{1 - 0}{n} = \frac{1}{n}$$

Georg Friedrich Bernhard Riemann (1826–1866) was born at Breselenz, Germany. His father was a Lutheran pastor, and the family, which included six children, lived in poverty. Riemann, a timid child with a horror of speaking in public, received his early education from his father and later at the Gymnasiums at Hanover and Lüneburg. Riemann's interest and ability in the theory of numbers were demonstrated at Lüneburg. Schmalfuss, director of the Gymnasium, gave young Riemann Legendre's *Théorie des Nombres* and Riemann apparently mastered the 859-page book in six days.

At the age of 19, Riemann entered the University of Göttingen, where he first studied philology and theology. A year later he transferred to the University of Berlin, where he learned much from the great mathematicians Dirichlet, Jacobi, Steiner, and Eisenstein. After two years in Berlin, he returned to the University of Göttingen to obtain his doctorate. Riemann's doctoral dissertation: "Foundations for a general theory of functions of a complex variable" impressed even Gauss. Riemann joined the faculty at the University of Göttingen as an unpaid lecturer, and in 1859 succeeded Dirichlet as full professor.

Riemann was honored by many of the learned societies including the French Academy of Sciences and the Royal Society of London. His methods and ideas have greatly influenced research in mathematics and mathematical physics. He died of tuberculosis at the age of 40.

Georg Friedrich Bernhard Riemann

Figure 7.9

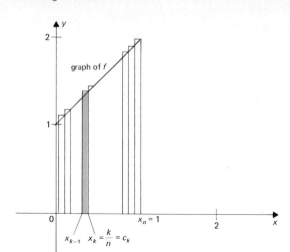

Step 2   Select c_k. In this case let c_k be the right-hand endpoint of the kth subinterval, that is,

$$c_1 = \frac{1}{n}, \; c_2 = \frac{2}{n}, \; c_3 = \frac{3}{n}, \; \cdots, \; c_k = \frac{k}{n}, \; \cdots, \; c_n = 1$$

Step 3   Calculate $\Sigma_{k=1}^{n} f(c_k) \, dx_k$. Here we have

$$\sum_{k=1}^{n} f(c_k) \, dx_k = \sum_{k=1}^{n} f\left(\frac{k}{n}\right) dx_k = \sum_{k=1}^{n} \left(\frac{k}{n} + 1\right)\frac{1}{n} = \sum_{k=1}^{n} \left(\frac{k}{n^2} + \frac{1}{n}\right)$$

$$= \frac{1}{n^2} \sum_{k=1}^{n} k + \frac{1}{n} \sum_{k=1}^{n} 1 = \frac{1}{n^2}\left[\frac{n(n+1)}{2}\right] + \frac{1}{n} \cdot n$$

$$= \frac{1}{2} + \frac{1}{2n} + 1 = \frac{3}{2} + \frac{1}{2n}$$

Step 4   Calculate $\int_0^1 (x + 1) \, dx$. We find that

$$\int_0^1 (x + 1) \, dx = \lim_{n \to \infty} \sum_{k=1}^{n} f(c_k) \, dx_k = \lim_{n \to \infty} \left(\frac{3}{2} + \frac{1}{2n}\right) = \frac{3}{2}$$

Note that the area of the region bounded by the graph of f and the lines $x = 1$, $x = 0$, and $y = 0$ (see Figure 7.10) is given by

$$A(R) = A(R_1) + A(R_2)$$

where

$$A(R_1) = (1)(1) = 1 \quad \text{and} \quad A(R_2) = \tfrac{1}{2}\,(1)(1) = \tfrac{1}{2}$$

Hence

$$A(R) = 1 + \tfrac{1}{2} = \tfrac{3}{2} = \int_0^1 (x + 1) \, dx$$

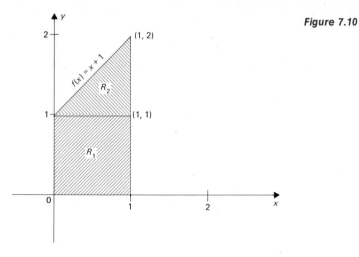

Figure 7.10

Example 7.5   Calculate $\int_1^3 (16 - x^2)\, dx$.

SOLUTION   The region is indicated in Figure 7.11. The region is bounded by the graph of $f(x) = 16 - x^2$ and the lines $x = 1$, $x = 3$, and $y = 0$.

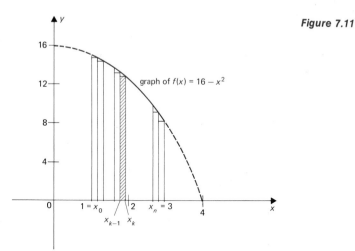

Figure 7.11

Step 1   Divide the interval $[1, 3]$ into n equal subintervals. Hence each subinterval is of length

$$dx_k = \frac{3 - 1}{n} = \frac{2}{n}$$

Step 2   Select c_k to be the right-hand endpoint of the kth subinterval. Then

$$c_1 = 1 + \frac{2}{n}, \quad c_2 = 1 + \frac{4}{n}, \quad \cdots, \quad c_k = 1 + \frac{2k}{n}, \quad \cdots, \quad c_n = 1 + \frac{2n}{n} = 3$$

Step 3   Calculate $\sum_{k=1}^{n} f(c_k) \, dx_k$. We have

$$\sum_{k=1}^{n} f(c_k) \, dx_k = \sum_{k=1}^{n} f\left(1 + \frac{2k}{n}\right) dx_k = \sum_{k=1}^{n} \left[16 - \left(1 + \frac{2k}{n}\right)^2 \right] \frac{2}{n}$$

$$= \sum_{k=1}^{n} \left(\frac{30}{n} - \frac{8k}{n^2} - \frac{8k^2}{n^3} \right)$$

$$= \frac{30}{n} \sum_{k=1}^{n} 1 - \frac{8}{n^2} \sum_{k=1}^{n} k - \frac{8}{n^3} \sum_{k=1}^{n} k^2$$

$$= \frac{30}{n} \cdot n - \frac{8}{n^2} \left[\frac{n(n+1)}{2} \right] - \frac{8}{n^3} \left[\frac{n(n+1)(2n+1)}{6} \right]$$

$$= \frac{70}{3} - \frac{8}{n} - \frac{4}{3n^2}$$

Step 4   Calculate $\int_1^3 (16 - x^2) \, dx$. Here we have

$$\int_1^3 (16 - x^2) \, dx = \lim_{n \to \infty} \sum_{k=1}^{n} f(c_k) \, dx_k$$

$$= \lim_{n \to \infty} \left(\frac{70}{3} - \frac{8}{n} - \frac{4}{3n^2} \right) = \frac{70}{3}$$

Therefore the area of the region in Figure 7.11 is $\frac{70}{3}$.

Exercises 7.2   In Problems 1 through 6, divide the interval $[a, b]$ into eight equal parts and evaluate s_8 and S_8, such that

$$s_8 < \int_a^b f(x) \, dx < S_8$$

1. $\displaystyle\int_1^2 (1 + x) \, dx$ 2. $\displaystyle\int_1^3 \frac{x}{2} \, dx$

3. $\displaystyle\int_0^4 x^2 \, dx$ 4. $\displaystyle\int_{-1}^1 x^2 \, dx$

5. $\displaystyle\int_1^2 \frac{1}{x} \, dx$ 6. $\displaystyle\int_1^4 \sqrt{x + 5} \, dx$

7.   Use mathematical induction to establish formulas (7.2a) and (7.2b).

8.   Repeat Example 7.4 by selecting c_k to be the left-hand endpoint of the kth subinterval.

9.   Repeat Example 7.4 by dividing the interval $[0, 1]$ into $2n$ equal subintervals.

10.   Repeat Example 7.5 by dividing the interval $[1, 3]$ into $2n$ equal subintervals.

In Problems 11 through 20, evaluate the indicated definite integral. Sketch the region involved in each case and show the kth rectangle in the approximating sum.

11. $\displaystyle\int_0^2 x\,dx$ 12. $\displaystyle\int_1^2 (2+x)\,dx$

13. $\displaystyle\int_0^1 x^2\,dx$ 14. $\displaystyle\int_1^3 (1+x^2)\,dx$

15. $\displaystyle\int_2^5 (x-1)\,dx$ 16. $\displaystyle\int_{-1}^3 (4-x)\,dx$

17. $\displaystyle\int_{-1}^2 (1+2x^2)\,dx$ 18. $\displaystyle\int_{-1}^1 (2-x)^2\,dx$

19. $\displaystyle\int_1^4 (2-x)^2\,dx$ 20. $\displaystyle\int_0^1 x^3\,dx$

21. Show that $\displaystyle\int_a^b x\,dx = \frac{b^2 - a^2}{2}$

*22. Show that $\displaystyle\int_a^b x^2\,dx = \frac{b^3 - a^3}{3}$

23. If $f(x) = x^2$ and $a < c < b$, show that

(a) $\displaystyle\int_a^b f(x)\,dx = \int_a^c f(x)\,dx + \int_c^b f(x)\,dx$

(b) $\displaystyle\int_a^b mf(x)\,dx = m\int_a^b f(x)\,dx$ (m is a constant)

24. Consider the region bounded by the graph of $f(x) = x^3$ and the lines $x = -1$, $x = 1$, and $y = 0$. Divide the interval $[-1, 1]$ into four equal sub-intervals and construct the approximating four rectangles.
(a) Find the area of the four rectangles.
(b) Using formula (7.4), evaluate $\int_{-1}^1 x^3\,dx$.
(c) Does $\int_{-1}^1 x^3\,dx$ represent the area of the region under consideration? Explain.

7.3 The Fundamental Theorem of Calculus

Obviously, the evaluation of a definite integral can be tedious. Fortunately, there is a technique available to help. It is given in the fundamental theorem of calculus, which relates the definite integral to the antiderivative of a given function. This provides a simple method for evaluating the definite integral. We merely state this theorem; for a proof, the student should consult more advanced calculus texts.

◀ *The fundamental theorem of calculus*

Theorem 7.3 *If f is a function continuous on $[a, b]$ and F is an anti-derivative of f, then*

$$\int_a^b f = F(b) - F(a) \qquad (7.5)$$

This basic relationship between the area under the curve $y = f(x)$ and the antiderivative of f was noted first by Newton's teacher, Isaac Barrow (1630–1677). Barrow was a mathematician and a theologian. He held a mathematics chair at Cambridge and in 1669 he relinquished his post to the young Newton.

Another notation commonly used is

$$\int_a^b f(x)\, dx = F(x)\,\Big|_a^b = F(b) - F(a)$$

The symbol $\int_a^b f(x)\, dx$ reminds us that the definite integral is the limit of a sequence of sums. We note that $F(b) - F(a)$ does not involve the variable x, and so we may write

$$\int_a^b f(x)\, dx = \int_a^b f(r)\, dr = \int_a^b f(t)\, dt$$

where the letters x, r, and t are called *dummy variables*.

Now if G is also an antiderivative of f, then by Theorem 7.1,

$$G = F + C$$

and so

$$\int_a^b f = G(b) - G(a)$$

$$= F(b) + C - F(a) - C = F(b) - F(a)$$

Therefore for a given function f, we can compute the definite integral of f over $[a, b]$ by finding *any* antiderivative of f, evaluating this antiderivative at b and at a, and then subtracting the second value from the first.

Example 7.6   Evaluate $\int_{-1}^2 2x^3\, dx$.

SOLUTION   Here $f(x) = 2x^3$ and an antiderivative of f is

$$F(x) = \frac{x^4}{2}$$

Evaluating F at $x = 2$ and at $x = -1$ and subtracting, we obtain

$$F(2) - F(-1) = \frac{x^4}{2}\Big|_{-1}^2 = \frac{(2)^4}{2} - \frac{(-1)^4}{2} = \frac{15}{2}$$

Therefore

$$\int_{-1}^2 2x^3\, dx = \tfrac{15}{2}$$

Example 7.7   Evaluate $\displaystyle\int_1^4 \left(x^2 - \sqrt{x} + \frac{1}{x^2}\right) dx$.

SOLUTION   An antiderivative of f, where $f(x) = x^2 - \sqrt{x} + (1/x^2)$, is

$$F(x) = \frac{x^3}{3} - \frac{2}{3}x^{3/2} - \frac{1}{x}$$

Therefore

$$\int_1^4 \left(x^2 - \sqrt{x} + \frac{1}{x^2} \right) dx = \left(\frac{x^3}{3} - \frac{2}{3} x^{3/2} - \frac{1}{x} \right) \bigg|_1^4$$

$$= \left[\frac{(4)^3}{3} - \frac{2}{3} (4)^{3/2} - \frac{1}{4} \right] - \left[\frac{(1)^3}{3} - \frac{2}{3} (1)^{3/2} - \frac{1}{1} \right]$$

This is $F(4) - F(1)$.

$$= \frac{205}{12}.$$

We make the following definitions:

$$\int_a^a f(x) \, dx = 0 \qquad\qquad (7.6)$$

$$\int_a^b f(x) \, dx = - \int_b^a f(x) \, dx \qquad\qquad (7.7)$$

If F is an antiderivative of f, then

$$\int_a^a f(x) \, dx = F(a) - F(a) = 0$$

The following property of the definite integral can be deduced readily by application of Theorem 7.3.

$$\int_a^b f(x) \, dx = \int_a^c f(x) \, dx + \int_c^b f(x) \, dx \qquad (a < c < b) \qquad (7.8)$$

The student is encouraged to establish properties (7.7) and (7.8).

In the previous section we indicated that if f is nonnegative on $[a, b]$, then the definite integral of f from a to b defines the area of the region bounded by the graph of f, the x-axis, and the lines $x = a$ and $x = b$. In Theorem 7.3, we did not restrict f to be positive or negative on $[a, b]$. The question then arises as to what $\int_a^b f$ does represent if $f(x) \le 0$.

Theorem 7.4   *If f is a continuous function on $[a, b]$ and $f(x) \le 0$ on $[a, b]$, then*   ◀

$$\int_a^b f = -A(R)$$

where $A(R)$ is the area of the region R bounded by the graph of f, the x-axis, and the lines $x = a$ and $x = b$ (see Figure 7.12), that is,

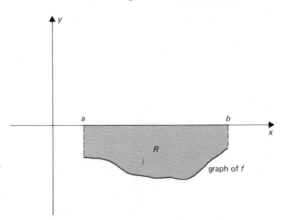

Figure 7.12

$$A(R) = -\int_a^b f \qquad (7.9)$$

Thus, in general, if f is a continuous function on $[a, b]$ whose graph is as shown in Figure 7.13, then

Figure 7.13

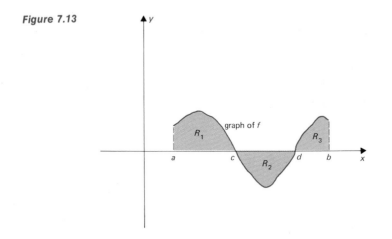

$$\int_a^b f = A(R_1) - A(R_2) + A(R_3)$$

In this case, we find that

By Theorem 7.4.
$$A(R_1) = \int_a^c f, \quad A(R_2) = -\int_c^d f, \quad \text{and} \quad A(R_3) = \int_d^b f$$

Thus

Application of property (7.8).
$$A(R_1) - A(R_2) + A(R_3) = \int_a^c f + \int_c^d f + \int_d^b f$$

$$= \int_a^d f + \int_d^b f = \int_a^b f$$

Example 7.8 Find the area of the region bounded by the graphs of functions f and g, where $f(x) = x$ and $g(x) = \frac{1}{2} x^2$.

SOLUTION

Step 1 Sketch a graph of the region R as shown in Figure 7.14.

Step 2 Find the points of intersection of the graphs of f and g, that is, find

$$\left\{ (x, y) \,|\, y = x \text{ and } y = \tfrac{1}{2} x^2 \right\}$$

Here we have to solve the equation

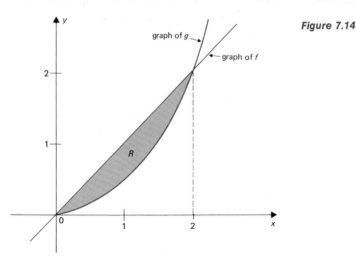

Figure 7.14

$$x = \tfrac{1}{2} x^2 \quad \text{or} \quad 2x - x^2 = 0$$

Thus $x = 0$ and $x = 2$, and the points of intersection are $(0, 0)$ and $(2, 2)$. Note that the points of intersection will define the region R.

Step 3 We observe that the area $A(R)$ of the region R is given by

$$A(R) = A(R_1) - A(R_2)$$

where $A(R_1)$ is the area of the region R_1 bounded by the graph of f, the x-axis, and the line $x = 2$; and $A(R_2)$ is the area of the region R_2 bounded by the graph of g, the x-axis, and the line $x = 2$. Now

$$A(R_1) = \int_0^2 f(x)\, dx = \int_0^2 x\, dx = \frac{x^2}{2}\Big|_0^2 = \frac{(2)^2}{2} - 0 = 2$$

and

$$A(R_2) = \int_0^2 g(x)\, dx = \int_0^2 \frac{1}{2} x^2\, dx = \frac{1}{2}\frac{x^3}{3}\Big|_0^2 = \frac{(2)^3}{6} - 0 = \frac{4}{3}$$

Therefore the area of the region in question is

$$A(R) = \int_0^2 x\, dx - \int_0^2 \tfrac{1}{2} x^2\, dx = 2 - \tfrac{4}{3} = \tfrac{2}{3}$$

In the previous section we indicated that

$$A(R) = \int_a^b f(x)\, dx$$

is the area of the region R bounded by the graph of f, the x-axis, and the lines $x = a$ and $x = b$ provided that f is nonnegative on $[a, b]$. Now we generalize this definition.

If f and g are continuous functions on $[a, b]$ and $f(x) \geq g(x)$ for each point on $[a, b]$, then the area of the region R bounded by the graphs of f and g and the lines $x = a$ and $x = b$ is defined by

$$A(R) = \int_a^b [f(x) - g(x)] \, dx \qquad (7.10)$$

Such a region is illustrated in Figure 7.15.

Figure 7.15

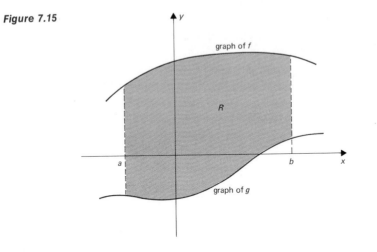

graph of f

R

graph of g

a   b

Example 7.9   Compute the area of the region for which $x \in [-1, 2]$ and which is bounded by the graphs of the function f and g, where

$$f(x) = -\tfrac{1}{4} x^2 + 1 \quad \text{and} \quad g(x) = \tfrac{1}{2} x^2 - 3$$

SOLUTION   The region R is illustrated in Figure 7.16. Since $f(x) \geq g(x)$ for each x on $[-1, 2]$, we have

Figure 7.16

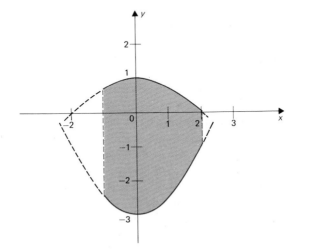

$$A(R) = \int_{-1}^{2} [f(x) - g(x)] \, dx = \int_{-1}^{2} \left[\left(-\frac{1}{4} x^2 + 1 \right) - \left(\frac{1}{2} x^2 - 3 \right) \right] dx$$

$$= \int_{-1}^{2} \left(-\frac{3}{4} x^2 + 4 \right) dx = \left(-\frac{3}{4} \frac{x^3}{3} + 4x \right) \Big|_{-1}^{2}$$

$$= \left[-\frac{(2)^3}{4} + 4(2) \right] - \left[-\frac{(-1)^3}{4} + 4(-1) \right] = \frac{39}{4}$$

Example 7.10   Compute the area of the region R bounded by the graphs of the functions f and g, where

$$f(x) = x \quad \text{and} \quad g(x) = x^3$$

SOLUTION   We follow the steps indicated in Example 7.8. First we sketch a graph of the region R as shown in Figure 7.17. We find the

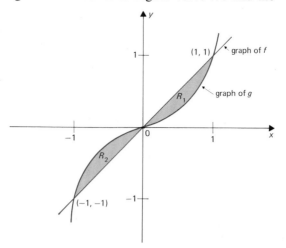

Figure 7.17

points of intersection of the graphs of f and g by solving the equation

$$x = x^3$$

from which we obtain $x = 0$, $x = 1$, and $x = -1$. Thus the points of intersection are $(0, 0)$, $(1, 1)$, and $(-1, -1)$.

Here we see that there are two regions to be considered:

1. In R_1, $x \in [0, 1]$ and $f(x) \geq g(x)$ for each x.
2. In R_2, $x \in [-1, 0]$ and $g(x) \geq f(x)$ for each x.

Hence we treat the problem in two parts. First, we find $A(R_1)$:

$$A(R_1) = \int_0^1 [f(x) - g(x)] \, dx = \int_0^1 (x - x^3) \, dx = \left(\frac{x^2}{2} - \frac{x^4}{4} \right) \Big|_0^1$$

$$= \left[\frac{(1)^2}{2} - \frac{(1)^4}{4} \right] - \left[\frac{0}{2} - \frac{0}{4} \right] = \frac{1}{4}$$

Next we find $A(R_2)$.

$$A(R_2) = \int_{-1}^0 [g(x) - f(x)] \, dx = \int_{-1}^0 (x^3 - x) \, dx = \left(\frac{x^4}{4} - \frac{x^2}{2} \right) \Big|_{-1}^0$$

$$= \left[\frac{0}{4} - \frac{0}{2} \right] - \left[\frac{(-1)^4}{4} - \frac{(-1)^2}{2} \right] = \frac{1}{4}$$

The area $A(R) = A(R_1) + A(R_2)$. Therefore

$$A(R) = \frac{1}{4} + \frac{1}{4} = \frac{1}{2}$$

Example 7.11 Suppose that $F(x) = \int_1^x t^3 \, dt$. Find $F'(x)$.

SOLUTION

$$F(x) = \frac{t^4}{4}\Big|_1^x = \frac{x^4}{4} - \frac{1}{4}$$

Thus

$$F'(x) = \frac{d}{dx}\left(\frac{x^4}{4} - \frac{1}{4}\right) = x^3$$

Exercises 7.3

In Problems 1 through 12, evaluate the given definite integrals.

1. $\displaystyle\int_1^3 (x^2 - 3) \, dx$

2. $\displaystyle\int_{-1}^2 (2x^2 + 4x) \, dx$

3. $\displaystyle\int_{-1}^2 (r + 1)^2 \, dr$

4. $\displaystyle\int_1^2 (x + 2)(x - 2) \, dx$

5. $\displaystyle\int_4^9 3\sqrt{t} \, dt$

6. $\displaystyle\int_1^3 \frac{2}{s^2} \, ds$

7. $\displaystyle\int_{-2}^4 x(x + 1) \, dx$

8. $\displaystyle\int_2^5 \frac{1}{\sqrt{u}} \, du$

9. $\displaystyle\int_{-1}^1 (2x^2 + x^{1/3} - 6) \, dx$

10. $\displaystyle\int_0^2 \frac{1 + 2v^{1/3}}{v^{2/3}} \, dv$

11. $\displaystyle\int_1^3 \left(\frac{1}{u}\right)^4 \, du$

12. $\displaystyle\int_0^{16} 2x^{1/4} \, dx$

13. Establish formulas (7.7) and (7.8).

In Problems 14 through 22, find the area of the region in the plane bounded by the given curves. Sketch a graph of the region.

14. $y = 2x + 1$, $x = 1$, $x = 3$, and the x-axis

15. $y = 9 - x^2$ and the x-axis.

16. $y = x^2 + 1$, $x = 2$, $x = -3$, and the x-axis

17. $y = x^2$ and $y = 2x$

18. $y = x^2$ and $y = x + 2$

19. $y = \sqrt{x}$ and $y = x^2$

20. $y = x^2$ and $y = 2 - x^2$

21. $y = 4x - x^2$ and $y = 3$

22. $y = 4x - x^2$ and $y = -5$

In Problems 23 through 30, determine $F'(x)$.

23. $F(x) = \displaystyle\int_1^x t \, dt$

24. $F(x) = \displaystyle\int_{-2}^x u^2 \, du$

25. $F(x) = \displaystyle\int_0^x (r - 1)^2 \, dr$

26. $F(x) = \displaystyle\int_1^x 2\sqrt{t} \, dt$

27. $F(x) = \int_x^2 u^2 \, du$

28. $F(x) = \int_x^{3x} r^2 \, dr$

29. $F(x) = \int_x^{x+2} r^2 \, dr$

30. $F(x) = \int_x^{x+2} r^2 \, dr + \int_{x+2}^x r^2 \, dr$

In Problems 31 through 34, determine the values for x which satisfy the given equation.

31. $\int_0^x 3t^2 \, dt = 8$

32. $\int_0^x (4t - 5) \, dt = -2$

33. $\int_{-x}^x \frac{u^6}{2} \, du = \frac{128}{7}$

34. $\int_x^{2x^2} dt = 6$

7.4 Applications of the Integral

Now we show how the process of integration can be used in solving problems arising in the sciences. Many problems in the sciences are described as rates of change of a function. The simplest of such problems are of the form

$$y'(t) = f(t) \tag{7.11}$$

Such an equation is called a *differential equation*. In general, a differential equation is an equation involving a variable t (or x or s, etc.), an unknown function $y(t)$, and certain derivatives of y with respect to t. The problem is usually to determine a function $y(t)$ that satisfies the given equation. Such a function is called a *solution* of the differential equation. In many applications, equations such as (7.11) are usually associated with additional conditions that the function $y(t)$ is to satisfy. For instance, it may be required that

Differential equation

$$y(t_0) = k \tag{7.12}$$

at $t = t_0$. This requirement is called an *initial condition*. Equation (7.11) together with (7.12) form an *initial-value problem*. The problem is to find a function $y(t)$ that satisfies the differential equation (7.11) as well as the initial condition (7.12). The following examples illustrate the procedure for solving such problems.

Example 7.12 The Goldegg Chicken Farm produces $E(t)$ dozen eggs in t days. Its *production rate* is $E'(t)$ and this can be estimated by

$$E'(t) \approx \frac{E(t+h) - E(t)}{h}$$

If the production rate (in dozens/day) is given by

$$E'(t) = 150 + \tfrac{2}{5} t \tag{7.13}$$

find the following.

(a)   The number of eggs produced in t days, assuming that $E(0) = 0$
(b)   The number of eggs produced in one year (365 days)
(c)   The average daily production

SOLUTION

(a)   We shall use two methods for solving part (a).

First method. The production E in t days is

$$E(t) = \int E'(t)\,dt = \int \left(150 + \frac{2}{5}t\right)dt = 150t + \frac{t^2}{5} + C \quad (7.14)$$

The student can show that $E(t)$ satisfies Equation (7.13). We now satisfy the initial condition. From (7.14), we have

$$E(0) = (150)(0) + \frac{0^2}{5} + C = 0$$

which implies that $C = 0$. Therefore the solution to the initial-value problem is

$$E(t) = 150t + \frac{t^2}{5} \quad (7.15)$$

Second method. Integrating both sides of Equation (7.13) from 0 to t, we have

$$\int_0^t E'(s)\,ds = \int_0^t \left(150 + \frac{2}{5}s\right)ds \quad (7.16)$$

Note

(a)   *The lower limit of integration is the initial value for t.*

(b)   *The dummy variable s is used to avoid confusion with the upper limit of integration.*

This yields

$$E(s)\Big|_0^t = \left(150s + \frac{s^2}{5}\right)\Big|_0^t$$

$$E(t) - E(0) = \left[150t + \frac{t^2}{5}\right] - \left[(150)(0) + \frac{0^2}{5}\right]$$

or

Observe that E(0) = 0.

$$E(t) = 150t + \frac{t^2}{5}$$

as before.

(b)   To obtain the production in one year, we have

$$E(365) = (150)(365) + \frac{(365)^2}{5} = 81{,}395 \text{ dozen eggs} \quad (7.17)$$

(c)   The average daily production is

$$\frac{E(365)}{365} = \frac{81{,}395}{365} = 223 \text{ dozens/day} \quad (7.18)$$

Example 7.13   Consider an (idealized) experiment in which a colony of live bacteria is introduced into a limited food supply. (See Example 5.8.) Suppose that the rate of change in the number N of live bacteria with respect to time t is given by

$$N'(t) = 6000t^2 - 75t^4 \qquad (7.19)$$

Find the size $N(t)$ of the population of bacteria at time t if 1000 bacteria were introduced into the food supply initially.

SOLUTION   We have an initial-value problem involving the differential equation (7.19) with initial condition

$$N(0) = 1000 \qquad (7.20)$$

Integrating both sides of Equation (7.19), we have

$$\int_0^t N'(s)\, ds = \int_0^t (6000s^2 - 75s^4)\, ds$$

$$N(s) \Big|_0^t = (2000s^3 - 15s^5) \Big|_0^t$$

$$N(t) - N(0) = [2000t^3 - 15t^5] - [(2000)(0)^3 - (15)(0)^5]$$

that is,

$$N(t) = 2000t^3 - 15t^5 + 1000 \qquad (7.21) \qquad N(0) = 1000$$

The student is encouraged to show that the function defined by (7.21) satisfies Equations (7.19) and (7.20).

A rough sketch of the graph of N is shown in Figure 7.18, indicating

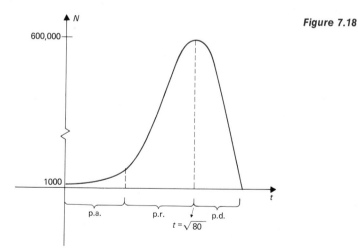

Figure 7.18

the period of adaptation (p.a.), period of reproduction (p.r.), and period of dying (p.d.). The student should solve this initial-value problem using the first method described in Example 7.12.

We now consider the motion of an object defined by

$$s = f(t), \tag{7.22}$$

where s, the distance, is a function of time t. We recall that the *velocity function*, $v(t)$, is defined by

$$v(t) = \frac{ds}{dt} \tag{7.23}$$

which is the rate of change of the distance with respect to time. In addition, the *acceleration function*, $a(t)$, is defined by

$$a(t) = \frac{dv}{dt} = \frac{d^2 s}{dt^2} \tag{7.24}$$

which is the rate of change of the velocity with respect to time.

We assume that air resistance or other forces do not act on the object.

Example 7.14 An object on the ground is projected vertically with initial velocity of 96 ft/sec. If the acceleration $a(t) = -32$ ft/sec^2, find the following.

Figure 7.19

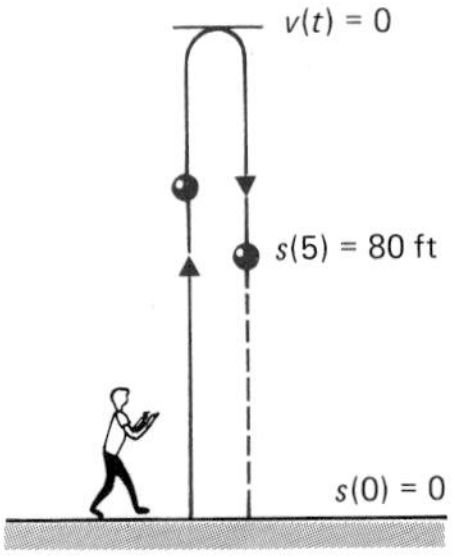

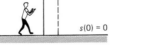

$v(t) = 0$

$s(5) = 80$ ft

$s(0) = 0$

(a) The velocity function
(b) The distance at time t
(c) The maximum height the object will attain
(d) The height of the object in 5 sec

SOLUTION

(a) Here we are dealing with the initial-value problem

$$v'(t) = -32 \tag{7.25}$$

$$v(0) = 96 \tag{7.26}$$

Integrating both sides of Equation (7.25), we obtain

$$\int_0^t v'(r)\,dr = -32 \int_0^t dr \quad \text{or} \quad v(r)\Big|_0^t = -32r\Big|_0^t \tag{7.27}$$

that is,

$$v(t) - v(0) = -32[t - 0]$$

Since $v(0) = 96$, we have

$$v(t) = -32t + 96 \qquad (7.28)$$

(b) We observe that the object is at ground level at $t = 0$: The initial condition is $s(0) = 0$. Since

$$s'(t) = v(t) = -32t + 96$$

we have

$$\int_0^t s'(r)\,dr = \int_0^t (-32r + 96)\,dr$$

$$s(r)\Big|_0^t = (-16r^2 + 96r)\Big|_0^t$$

$$s(t) - s(0) = [-16t^2 + 96t] - [(-16)(0)^2 + (96)(0)]$$

Since $s(0) = 0$, we have

$$s(t) = -16t^2 + 96t \qquad (7.29)$$

(c) The object will continue to move as long as $v(t) \neq 0$, and it will be at instantaneous rest when $v(t) = 0$. Therefore from Equation (7.28), we have

$$v(t) = -32t + 96 = 0$$

which is satisfied when $t = \frac{96}{32} = 3$. In other words, 3 sec after the object is projected upward, it comes to instantaneous rest and then begins its journey downward. To find the distance traveled by the object in 3 sec, we use Equation (7.29) and obtain

$$s(3) = (-16)(3)^2 + (96)(3) = 144$$

that is, the object will rise to a height of 144 ft above the ground.

(d) From Equation (7.29), we have

$$s(5) = (-16)(5)^2 + (96)(5) = 80$$

In 5 sec the object will be 80 ft above ground level. We observe that $s(6) = 0$. Hence the object will be back on the ground in 6 sec, and for $t > 6$, $v(t) \equiv 0$, and $s(t) \equiv 0$. (Why?)

Example 7.15 A water supply tank with a capacity of 1500 cu ft develops a leak. Suppose that the rate at which the water is flowing out of the tank is given by

$$V'(t) = -\tfrac{1}{30}\,t \qquad (7.30)$$

The negative sign indicates that the volume is decreasing.

where V, the volume of water, is a function of time t measured in minutes.

(a) Find the volume of the water in the tank at time t.

(b)   If the tank develops the leak at 9:05 P.M., at what time will the tank be empty?

SOLUTION

(a)   We assume that time is measured starting at 9:05 P.M. Hence $V = 1500$ at $t = 0$, and we have the initial-value problem

$$V'(t) = -\tfrac{1}{30} t, \qquad V(0) = 1500$$

Then

$$\int_0^t V'(s)\, ds = -\frac{1}{30} \int_0^t s\, ds$$

which yields

$$V(s)\Big|_0^t = -\frac{1}{30} \frac{s^2}{2}\Big|_0^t$$

$$V(t) - V(0) = -\frac{1}{60} t^2$$

$$V(t) = -\frac{1}{60} t^2 + 1500 \tag{7.31}$$

(b)   When the tank is empty, $V = 0$. From Equation (7.31), we have

$$-\tfrac{1}{60} t^2 + 1500 = 0$$

which implies that $t = \pm 300$. We disregard $t = -300$. Therefore

$$t = 300$$

This result means that the tank will be empty 300 min after the leak develops—at precisely 2:05 A.M.

It is important to note that the volume function defined by (7.31) has domain $\{t \mid 0 \le t \le 300 \text{ min}\}$, since we cannot have negative volume. For $t > 300$, we define $V(t) = 0$.

Exercises 7.4   In Problems 1 through 10, find a solution of the initial-value problem. Use two methods.

1. $y'(t) = 1 - 3t, \quad y(0) = 4$

2. $\dfrac{dy}{dx} = x^2 + 4x^3, \quad y(1) = -2$

3. $y'(t) = t^3 + t^{-2}, \quad y(-2) = 1$

4. $y'(t) = \sqrt{t} + 4, \quad y(4) = 3$

5. $\dfrac{dy}{dx} = \dfrac{3x^2 + x}{x^5}, \quad y(1) = -1$

6. $\dfrac{dy}{dx} = 2 - x^{3/2}, \quad y(9) = 2$

7. $y' = x^{1/3}(1 + 3x^{1/3}), \quad y(-8) = -\tfrac{1}{4}$ 8. $\sqrt{t}\, y'(t) = 2, \quad y(1) = 3$

9. $\dfrac{\sqrt{x}}{x^2 - 1}\dfrac{dy}{dx} + 1 = 0,\quad y(4) = -\tfrac{1}{2}$ 10. $\dfrac{dy}{dx} = 3x^{1/3}(1 - x^{1/3}),\quad y(8) = -1$

11. Suppose that the marginal cost in producing x units of an item is given by

$$C'(x) = 4x - 350$$

   (a) Find the total cost if there is a fixed cost of 500, i.e., $C(0) = 500$.
   (b) Find the profit if each item produced sells at $10.
   (c) How many units should be sold to produce maximum profit?

12. Suppose that the marginal cost of producing x gal of Slick oil is $10\sqrt{x}$ dollars. Find the cost function $C(x)$ if there is an overhead cost of $350. (Note that the overhead cost is independent of x.)

13. Suppose that the marginal revenue of producing and selling x units of a product is $20 + (x/50)$. If the revenue for the first 500 units is $12,500, find the revenue function and the price function.

14. A snowball in the form of a perfect sphere has radius 72 in. Suppose that it melts so that the radius decreases at the rate of 4 in./min. (Assume that the snowball retains its spherical shape until it is completely melted.)

   (a) Find the time t when the diameter is 72 in.
   (b) What is the diameter of the snowball after 12 min?
   (c) At what time will the snowball be completely melted?

15. Three astronauts are marooned in a spacecraft. There are 350 units of oxygen available, and the rate of consumption in units per minute is

$$A'(t) = -0.03\sqrt{t}$$

   (a) Determine the amount A of oxygen in the spacecraft as a function of time.
   (b) Find the time at which the oxygen supply is exhausted.

16. A driver sees a dog crossing the highway ahead, and applies the brakes when the car is 100 ft from the dog. If the car's velocity after application of the brakes is $(70 - 26t)$ ft/sec, will the driver stop in time?

17. Find the velocity and distance functions under the following conditions:

   (a) $a(t) = t + 1,\quad v(0) = 2,\quad s(0) = 1$
   (b) $a(t) = t^2 - 5,\quad v(0) = -1,\quad s(0) = 4$

In the following problems, assume that $a(t) = 32$ ft/sec^2.

18. An object is projected vertically from the surface of the earth with initial velocity $v(0) = 288$ ft/sec.

   (a) Find the velocity and distance functions.
   (b) How high will the object rise?

19. An object projected from the surface of the Earth rises 228 ft before starting to fall. Find the time it took the object to rise to this height.

20. An object is thrown upward from the roof of a building 64 ft above the ground. If the initial velocity is 48 ft/sec, determine when the object will hit the ground. (See Figure 7.20, p. 282.)

Figure 7.20

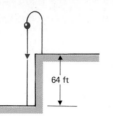

64 ft

7.5   Integration by Substitution

Suppose we wish to find $\int (2x^3 + 5)^{137} 6x^2\,dx$. We could expand $(2x^3 + 5)^{137}$ and multiply the result by $6x^2$. Then the integral of the expression obtained can be evaluated. Undoubtedly, the reader will agree that this approach is tedious, to say the least. To avoid such a cumbersome task, we use the chain rule.

Consider a function F expressible as a composite function

$$F(x) = G\big(u(x)\big)$$

By the chain rule,

$$F'(x) = G'\big(u(x)\big)u'(x)$$

and thus

$$\int F'(x)\,dx = \int G'\big(u(x)\big)u'(x)\,dx$$

The notation dx or du has an interesting history. Originally, Leibniz defined these quantities as "infinitesimals" or "very small quantities greater than zero but less than any positive number." This description led to much confusion since there are no such numbers.

If dx denotes a positive real number, Cauchy defined

$$du = u'(x)\,dx$$

from which we obtain

$$\int F'(x)\,dx = \int G'(u)\,du = G\big(u(x)\big) + C = F(x) + C \quad (7.32)$$

We illustrate this procedure in the following examples.

Example 7.16   Evaluate $\int (2x^3 + 5)^{137}\,6x^2\,dx$.

    SOLUTION   Let $u = 2x^3 + 5$. Then

$$du = u'(x)\,dx = 6x^2\,dx$$

Using these substitutions, we obtain

Application of formula (7.1)

$$\int (2x^3 + 5)^{137}\,6x^2\,dx = \int u^{137}\,du = \frac{u^{138}}{138} + C = \frac{(2x^3 + 5)^{138}}{138} + C$$

Example 7.17   Evaluate $\int \sqrt{2x + 1}\,dx$.

SOLUTION   Let $u = 2x + 1$. Then

$$du = u'(x)\ dx = 2\ dx$$

Using these substitutions, we obtain

$$\int \sqrt{2x+1}\ dx = \frac{1}{2} \int \sqrt{2x+1}\ 2\ dx = \frac{1}{2} \int u^{1/2}\ du = \frac{1}{2} \frac{u^{3/2}}{3/2} + C$$

$$= \frac{1}{3} u^{3/2} + C = \frac{1}{3} (2x+1)^{3/2} + C$$

Multiply and divide by 2. We have used the fact that $c \int f(x)\ dx = \int cf(x)\ dx$ ($c = $ constant).

Example 7.18   Evaluate $\int (2x+3)(x^2+3x+5)^{-3/2}\ dx$.

SOLUTION   Let $u = x^2 + 3x + 5$. Then

$$du = u'(x)\ dx = (2x+3)\ dx$$

Thus we have

$$\int (2x+3)(x^2+3x+5)^{-3/2}\ dx = \int u^{-3/2}\ du$$

$$= -2u^{-1/2} + C = -2(x^2+3x+5)^{-1/2} + C$$

In the preceding examples it is important to observe how the function u is selected. In each case we choose u so as to change the form of the integrand to u^n; that is, $\int f(x)\ dx$ becomes $\int u^n\ du$, $n \neq -1$.

We now consider the evaluation of definite integrals. The method involves first finding an antiderivative of the integrand and then evaluating this antiderivative at the limits of integration. There are two possible ways of doing this and we shall illustrate both methods.

Example 7.19   Evaluate $\int_1^2 (1-x^{-1})^3\ x^{-2}\ dx$.

SOLUTION 1   We first evaluate the indefinite integral

$$\int (1-x^{-1})^3\ x^{-2}\ dx$$

We let

$$u = 1 - x^{-1}, \qquad du = u'(x)\ dx = x^{-2}\ dx$$

Substitution yields

$$\int (1-x^{-1})^3\ x^{-2}\ dx = \int u^3\ du = \tfrac{1}{4} u^4 + C = \tfrac{1}{4}(1-x^{-1})^4 + C$$

so that

$$\int_1^2 (1-x^{-1})^3\ x^{-2}\ dx = \tfrac{1}{4}(1-x^{-1})^4 \Big|_1^2$$

$$= \left[\tfrac{1}{4}\left(1 - \tfrac{1}{2}\right)^4 \right] - \left[\tfrac{1}{4}\left(1-1\right)^4 \right] = \tfrac{1}{64}$$

SOLUTION 2   Let

$$u = 1 - x^{-1}, \qquad du = x^{-2} \, dx$$

We note that the limits of the definite integral for values of x are $x = 1$ and $x = 2$. We find the corresponding values of u:

$$\text{when} \quad x = 1, \qquad u = 1 - 1 = 0$$

$$\text{when} \quad x = 2, \qquad u = 1 - \tfrac{1}{2} = \tfrac{1}{2}$$

In Solution 2 we are using the formula

$$\int_a^b g(u(x)) \, u'(x) \, dx$$

$$= \int_{u(a)}^{u(b)} g(u) \, du$$

Under certain conditions on u(x) and g(x), this formula is valid. The conditions will be satisfied for the functions considered in this text.

Be careful!

$$\int (2x^3 + 5)^{137} \, dx$$

$$\neq \frac{1}{6x^2} \int (2x^3 + 5)^{137} \, 6x^2 \, dx$$

In general,

$$\int f(x) \, dx = \frac{1}{g(x)} \int f(x) \, g(x) \, dx$$

if and only if g is a constant function.

Therefore by substitution, we have

$$\int_1^2 (1 - x^{-1})^3 \, x^{-2} \, dx = \int_0^{1/2} u^3 \, du$$

$$= \tfrac{1}{4} u^4 \Big|_0^{1/2} = \left[\tfrac{1}{4} \left(\tfrac{1}{2} \right)^4 \right] - \left[\tfrac{1}{4} (0)^4 \right] = \tfrac{1}{64}$$

Reviewing the previous examples, we become aware that the method of substitution of variables is not always applicable. For instance, in Example 7.16 the factor $6x^2$ plays an important role. To evaluate

$$\int (2x^3 + 5)^{137} \, dx$$

we will have to use the binomial theorem to expand the integrand and then find the antiderivative. Note that if we let

$$u = 2x^3 + 5 \quad \text{then} \quad du = 6x^2 \, dx$$

and

$$\int (2x^3 + 5)^{137} \, dx = \int u^{137} \, \frac{1}{6x^2} \, du$$

The factor $1/6x^2$ makes it impossible to evaluate the integral on the right-hand side. The method of substitution of variables does not work in this case.

Exercises 7.5

In Problems 1 through 14, use the method of substitution of variables to evaluate the given integral.

1. $\displaystyle\int (x + 4)^2 \, dx$

2. $\displaystyle\int (x - 5)^6 \, dx$

3. $\displaystyle\int 3\sqrt{3x + 4} \, dx$

4. $\displaystyle\int \frac{1}{\sqrt{3x + 4}} \, dx$

5. $\displaystyle\int (4 - x)^3 \, dx$

6. $\displaystyle\int \left(\frac{x - 3}{2} \right)^2 \, dx$

7. $\displaystyle\int (4x - 1)^{-2} \, dx$

8. $\displaystyle\int (2x + 3)(x^2 + 3x - 1)^{-3} \, dx$

9. $\displaystyle\int x^4 \, (4x^5 + 6)^{100} \, dx$

10. $\displaystyle\int (x + 1)(x^2 + 2x - 5)^3 \, dx$

11. $\displaystyle\int x\sqrt{x^2 - 4} \, dx$

12. $\displaystyle\int x\sqrt{4 - x^2} \, dx$

13. $\displaystyle\int x^{-2}\sqrt{1-x^{-1}}\,dx$ 14. $\displaystyle\int x^{2/3}(2+x^{5/3})^{1/2}\,dx$

In Problems 15 through 20, evaluate the indicated definite integrals.

15. $\displaystyle\int_0^1 (3x-1)\,dx$ 16. $\displaystyle\int_{-2}^1 \sqrt{1-x}\,dx$

17. $\displaystyle\int_3^7 (x-1)^{-1/2}\,dx$ 18. $\displaystyle\int_{-1}^0 (1-3x)^{3/2}\,dx$

19. $\displaystyle\int_0^1 x(x^2+1)^{-2}\,dx$ 20. $\displaystyle\int_0^1 x(1-x^2)^{-1/2}\,dx$

7.6 Integration Involving e^x and In x

We can easily obtain the two basic formulas for integrating e^x and $1/x$ by inspecting the derivatives of e^x and In x developed in the previous chapter. We recall that

$$\int f(x)\,dx = F(x) + C$$

where C is a constant and $F'(x) = f(x)$.

Theorem 7.5 ◀

$$\int e^x\,dx = e^x + C \tag{7.33}$$

PROOF   Since

$$\frac{d}{dx}(e^x) = e^x \quad \text{it follows that} \quad \int e^x = e^x + C$$

Theorem 7.6 ◀

$$\int \frac{dx}{x} = \ln|x| + C \tag{7.34}$$

PROOF

Case 1   If $x > 0$, then $|x| = x$, and

$$\frac{d}{dx}\ln|x| = \frac{d}{dx}\ln x = \frac{1}{x}$$

Case 2   If $x < 0$, then $|x| = -x$, and

$$\frac{d}{dx}\ln|x| = \frac{d}{dx}\ln(-x) = \frac{1}{-x}(-1) = \frac{1}{x}$$

Therefore $\dfrac{d}{dx}\ln|x| = 1/x$ in both cases, and consequently

$$\int \frac{dx}{x} = \ln |x| + C$$

By using these two theorems, we can extend our techniques of integration to more complicated functions.

Example 7.20   Find $\int e^{3x}\ dx$.

SOLUTION   Let $u = 3x$; then $du = 3\ dx$. Using these substitutions, we have

$$\int e^{3x}\ dx = \tfrac{1}{3} \int e^{3x} \cdot 3\ dx = \tfrac{1}{3} \int e^{u}\ du = \tfrac{1}{3}\ (e^{u} + C_1)$$

$$= \tfrac{1}{3}\ e^{u} + C \qquad (\text{where } C = \tfrac{1}{3}\ C_1)$$

$$= \tfrac{1}{3}\ e^{3x} + C$$

Example 7.21   Find $\int xe^{2x^2+1}\ dx$.

SOLUTION   Letting $u = 2x^2 + 1$, we have $du = 4x\ dx$. Hence

$$\int xe^{2x^2+1}\ dx = \tfrac{1}{4} \int e^{2x^2+1}\ (4x)\ dx = \tfrac{1}{4} \int e^{u}\ du = \tfrac{1}{4}\ e^{u} + C$$

$$= \tfrac{1}{4}\ e^{2x^2+1} + C$$

It follows from Theorem 7.6 that $d/dx\ (\ln|x|) = 1/x,\ x \neq 0$, so that we can generalize formula (7.34) and obtain

$$\int \frac{1}{u}\frac{du}{dx}\ dx = \int \frac{1}{u}\ du = \ln |u| + C, \qquad u \neq 0 \qquad\qquad (7.35)$$

Example 7.22   Evaluate $\int \dfrac{2x}{x^2 + 4}\ dx$

SOLUTION   Let $u = x^2 + 4$. Then $du/dx = 2x$. Since $u(x) > 0$ for all x, we have

$$\int \frac{2x}{x^2 + 4}\ dx = \int \frac{1}{u}\frac{du}{dx}\ dx = \ln u + C = \ln(x^2 + 4) + C$$

Example 7.23   Evaluate $\int \dfrac{2x^2 + 2x + 1}{x + 1}\ dx, \quad x \neq -1$

SOLUTION   We first divide the integrand to obtain

$$\int \frac{2x^2 + 2x + 1}{x + 1} \, dx = \int \left(2x + \frac{1}{x + 1}\right) dx \qquad\qquad \frac{2x^2 + 2x + 1}{x + 1} = 2x + \frac{1}{x + 1}$$

$$= \int 2x \, dx + \int \left(\frac{1}{x + 1}\right) dx = x^2 + \ln|x + 1| + C$$

In Problems 1 through 20, calculate the given integrals.   Exercises 7.6

1. $\displaystyle\int e^{4x} \, dx$ 2. $\displaystyle\int e^{-5x} \, dx$

3. $\displaystyle\int e^{(1/2)x} \, dx$ 4. $\displaystyle\int e^{(-3/4)x} \, dx$

5. $\displaystyle\int x e^{x^2} \, dx$ 6. $\displaystyle\int_{1}^{2} e^{3x + 4} \, dx$

7. $\displaystyle\int_{0}^{1} x e^{x^2} \, dx$ 8. $\displaystyle\int_{1}^{2} 3x e^{x^2} \, dx$

9. $\displaystyle\int \frac{2}{x^2} e^{1/x} \, dx$ 10. $\displaystyle\int (e^x + e^{-x}) \, dx$

11. $\displaystyle\int \left(e^{-x} + \frac{1}{x}\right) dx$ 12. $\displaystyle\int \frac{1}{x + 1} \, dx$

13. $\displaystyle\int \left(\frac{e^{\sqrt{x}}}{\sqrt{x}} + \frac{1}{x}\right) dx$ 14. $\displaystyle\int \frac{e^x}{1 + e^x} \, dx$

15. $\displaystyle\int \frac{1}{x - 2} \, dx$ 16. $\displaystyle\int \frac{x}{x^2 + 4} \, dx$

17. $\displaystyle\int_{1}^{2} \frac{(\ln x)^2}{x} \, dx$ 18. $\displaystyle\int_{2}^{3} \frac{x^2 - x - 1}{x - 1} \, dx$

19. $\displaystyle\int_{0}^{1} \frac{6x^2 + 3x - 2}{2x + 1} \, dx$ 20. $\displaystyle\int \frac{x^3 + 5x^2 + 10x - 8}{x + 2} \, dx$

7.7 Integration by Parts

Suppose f and g are differentiable functions of x. Then by the rule of the derivative of a product, we have

$$(f \cdot g)' = f \cdot g' + f' \cdot g \qquad (7.36)$$

Integrating both sides of Equation (7.36) with respect to x, we obtain

$$\int \frac{d}{dx} [f(x)g(x)] \, dx = \int [f(x) \cdot g'(x) + f'(x)g(x)] \, dx \qquad (7.37)$$

or

$$f(x)g(x) = \int f(x)g'(x) \, dx + \int g(x)f'(x) \, dx \qquad (7.38)$$

From this, we have

$$\int f(x)g'(x) \, dx = f(x)g(x) - \int g(x)f'(x) \, dx \qquad (7.39)$$

Formula (7.39) is called *integration by parts*.

For the definite integral, we have

$$\int_a^b f(x)g'(x)\,dx = f(x)g(x)\Big|_a^b - \int_a^b g(x)f'(x)\,dx \qquad (7.40)$$

At first glance, it might appear that we are complicating the problem. However, integration by parts facilitates the evaluation of antiderivatives of a large class of functions.

Example 7.24 Evaluate $\int x\sqrt{x-1}\,dx$.

SOLUTION There are two choices we can make for f and g'. First we let

$$f(x) = \sqrt{x-1} \quad\text{and}\quad g'(x) = x$$

then

$$f'(x) = \tfrac{1}{2}(x-1)^{-1/2} \quad\text{and}\quad g(x) = \tfrac{1}{2}x^2$$

and integration by parts yields

$$\int x\sqrt{x-1}\,dx = \tfrac{1}{2}x^2\sqrt{x-1} - \int \tfrac{1}{4}x^2(x-1)^{-1/2}\,dx$$

We note that the problem is getting more complicated. Therefore this choice of f and g' is discarded.

A second choice is

$$f(x) = x \quad\text{and}\quad g'(x) = \sqrt{x-1}$$

Then

$$f'(x) = 1 \quad\text{and}\quad g(x) = \tfrac{2}{3}(x-1)^{3/2}$$

Substitution into formula (7.39) yields

To evaluate $\int (2/3)(x-1)^{3/2}\,dx$, we use the method of substitution of variables with $u = x-1$ and $du = dx$.

$$\int x\sqrt{x-1}\,dx = \tfrac{2}{3}x(x-1)^{3/2} - \int \tfrac{2}{3}(x-1)^{3/2}\,dx$$
$$= \tfrac{2}{3}x(x-1)^{3/2} - \tfrac{2}{3}\cdot\tfrac{2}{5}(x-1)^{5/2} + C$$
$$= \tfrac{2}{3}x(x-1)^{3/2} - \tfrac{4}{15}(x-1)^{5/2} + C$$

Example 7.25 Evaluate $\int_0^1 x^3\sqrt{1+x^2}\,dx$.

We choose g' to be the most complicated factor in the integrand that can be integrated readily.

SOLUTION Let

$$f(x) = x^2 \quad\text{and}\quad g'(x) = x\sqrt{1+x^2}$$

Then

To find g we use the method of substitution of variables with $u = 1+x^2$ and $du = 2x\,dx$.

$$f'(x) = 2x \quad\text{and}\quad g(x) = \tfrac{1}{3}(1+x^2)^{3/2}$$

Therefore

$$\int_0^1 x^3 \sqrt{1 + x^2} \, dx = \int_0^1 x^2 (x \sqrt{1 + x^2}) \, dx$$

$$= \frac{1}{3} x^2 (1 + x^2)^{3/2} \Big|_0^1 - \int_0^1 \frac{2}{3} x (1 + x^2)^{3/2} \, dx$$

$$= \frac{1}{3} x^2 (1 + x^2)^{3/2} \Big|_0^1 - \frac{1}{3} \cdot \frac{2}{5} (1 + x^2)^{5/2} \Big|_0^1$$

$$= \left[\frac{2\sqrt{2}}{3} - 0 \right] - \frac{2}{15} [4\sqrt{2} - 1] = \frac{2}{15} (\sqrt{2} + 1)$$

Integration by substitution of variables

Example 7.26   Find $\int x^2 e^x \, dx$.

SOLUTION   We use integration by parts. Set

$$f(x) = x^2, \quad g'(x) = e^x \quad \text{and} \quad f'(x) = 2x, \quad g(x) = e^x$$

Using formula (7.39), we have

$$\int x^2 e^x \, dx = x^2 e^x - \int e^x 2x \, dx = x^2 e^x - 2 \int x e^x \, dx$$

We find $\int x e^x \, dx$ by parts. Again, setting

$$f(x) = x, \quad g'(x) = e^x \quad \text{and} \quad f'(x) = 1, \quad g(x) = e^x$$

and using (7.39), we have

$$\int x e^x \, dx = x e^x - \int e^x \, dx = x e^x - e^x + C_1$$

Consequently,

$$\int x^2 e^x \, dx = x^2 e^x - 2 \int x e^x \, dx = x^2 e^x - 2(x e^x - e^x) + C$$
$$= x^2 e^x - 2x e^x + 2 e^x + C$$

Example 7.27   Find $\int \ln x \, dx$.

SOLUTION   Setting

$$f(x) = \ln x \text{ and } g'(x) = 1, \text{ then } f'(x) = \frac{1}{x} \text{ and } g(x) = x.$$

Using (7.39), we have

$$\int \ln x \, dx = x \ln x - \int x \cdot \frac{1}{x} \, dx = x \ln x - \int dx = x \ln x - x + C$$

Use integration by parts to find the following integrals.

Exercises 7.7

1. $\int x(2x - 1) \, dx$

2. $\int x \sqrt{1 + x} \, dx$

3. $\int x^3(1+x^2)^{3/2}\,dx$ 4. $\int_0^8 x^2(1+x)^{-1/2}\,dx$

5. $\int xe^x\,dx$ 6. $\int x^3 e^x\,dx$

7. $\int x\ln x\,dx$ 8. $\int x^2\ln x\,dx$

9. $\int_2^4 5x^4\ln x\,dx$ 10. $\int \sqrt{x}\ln x\,dx$

11. $\int (1-x)\,e^x\,dx$ 12. $\int (2+x^2)\,e^x\,dx$

13. $\int xe^{2x}\,dx$ 14. $\int 3xe^{-2x+4}\,dx$

15. $\int \ln(2x+3)\,dx$ 16. $\int 2x\ln(4x-3)\,dx$

17. $\int (1+x)\,e^{-x}\,dx$ 18. $\int (1-x^2)\,e^{-x}\,dx$

19. $\int \dfrac{\ln(\ln x)}{x}\,dx$ 20. $\int e^x\ln e^{-x}\,dx$

7.8 Improper Integrals

The definite integrals considered earlier all had the form

$$\int_a^b f(x)\,dx$$

We assumed that the function f was continuous on the finite closed interval $[a, b]$. We shall now relax these assumptions and extend the definition of the integral to consider

Note *Rigorous definition of improper integrals is given in a higher level calculus course.*

(i) Those integrals in which the function f does not exist at some points in the interval $[a, b]$

(ii) An infinite interval of integration, such as $[a, \infty)$, $(-\infty, a]$, or $(-\infty, \infty)$

The integrals having properties (i) or (ii) are called *improper integrals*. First we consider the case where f is defined and continuous on an interval $(a\ \ b]$, but not on $[a, b]$. For example, let f be the function defined by

$$y = \frac{1}{\sqrt{x}}$$

and consider the integral

$$\int_0^1 \frac{1}{\sqrt{x}}\,dx \tag{7.41}$$

Here f is not defined at $x = 0$ and $f(x) \to \infty$ as $x \to 0^+$ (see Figure 7.21). Suppose that we take $0 < a < 1$. Then

$$\int_a^1 \frac{1}{\sqrt{x}}\, dx = 2\sqrt{x}\,\Big|_a^1 = 2 - 2\sqrt{a}$$

and we find that $(2 - 2\sqrt{a}) \to 2$ as $a \to 0^+$. We make the following definition.

$$\int_0^1 \frac{1}{x}\, dx = \lim_{a \to 0^+} \int_a^1 \frac{1}{\sqrt{x}}\, dx = \lim_{a \to 0^+} (2 - 2\sqrt{a}) = 2$$

Therefore we assign 2 to the area of the region shown in Figure 7.21.

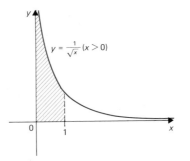

Figure 7.21

The integral in (7.41) is an example of an *improper integral* which is said to be *convergent*.

Definition 7.2   If f is continuous on $(a, b]$ and $f(a)$ does not exist,   ◀
then

$$\int_a^b f(x)\, dx = \lim_{\epsilon \to 0^+} \int_{a+\epsilon}^b f(x)\, dx$$

provided this limit exists.

Another example of an improper integral is

$$\int_{-2}^0 \frac{1}{x^2}\, dx \qquad\qquad (7.42)$$

Here the function $y = 1/x^2$ is continuous on $[-2, 0)$ but not on $[-2, 0]$. To evaluate this integral, we compute for $-2 < b < 0$

$$\int_{-2}^b \frac{1}{x^2}\, dx = -\frac{1}{x}\,\Big|_{-2}^b = -\frac{1}{b} - \frac{1}{2}$$

In this case, we find that as $b \to 0^-$, the expression $[-(1/b) - \frac{1}{2}] \to \infty$. Hence no limit exists, and no number can be assigned to the area of the

Figure 7.22

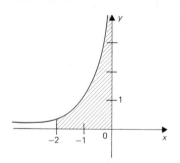

region shown in Figure 7.22. The integral in (7.42) is an example of an *improper integral* that is said to be *divergent*.

▶ **Definition 7.3** *If f is continuous on [a, b) and f(b) does not exist, then*

$$\int_a^b f(x)\, dx = \lim_{\epsilon \to 0^+} \int_a^{b-\epsilon} f(x)\, dx$$

provided this limit exists.

If the integrand does not exist at an interior point of the interval of integration, we have the following definition.

▶ **Definition 7.4** *If f is continuous at all x in the interval [a, b] except at c, where a < c < b and f(c) does not exist, then*

$$\int_a^b f(x)\, dx = \lim_{\epsilon_1 \to 0^+} \int_a^{c-\epsilon_1} f(x)\, dx + \lim_{\epsilon_2 \to 0^+} \int_{c+\epsilon_2}^b f(x)\, dx$$

provided both *of these limits exist.*

If the limits in Definitions 7.2, 7.3, and 7.4 exist, we say that the improper integral is *convergent*; otherwise it is *divergent*.

Note *It is important for the student to observe the points of discontinuity of the integrand. For instance, a careless evaluation of (7.43) may lead to the following erroneous answer:*

$$\int_{-2}^3 \frac{1}{x^2}\, dx = -\frac{1}{x}\bigg|_{-2}^3 = -\frac{1}{3} + \frac{1}{2} = \frac{1}{6}$$

This example represents incorrect application of the definition of the definite integral.

Example 7.28 Evaluate the following improper integral if it exists:

$$\int_{-2}^3 \frac{1}{x^2}\, dx \tag{7.43}$$

SOLUTION In this case, $y = 1/x^2$ is continuous at every x in $[-2, 3]$ except at $x = 0$, and $f(0)$ does not exist. To evaluate the integral, we use Definition 7.4, and write

$$\int_{-2}^3 \frac{1}{x^2}\, dx = \lim_{\epsilon_1 \to 0^+} \int_{-2}^{0-\epsilon_1} \frac{1}{x^2}\, dx + \lim_{\epsilon_2 \to 0^+} \int_{0+\epsilon_2}^3 \frac{1}{x^2}\, dx$$

$$= \lim_{\epsilon_1 \to 0^+} -\frac{1}{x}\Big|_{-2}^{-\epsilon_1} + \lim_{\epsilon_2 \to 0^+} -\frac{1}{x}\Big|_{\epsilon_2}^{3}$$

$$= \lim_{\epsilon_1 \to 0^+} \left(\frac{1}{\epsilon_1} - \frac{1}{2}\right) + \lim_{\epsilon_2 \to 0^+} \left(-\frac{1}{3} + \frac{1}{\epsilon_2}\right)$$

Since neither of the limits exist, the improper integral (7.43) is divergent.

Example 7.29 Evaluate the following improper integral if it exists.

$$\int_{-1}^{2} \frac{1}{x^{1/3}} \, dx \tag{7.44}$$

SOLUTION In this case, $y = 1/x^{1/3}$ is continuous at every x in $[-1, 2]$ except at $x = 0$, and $f(0)$ does not exist. Using Definition 7.4, we find that

$$\int_{-1}^{2} \frac{1}{x^{1/3}} \, dx = \lim_{\epsilon_1 \to 0^+} \int_{-1}^{0-\epsilon_1} \frac{1}{x^{1/3}} \, dx + \lim_{\epsilon_2 \to 0^+} \int_{0+\epsilon_2}^{2} \frac{1}{x^{1/3}} \, dx$$

$$= \lim_{\epsilon_1 \to 0^+} \tfrac{3}{2} x^{2/3}\Big|_{-1}^{-\epsilon_1} + \lim_{\epsilon_2 \to 0^+} \tfrac{3}{2} x^{2/3}\Big|_{\epsilon_2}^{2}$$

$$= -\tfrac{3}{2} + \tfrac{3}{2} (2)^{2/3} = \tfrac{3}{2} (\sqrt[3]{4} - 1)$$

Next we consider the case of a continuous function that is to be integrated over an infinite interval. An interesting application of such improper integrals arises in the field of economics. For example, if the number of dollars in the annual income is $f(t)$ at t years and r is the annual rate of interest compounded continuously, then the present value of all future income is given by the improper integral

$$\int_{0}^{\infty} f(t) e^{-rt} \, dt$$

Improper integrals such as

$$\int_{a}^{\infty} f(x) \, dx, \quad \int_{-\infty}^{b} f(x) \, dx, \quad \text{and} \quad \int_{-\infty}^{\infty} f(x) \, dx$$

are very common and useful in applications. Such integrals arise in the theory of probability, which provides very powerful mathematical tools for predicting events in most branches of the sciences. A biologist studying heredity or a psychologist conducting learning experiments would encounter probability theory frequently. The procedure for evaluating improper integrals of this type is similar to the procedure considered in earlier examples. We have the following definition.

Definition 7.5 *If f is continuous for all $x \geq a$, then* ◀

$$\int_a^\infty f(x)\,dx = \lim_{b\to\infty} \int_a^b f(x)\,dx$$

provided this limit exists.

Example 7.30   Evaluate the following improper integral if it exists:

$$\int_0^\infty e^{-x}\,dx \tag{7.45}$$

SOLUTION   By definition, we have

$$\int_0^\infty e^{-x}\,dx = \lim_{b\to\infty} \int_0^b e^{-x}\,dx = \lim_{b\to\infty} -e^{-x}\Big|_0^b = \lim_{b\to\infty}(1 - e^{-b}) = 1$$

Therefore the improper integral (7.45) is convergent and its value is 1.

▶ **Definition 7.6**   *If f is continuous for all $x \le a$, then*

$$\int_{-\infty}^a f(x)\,dx = \lim_{b\to -\infty} \int_b^a f(x)\,dx$$

provided this limit exists.

If a function f is continuous on the interval $(-\infty, \infty)$, we make the following definition.

▶ **Definition 7.7**   *If f is continuous for all values of x, then*

$$\int_{-\infty}^\infty f(x)\,dx = \lim_{a\to -\infty} \int_a^0 f(x)\,dx + \lim_{b\to\infty} \int_0^b f(x)\,dx$$

provided **both** *limits exist.*

Example 7.31   Evaluate the following improper integrals if they exist.

(a) $\displaystyle\int_{-\infty}^0 \frac{1}{(2-x)^2}\,dx$ (b) $\displaystyle\int_{-\infty}^\infty x\,dx$

SOLUTION

(a)   By definition, we have

$$\int_{-\infty}^0 \frac{1}{(2-x)^2}\,dx = \lim_{b\to -\infty} \int_b^0 \frac{1}{(2-x)^2}\,dx = \lim_{b\to -\infty}\left(\frac{1}{2-x}\right)\Big|_b^0$$

$$= \lim_{b\to -\infty}\left(\frac{1}{2} - \frac{1}{2-b}\right) = \tfrac{1}{2} + 0 = \tfrac{1}{2}$$

Therefore the improper integral in (a) is convergent and has the volume $\tfrac{1}{2}$.

(b) In this case, we have

$$\int_{-\infty}^{\infty} x \, dx = \lim_{b \to -\infty} \int_{b}^{0} x \, dx + \lim_{a \to \infty} \int_{0}^{a} x \, dx$$

$$= \lim_{b \to -\infty} \frac{x^2}{2} \Big|_{b}^{0} + \lim_{a \to \infty} \frac{x^2}{2} \Big|_{0}^{a} = \lim_{b \to -\infty} \left(-\frac{b^2}{2} \right) + \lim_{a \to \infty} \frac{a^2}{2}$$

Since neither of these two limits exist, the improper integral in (b) is divergent.

A common mistake is to write

$$\int_{-\infty}^{\infty} f(x) \, dx = \lim_{b \to \infty} \int_{-b}^{b} f(x) \, dx$$

which is not the definition of the improper integral on the left-hand side of this equation. For example,

$$\lim_{b \to \infty} \int_{-b}^{b} x \, dx = \lim_{b \to \infty} \frac{x^2}{2} \Big|_{-b}^{b} = \lim_{b \to \infty} \left(\frac{b^2}{2} - \frac{b^2}{2} \right) = \lim_{b \to \infty} 0 = 0$$

whereas the improper integral $\int_{-\infty}^{\infty} x \, dx$ is divergent.

In each of Problems 1 through 26, evaluate the given improper integral if it exists.

Exercises 7.8

1. $\displaystyle\int_{0}^{4} \frac{1}{\sqrt{x}} \, dx$

2. $\displaystyle\int_{-1}^{0} \frac{1}{x^2} \, dx$

3. $\displaystyle\int_{0}^{1} \frac{1}{x^{1/3}} \, dx$

4. $\displaystyle\int_{0}^{9} \frac{1}{x^{3/2}} \, dx$

5. $\displaystyle\int_{2}^{4} \frac{1}{\sqrt{x-2}} \, dx$

6. $\displaystyle\int_{-5}^{1} \frac{1}{\sqrt{1-x}} \, dx$

7. $\displaystyle\int_{-2}^{2} \frac{1}{x^2} \, dx$

8. $\displaystyle\int_{2}^{10} \frac{1}{(x-2)^{2/3}} \, dx$

9. $\displaystyle\int_{0}^{2} \frac{1}{(x-1)^2} \, dx$

10. $\displaystyle\int_{0}^{2} \frac{1}{(x-1)^{2/3}} \, dx$

11. $\displaystyle\int_{0}^{2} \frac{x^2}{(x^3-1)^2} \, dx$

12. $\displaystyle\int_{-1}^{2} \frac{x}{\sqrt{x^2-1}} \, dx$

13. $\displaystyle\int_{1}^{\infty} \frac{1}{x^2} \, dx$

14. $\displaystyle\int_{1}^{\infty} \frac{1}{x} \, dx$

15. $\displaystyle\int_{0}^{\infty} e^{-3x} \, dx$

16. $\displaystyle\int_{2}^{\infty} \frac{1}{x^2} \, dx$

17. $\displaystyle\int_{1}^{\infty} \frac{1}{x^4} \, dx$

18. $\displaystyle\int_{-\infty}^{0} e^{3x} \, dx$

19. $\displaystyle\int_{-1}^{\infty} \frac{dx}{(x+3)^2}$

20. $\displaystyle\int_{-\infty}^{-1} \frac{1}{x^3} \, dx$

21. $\displaystyle\int_{5}^{\infty} \frac{1}{\sqrt{x-1}} \, dx$

22. $\displaystyle\int_{1}^{\infty} \frac{1}{x \ln x} \, dx$

23. $\displaystyle\int_0^\infty \frac{1}{\sqrt{e^x}}\, dx$ 24. $\displaystyle\int_e^\infty \frac{1}{x(\ln x)^2}\, dx$

25. $\displaystyle\int_{-\infty}^\infty \frac{x}{1+2x^2}\, dx$ 26. $\displaystyle\int_{-\infty}^\infty xe^{-x^2}\, dx$

27. If $a > 0$, show that $\displaystyle\int_0^\infty e^{-ax}\, dx = 1/a$.

28. If $a > 0$, show that $\displaystyle\int_0^\infty xe^{-ax}\, dx = 1/a^2$ [assume that $\displaystyle\lim_{c\to\infty}(c/e^{ac}) = 0$].

29. If $a > 1$, show that $\displaystyle\int_1^\infty \frac{1}{x^a}\, dx = \frac{1}{a-1}$

30. The formula for finding the present value V of all future income of D dollars per year and at an interest rate of $r\%$ compounded continuously is given by

$$V = \int_0^\infty De^{-rt}\, dt$$

where t is time.

(a) If the perpetual cash flow $D = \$4000$ per year and $r = 8\%$, what is the present value V?

(b) Show that for constant D, we have $V = D/r$.

Chapter 7 Review   Part 1 *(Oral)*

Define or discuss the following:

1. Antiderivative

2. The definite integral $\int_a^b f(x)\, dx$.

3. Riemann sum

4. Fundamental theorem of calculus

5. When $f(x) \geq 0$, what is the geometric interpretation of $\int_a^b f(x)\, dx$?

6. When $f(x) \leq 0$, what is the geometric interpretation of $\int_a^b f(x)\, dx$?

7. Differential equation

8. Initial-value problem

9. Integration by parts

10. Improper integrals

Part 2 *(Written)*

In Problems 1 through 42, find the given antiderivative.

1. $\displaystyle\int x^5\, dx$ 2. $\displaystyle\int x^{1/2}\, dx$

3. $\displaystyle\int \frac{1}{x^3}\, dx$

4. $\displaystyle\int 6x^5\, dx$

5. $\displaystyle\int x^{2/3}\, dx$

6. $\displaystyle\int \frac{1}{x^n}\, dx$

7. $\displaystyle\int (x^5 + 3x^4 + 1)\, dx$

8. $\displaystyle\int \left(3x + 2 + \frac{5}{x}\right) dx$

9. $\displaystyle\int \frac{x^3 + 3x^2 + 4}{\sqrt{x}}\, dx$

10. $\displaystyle\int x(x^2 + 3)\, dx$

11. $\displaystyle\int x(x + 1)(x - 1)\, dx$

12. $\displaystyle\int \sqrt{5x}\, dx$

13. $\displaystyle\int (x + 2)^3\, dx$

14. $\displaystyle\int (2x + 3)^4\, dx$

15. $\displaystyle\int 3\sqrt{3x - 5}\, dx$

16. $\displaystyle\int [(x + 1)^2 + (x - 2)^3]\, dx$

17. $\displaystyle\int \frac{1}{(x - 1)^3}\, dx$

18. $\displaystyle\int \left(\sqrt{x + 1} + \frac{1}{\sqrt{x + 1}}\right) dx$

19. $\displaystyle\int \left(2x + \frac{1}{x}\right) dx$

20. $\displaystyle\int \left(2x + \frac{1}{x}\right)^2 dx$

21. $\displaystyle\int \left(2x + \frac{1}{x}\right)^3 dx$

22. $\displaystyle\int \frac{1 + x}{x}\, dx$

23. $\displaystyle\int \frac{x}{1 + x}\, dx$

24. $\displaystyle\int \frac{x}{1 - x}\, dx$

25. $\displaystyle\int \frac{1 - x}{1 + x}\, dx$

26. $\displaystyle\int \left(\frac{1}{1 + x} + \frac{1}{1 - x}\right) dx$

27. $\displaystyle\int \frac{x}{(x + 2)^2}\, dx$

28. $\displaystyle\int x(1 + 2x^2)^7\, dx$

29. $\displaystyle\int \frac{x}{\sqrt{1 - x^2}}\, dx$

30. $\displaystyle\int x(4 + x^2)^{4/7}\, dx$

31. $\displaystyle\int 3x^2 e^{x^3 + 1}\, dx$

32. $\displaystyle\int xe^{-x^2 - 4}\, dx$

33. $\displaystyle\int e^{-ax}\, dx$

34. $\displaystyle\int \frac{dx}{3x - 4}$

35. $\displaystyle\int \frac{dx}{3 - 7x}$

36. $\displaystyle\int \frac{2x}{(5x^2 + 2)^5}\, dx$

37. $\displaystyle\int xe^{2x}\, dx$

38. $\displaystyle\int 2x \ln x\, dx$

39. $\displaystyle\int -\ln x\, dx$

40. $\displaystyle\int x^2 e^{-x}\, dx$

41. $\displaystyle\int (1 + x^2)\, e^x\, dx$

42. $\displaystyle\int (1 + x^2) \ln x\, dx$

43. Find the equation of the curve whose tangent line at x has a slope of

$$f'(x) = 2x - 3$$

and the curve passes through the point (1, 2).

44. For a certain curve $y = f(x)$, we have

$$\frac{d^2y}{dx^2} = 3x^2$$

If the curve passes through the point (1, 2) and the slope of the tangent line to the curve at (1, 2) is 3, find the equation of the curve.

45. An object is thrown vertically upward with a velocity of 256 ft/sec. Find the greatest height attained by the particle and the time required to attain it. Assume the acceleration due to gravity is -32 ft/sec^2.

46. The volume of a spherical drop of liquid increases at a rate proportional to its surface area. If a is the initial radius of the drop, find its radius at time t.

47. Find the area of the region whose boundaries are given:

(a) The x-axis, the curve $y = x^3$, and the line $x = 3$

(b) The x-axis, the curve $y = x^2$, and the lines $x = 1$ and $x = 3$

(c) The x-axis, the curve $y = x^2 - 4$, and the lines $x = 3$ and $x = 5$

(d) The curve $y = \ln x$, the x-axis, and the lines $x = 1$ and $x = e$

(e) The curve $y = x^2 + 2x - 8$ and the line $y = 4x$

(f) The curve $y = -x^2 + 2x + 8$ and the line $y = 2x - 1$

In Problems 48 through 51, evaluate the given improper integral if it exists.

48. $\displaystyle\int_0^\infty e^{-(1/2)x} \, dx$

49. $\displaystyle\int_1^\infty \frac{1}{x^{1/2}} \, dx$

50. $\displaystyle\int_1^3 \frac{1}{(x-2)^2} \, dx$

51. $\displaystyle\int_{-3}^5 \frac{1}{(3x-4)^{1/2}} \, dx$

Numerical Applications Using Calculators

The definite integral of a function can be approximated by its definition. To obtain a reasonable approximation, enough terms should be added to the sum. A calculator can be useful in the computations. We shall demonstrate by considering the following example:

Example 7.32   Approximate $\int_2^4 (x^2 + 2)\, dx$ by definition.

Recall that

$$\int_a^b f(x)\, dx \approx \sum_{i=1}^n f(a + ih) \cdot h$$

where $h = (b - a)/n$ and n is a positive integer to be chosen. Note that the larger the n is chosen, the better approximation we obtain. Here we shall choose n to be 5 and 10. See Tables 7.1 and 7.2.

i	ih	$a + ih$	$f(a + ih) = (a + ih)^2 + 2$	$f(a + ih) \cdot h$
1	0.4	2.4	7.76	3.104
2	0.8	2.8	9.84	3.936
3	1.2	3.2	12.24	4.896
4	1.6	3.6	14.96	5.984
5	2.0	4.0	18	7.2

$\sum_{i=1}^5 f(a + ih) \cdot h = 25.12$

Table 7.1
$n = 5, h = (4 - 2)/5 = 2/5 = 0.4$

i	ih	$a + ih$	$f(a + ih) = (a + ih)^2 + 2$	$f(a + ih) \cdot h$
1	0.2	2.2	6.84	1.368
2	0.4	2.4	7.76	1.552
3	0.6	2.6	8.76	1.752
4	0.8	2.8	9.84	1.968
5	1.0	3.0	11	2.2
6	1.2	3.2	12.24	2.448
7	1.4	3.4	13.56	2.712
8	1.6	3.6	14.96	2.992
9	1.8	3.8	16.44	3.228
10	2.0	4.0	18.	3.6

$\sum_{i=1}^{10} f(a + ih) \cdot h = 23.88$

Table 7.2
$n = 10, h = 2/10 = 0.2$

By the fundamental theorem of calculus, we have

$$\int_2^4 (x^2 + 2)\, dx = \frac{x^3}{3} + 2x \Big|_2^4 = \frac{4^3}{3} + 2 \cdot 4 - \left(\frac{2^3}{3} + 2 \cdot 2\right)$$

$$\approx 22.66666666$$

Example 7.33   Compute $\int_5^{10} (1/x)\, dx$ from definition.

We shall choose $n = 10$. See Table 7.3.

Table 7.3

$n = 10, h = (10 - 5)/10 = 0.5$

i	ih	$a + ih$	$f(a + ih) = \dfrac{1}{a + ih}$	$f(a + ih) \cdot h$
1	0.5	5.5	0.1818181818	0.0909090909
2	1.0	6.0	0.1666666667	0.0833333333
3	1.5	6.5	0.1538461538	0.0769230769
4	2.0	7.0	0.1428571429	0.0714285714
5	2.5	7.5	0.1333333333	0.0666666666
6	3.0	8.0	0.125	0.0625
7	3.5	8.5	0.1176470588	0.0588235294
8	4.0	9.0	0.1111111111	0.0555555555
9	4.5	9.5	0.1052631579	0.0526315789
10	5.0	10.0	0.1	0.05

$$\sum_{i=1}^{10} f(a + ih) \cdot h = 0.668830221$$

Again, by the fundamental theorem of calculus, we have

$$\int_5^{10} \frac{1}{x}\, dx = \ln x \Big|_5^{10} = \ln 10 - \ln 5 \approx 2.302585093 - 1.609437912$$

$$= 0.693147181$$

Exercises

In Problems 1 through 10, approximate the given definite integral for $n = 5$ and 10. Compare your results with those obtained from the fundamental theorem of calculus.

1. $\displaystyle\int_5^7 (x^3 + x + 2)\, dx$

2. $\displaystyle\int_5^{10} \frac{1}{x+1}\, dx$

3. $\displaystyle\int_2^5 xe^x\, dx$

4. $\displaystyle\int_1^{10} (\sqrt{x} + 3)\, dx$

5. $\displaystyle\int_7^{10} \frac{x+1}{x-2}\, dx$

6. $\displaystyle\int_3^5 2xe^{x^2}\, dx$

7. $\displaystyle\int_{10}^{20} \frac{\ln x}{x}\, dx$

8. $\displaystyle\int_5^7 \frac{x^2 + 2x + 2}{x^2 + 2x + 1}\, dx$

9. $\displaystyle\int_{10}^{15} \frac{\sqrt{x}+1}{\sqrt{x}}\, dx$

10. $\displaystyle\int_1^5 \ln x\, dx$

11. Approximate the definite integral $\displaystyle\int_0^1 \frac{1}{x^3 + 1}\, dx$

Differential Equations

CHAPTER 8

8.1 Introduction

The subject of differential equations is an important aspect of modern mathematics. Numerous problems encountered in the various branches of the physical and biological sciences, the social sciences, economics, and engineering lead to differential equations. Such equations arise in the study of the motion of celestial objects and atomic particles, the rate of growth of populations, the decomposition of radioactive substances, and problems involved with social diffusion.

In this chapter we define and classify differential equations. We indicate in some cases how the differential equations arise and discuss simple methods of solution.

First let us consider a problem involving the diffusion of some information through the inhabitants of a country.

In the seventeenth century Galileo and Hooke used differential equations to study the behavior of beams under stress and strain. Such problems arose in the building of some of the famous cathedrals. Newton and others used differential equations in their study of such problems as the shape of the Earth and gravitational attraction, and it was Leibniz who coined the term "differential equation."

Example 8.1 A company is using the television media to advertise its latest product. Suppose the information concerning the product spreads from television commercials only. It is found that a fixed percentage (say, 5%) of the population is reached per day and that the rate of spread of information is proportional to the number of people who have not heard of the product. Describe this situation mathematically.

SOLUTION Let $y(t)$ denote the number of persons who have heard of the product at time t. Suppose A represents the number of persons in the country. Then, $A - y(t)$ denotes the number of persons who have not yet heard of the product. Since the rate of spread of information (that is, dy/dt) is proportional to the number of persons who have not yet heard of the product, we obtain

$$\frac{dy}{dt} = k(A - y) \tag{8.1}$$

where k is the constant of proportionality. Initially, before airing of the commercials, the number of persons who had heard of the product was zero. We denote this by

$$y(0) = 0 \tag{8.2}$$

Thus this situation is described by an initial-value problem. The problem involves an unknown function $y(t)$ to be determined so that Equations (8.1) and (8.2) are satisfied.

Note We are excluding the employees of the company.

Equation (8.1) is called a *differential equation*. In general, an ordinary differential equation is an equation involving an independent variable t (or x, or u, etc.); a dependent variable y; and certain derivatives of y with respect to t. We may use the expression

$$F(t, y, y', y'', \ldots, y^{(n)}) = 0 \tag{8.3}$$

The relationship (8.3) must involve some derivative of the unknown function y. For instance,

$$y'' + ty = 0 \tag{8.4}$$

$$y' + ty = e^t \tag{8.5}$$

$$y''' + y'y = t^2 + 1 \tag{8.6}$$

are examples of differential equations. However, the equation

$$x^2 + ye^x + y^2 = 0$$

is *not* a differential equation because no derivatives are involved. Also, identities of the form

$$\frac{d}{dx}(e^{-x}) + e^x = 0$$

are *not* differential equations because no unknown function is involved.

A differential equation in which the highest derivative is the nth derivative is called a differential equation of *order n*. Thus Equations (8.4), (8.5), and (8.6) are of order two, one, and three, respectively.

Linear differential equation

Equations (8.1), (8.4), and (8.5) are called *linear differential equations*. In general, a linear differential equation is of the form

$$a_n(t)y^{(n)} + a_{n-1}(t)y^{(n-1)} + \cdots + a_0(t)y = f(t) \tag{8.7}$$

Nonlinear differential equation

where $a_0(t), a_1(t), \ldots, a_n(t)$ and $f(t)$ are known functions of t. A differential equation that is not linear is called a *nonlinear differential equation*. Equation (8.6) cannot be written in the form (8.7). Thus

Equation (8.6) is a nonlinear differential equation. Other examples of nonlinear differential equations are

$$y'' + (y')^3 = x + 1 \tag{8.8}$$

$$\frac{dy}{dx} + xe^y = 0 \tag{8.9}$$

Note *Equation (8.6) is nonlinear due to the term $y'y$. Equation (8.8) is nonlinear due to the term $(y')^3$. Equation (8.9) is nonlinear due to the term e^y.*

Definition 8.1 *A solution of a differential equation is a function defined on some interval such that the function and its respective derivatives on that interval satisfy the given equation.*

◄

Example 8.2 Show that $y = e^t$ is a solution of the equation

$$y'' - y = 0 \tag{8.10}$$

SOLUTION For $t \in R$, we find that $y'' = e^t$, and substitution into Equation (8.10) yields

$$y'' - y = e^t - e^t = 0$$

Thus $y = e^t$ is a solution of the given equation for $t \in R$.

The reader can also show that $y = e^{-t}$ is a solution of Equation (8.10). In fact,

$$y = c_1 e^t + c_2 e^{-t} \tag{8.11}$$

where c_1 and c_2 are arbitrary constants, is also a solution of Equation (8.10) for $t \in R$. The function in (8.11) is called the *general solution* of Equation (8.10).

A solution of an nth-order differential equation that contains n essentially arbitrary constants is called the *general solution*. A solution that is obtained from the general solution by giving particular values to one or more of the n essentially arbitrary constants is called a *particular solution* of the equation.

Note *The function*

$$y = (c_1 + c_2)\, e^t + c_3 e^{-t}$$

has essentially two arbitrary constants and not three since $c_1 + c_2$ may be replaced by a single arbitrary constant.

Example 8.3 Show that $y = c_1 e^t + c_2 e^{3t}$ is the general solution of the equation

$$y'' - 4y' + 3y = 0 \tag{8.12}$$

SOLUTION From the function

$$y = c_1 e^t + c_2 e^{3t}$$

Remark *The general solution of a given differential equation contains all possible solutions of the equation.*

we find that

$$y' = c_1 e^t + 3c_2 e^{3t} \quad \text{and} \quad y'' = c_1 e^t + 9c_2 e^{3t}$$

Substituting y, y', and y'' into Equation (8.12), we find that for $t \in R$,

$$y'' - 4y' + 3y = c_1e^t + 9c_2e^{3t} - 4(c_1e^t + 3c_2e^{3t}) + 3(c_1e^t + c_2e^{3t}) = 0$$

Thus $y = c_1e^t + c_2e^{3t}$ is a solution of (8.12) for every pair of constants c_1 and c_2. Since Equation (8.12) is a second-order differential equation and $y = c_1e^t + c_2e^{3t}$ contains essentially two arbitrary constants, then it is the general solution.

Example 8.4   Find a particular solution of Equation (8.12) with the following initial conditions:

$$y(0) = 1 \quad \text{and} \quad y'(0) = 0 \tag{8.13}$$

SOLUTION   In this case we seek a function that is a solution of Equation (8.12) and satisfies the initial conditions (8.13). Using the general solution, we try to find values for c_1 and c_2 such that the resulting function satisfies not only Equation (8.12), but also (8.13). From the condition $y(0) = 1$ and the general solution, we find that

$$1 = c_1 + c_2 \tag{8.14}$$

and from $y'(0) = 0$, we get

$$0 = c_1 + 3c_2 \tag{8.15}$$

Solving Equations (8.14) and (8.15), we obtain

$$c_1 = \tfrac{3}{2} \quad \text{and} \quad c_2 = -\tfrac{1}{2}$$

Thus the function

$$y = \tfrac{3}{2} e^t - \tfrac{1}{2} e^{3t}$$

is a solution of (8.12) and (8.13). The reader is encouraged to verify this.

Example 8.5

(a)   Show that $y = c_1e^t + c_2e^{2t}$ is the general solution of the equation

$$y'' - 3y' + 2y = 0 \tag{8.16}$$

(b)   Find a solution satisfying the supplementary (or boundary) conditions

$$y(0) = 0 \quad \text{and} \quad y(1) = 2 \tag{8.17}$$

Note *A problem involving a differential equation associated with supplementary conditions which are related to two or more values of the independent variable is called a* boundary-value *problem.*

SOLUTION   Direct substitution of $y = c_1e^t + c_2e^{2t}$ and its derivatives y', y'' into Equation (8.16) yields

$$y'' - 3y' + 2y = (c_1e^t + 4c_2e^{2t}) - 3(c_1e^t + 2c_2e^{2t}) + 2(c_1e^t + c_2e^{2t}) = 0$$

Thus $y = c_1 e^t + c_2 e^{2t}$ is a solution of Equation (8.16). Since the differential equation is of order two and the solution contains two arbitrary constants, then it is the general solution. Now we try to solve for c_1 and c_2 so that the conditions in (8.17) are satisfied. The condition

$$y(0) = 0 \quad \text{implies} \quad 0 = c_1 + c_2$$

and the condition

$$y(1) = 2 \quad \text{implies} \quad 2 = c_1 e + c_2 e^2$$

Solving for c_1 and c_2, we obtain

$$c_1 = \frac{2}{e - e^2} \quad \text{and} \quad c_2 = \frac{2}{e^2 - e}$$

Thus a particular solution of Equation (8.16) satisfying the boundary conditions (8.17) is

$$y = \frac{2}{e^2 - e} \, (e^{2t} - e^t) \tag{8.18}$$

The reader is encouraged to verify that (8.18) is a solution of the boundary-value problem (8.16) and (8.17).

In Problems 1 through 12, give the order of the differential equation. State whether the equation is linear or nonlinear.

Exercises 8.1

1. $y'' + 3y' + y = t$

2. $y'' + 3tyy' = 0$

3. $x\dfrac{dy}{dx} + y - y^2 = 0$

4. $(y'')^2 + 1 + y^2 = 0$

5. $y^{(4)} = t^2 + 1$

6. $\dfrac{d^3 y}{dx^3} - 4\dfrac{dy}{dx} + 3y = e^x$

7. $x^2 \dfrac{dy}{dx} + y^2 = 0$

8. $\dfrac{d^3 y}{dx^3} + 2\left(\dfrac{dy}{dx}\right)^4 + 5y = 0$

9. $y'' + ye^x = 0$

10. $y'' + xe^y = 0$

11. $\left(\dfrac{dy}{dx}\right)^3 = \dfrac{d^2 y}{dx^2} + 1$

12. $(1 + t^2)\, y'' + ty' = 5$

In Problems 13 through 20, show that each of the functions f is a solution of the given differential equation for some interval on the real line.

13. $f(x) = 2e^{-x} + x, \quad \dfrac{dy}{dx} + y = 1 + x$

14. $f(x) = 3e^{-x} - 5e^x, \quad \dfrac{d^2 y}{dx^2} - y = 0$

15. $f(x) = 2e^x + e^{2x} + \frac{1}{6} e^{-x}, \quad \dfrac{d^3 y}{dx^3} - \dfrac{d^2 y}{dx^2} - 4\dfrac{dy}{dx} + 4y = e^{-x}$

16. $f(x) = 1 + 2e^x - 3e^{-x}, \quad \dfrac{d^3 y}{dx^3} - \dfrac{dy}{dx} = 0$

17. $f(x) = e^x + \frac{1}{2} xe^x$, $\quad \dfrac{d^2y}{dx^2} - y = e^x$

18. $f(x) = 2x - xe^{1/x}$, $\quad x^3 \dfrac{d^2y}{dx^2} + x \dfrac{dy}{dx} - y = 0$

19. $f(x) = \dfrac{1}{x^2}(1 - \ln x)$, $\quad x^2 \dfrac{d^2y}{dx^2} + 5x \dfrac{dy}{dx} + 4y = 0$

20. $f(x) = \dfrac{1}{2x+1}$, $\quad (2x+1)^2 \dfrac{d^2y}{dx^2} - (4x+2) \dfrac{dy}{dx} - 12y = 0$

In Problems 21 through 28, show that the given function f is the general solution of the given differential equation.

21. $f(t) = -16t^2 + c_1 t + c_2$, $\quad y'' = -32$

22. $f(x) = c_1 e^{3x} + c_2 e^{-3x}$, $\quad \dfrac{d^2y}{dx^2} - 9y = 0$

23. $f(t) = c_1 e^t + c_2 t e^t$, $\quad y'' - 2y' + y = 0$

24. $f(t) = c_1 e^t + c_2 e^{-2t} + t^2$, $\quad y'' + y' - 2y = 2(1 + t - t^2)$

25. $f(x) = c_1 x^2 + c_2 x^{-3}$, $\quad x^2 \dfrac{d^2y}{dx^2} + 2x \dfrac{dy}{dx} - 6y = 0$

26. $f(t) = c_1 e^t + c_2 e^{-t} + \frac{1}{2} t e^t$, $\quad y'' - y = e^t$

*27. $f(t) = c_1 + c_2 e^t + c_3 e^{-t} + \frac{1}{2}(t^2 - 3t)e^t$, $\quad y''' - y' = 2te^t$

*28. $f(t) = c_1 + c_2 e^{-2t} + c_3 t e^{-2t} + c_4 t^2 e^{-2t}$, $\quad y^{(4)} + 6y''' + 12y'' + 8y' = 0$

29. (a) Show that $y = ce^{-2t}$, where c is an arbitrary constant, is the general solution of the differential equation $y' + 2y = 0$.

(b) Find a particular solution satisfying the initial conditions $y(0) = -3$.

30. Consider the equation $y' = -ky$.

(a) Show that $y = ce^{-kt}$ is a solution of the equation.

(b) Find c given that $y(0) = 2$.

(c) Find k given that $y(1) = 1$.

(d) What is $y(2)$?

31. (a) Show that $y = c_1 e^{2x} + c_2 e^{3x}$, where c_1 and c_2 are arbitrary constants, is the general solution of the differential equation

$$\dfrac{d^2y}{dx^2} - 5 \dfrac{dy}{dx} + 6y = 0$$

(b) Find a particular solution satisfying the initial conditions $y(0) = 3$ and $y'(0) = 0$.

32. (a) Show that $y = c_1 e^{-x} + c_2 e^{2x}$ is the general solution of the equation

$$\dfrac{d^2y}{dx^2} - \dfrac{dy}{dx} - 2y = 0$$

(b) Find a particular solution satisfying the boundary conditions $y(0) = 1$ and $y(1) = 2$.

33. Show that $y = 1/(1 - x^2)$ is a solution of the initial-value problem

$$\frac{dy}{dx} - 2xy^2 = 0, \qquad y(0) = 1$$

34. (a) Show that $y = 1/(\sqrt{2x + c})$ is a solution of the equation

$$\frac{dy}{dx} + y^3 = 0$$

(b) Find the solution satisfying the initial condition $y(1) = 2$.

8.2 Some Methods of Solution

In this section we shall develop methods for solving two classes of first-order differential equations. First, we shall consider first-order differential equations in which the variables can be separated and then we shall consider the linear first-order differential equation for which the variables cannot be separated.

One of the earliest procedures used for solving differential equations is *separation of variables*. In certain cases a first-order differential equation may be written in the form

$$f(x) + g(y)\frac{dy}{dx} = 0 \tag{8.19}$$

Equation (8.19) can be integrated directly using the chain rule to obtain

$$\int \left[f(x) + g(y)\frac{dy}{dx} \right] dx = C, \qquad \int f(x)\, dx + \int g(y)\frac{dy}{dx}\, dx = C$$

or simply

$$\int f(x)\, dx + \int g(y)\, dy = C$$

where C is the constant of integration. We illustrate this procedure in the following examples.

Example 8.6 Find the general solution of the given equation.

$$2x\frac{dy}{dx} = 2x^3 - 1 \tag{8.20}$$

SOLUTION For $x \neq 0$, we separate the variables as follows

$$\frac{dy}{dx} = x^2 - \frac{1}{2x} \tag{8.21}$$

Direct integration with respect to x yields

$$\int \frac{dy}{dx}\, dx = \int \left(x^2 - \frac{1}{2x} \right) dx \quad \text{or} \quad y = \frac{x^3}{3} - \frac{1}{2}\ln |x| + C$$

Note Equations (8.20) and (8.21) are equivalent; that is, the general solution of (8.21) is also the general solution of (8.20).

where C is an arbitrary constant. The reader can verify that y is indeed the general solution of Equation (8.20) on any interval of the x-axis, where $x \neq 0$.

Example 8.7 Find the general solution of the given differential equation.

$$\frac{dy}{dx} = \frac{xy^2 - x}{xy^3 - y^3} \qquad (8.22)$$

Observe that we are assuming that no division by zero is allowed.

SOLUTION Separation of variables yields

$$\frac{y^3}{y^2 - 1}\frac{dy}{dx} = \frac{x}{x - 1}$$

Direct integration with respect to x yields

$$\int \frac{y^3}{y^2 - 1}\frac{dy}{dx}\,dx = \int \frac{x}{x - 1}\,dx$$

$$\int \left(y + \frac{y}{y^2 - 1}\right) dy = \int \left(1 + \frac{1}{x - 1}\right) dx$$

or

Note In this case we are unable to write $y = f(x)$. We say that (8.23) defines an implicit solution of (8.22).

$$\frac{y^2}{2} + \frac{1}{2}\ln |y^2 - 1| = x + \ln |x - 1| + C \qquad (8.23)$$

where C is an arbitrary constant.

Equation (8.23) is an *implicit* representation of the solution of the differential equation (8.24).

Example 8.8 Solve the given initial-value problem.

$$\frac{dy}{dx} = \frac{3x^2 - 1}{2(y - 1)}, \qquad y(0) = -2 \qquad (8.24)$$

SOLUTION Separation of variables and integration with respect to x yields

$$\int 2(y - 1)\,dy = \int (3x^2 - 1)\,dx$$

or

$$y^2 - 2y = x^3 - x + C \qquad (8.25)$$

where C is an arbitrary constant. To determine the particular solution satisfying the given initial condition, we set $x = 0$ and $y = -2$ in Equation (8.25) and obtain $C = 8$. Hence the desired particular solution is given implicitly by

$$y^2 - 2y = x^3 - x + 8 \qquad (8.26)$$

To obtain y in terms of x, we solve Equation (8.26) using the quadratic formula:

$$y = 1 \pm \sqrt{x^3 - x + 9} \qquad (8.27)$$

Thus Equation (8.27) provides two solutions of the differential equation; the one that satisfies the initial condition is

$$y = 1 - \sqrt{x^3 - x + 9} \qquad (8.28)$$

The reader should verify that Equation (8.28) gives the solution to the initial-value problem.

Unfortunately, not all first-order equations can be solved by the method we have just described. For example, the following equations are not separable:

$$xy \frac{dy}{dx} = x^2 + y^2 \quad \text{and} \quad 3x^2 y + y^3 + (x^2 + 3xy^2) \frac{dy}{dx} = 0$$

Methods for solving such equations are usually discussed in a first course on differential equations.

A very important class of first-order differential equations with applications in a wide variety of disciplines is the first-order linear differential equation

$$\frac{dy}{dx} + a(x)\, y = b(x) \qquad (8.29)$$

Except for special cases, such equations are not separable. We shall give a method for solving linear equations.

Consider the function v, where

$$v = e^{\int a(x)\,dx} \qquad (8.30)$$

If we multiply Equation (8.29) by v, we obtain

$$\frac{dy}{dx}\, e^{\int a(x)\,dx} + a(x)\, y e^{\int a(x)\,dx} = b(x)\, e^{\int a(x)\,dx} \qquad (8.31)$$

from which we get

$$\frac{d}{dx} \left(y e^{\int a(x)\,dx} \right) = b(x)\, e^{\int a(x)\,dx} \qquad (8.32)$$

The reader should verify the equality of the left-hand side of Equations (8.31) and (8.32).

Equation (8.32) can be integrated with respect to x and thus solved for y. The solution of Equation (8.32) is the solution of Equation (8.29). The function v defined by (8.30) is called an *integrating factor*. It should be noted that (8.30) is an integrating factor provided the differential equation is of the form (8.29). We illustrate this method in the next two examples.

Example 8.9 Solve the given differential equation.

$$\frac{dy}{dx} + 2y = x \qquad (8.33)$$

Note The method of separation of variables does not work in this case.

Other integrating factors are

$$v = e^{2x+C}$$

where C is any constant. We have simplified the situation by taking C = 0.

SOLUTION   First we find an integrating factor

$$v = e^{\int 2\,dx} = e^{2x} \tag{8.34}$$

Multiplying Equation (8.33) by e^{2x}, we get

$$e^{2x}\frac{dy}{dx} + 2e^{2x}y = xe^{2x}$$

or

$$\frac{d}{dx}(ye^{2x}) = xe^{2x} \tag{8.35}$$

Note that when the linear equation is multiplied by the integrating factor, the left-hand side is always the derivative of y times the integrating factor. Integrating equation (8.35) with respect to x, we have

$$\int \frac{d}{dx}(ye^{2x})\,dx = \int xe^{2x}\,dx$$

Using integration by parts

$$ye^{2x} = \tfrac{1}{2}xe^{2x} - \tfrac{1}{4}e^{2x} + C$$

$$y = \tfrac{1}{2}x - \tfrac{1}{4} + Ce^{-2x}$$

where C is an arbitrary constant. The reader should verify that y is indeed the general solution of (8.33).

Example 8.10   Solve the given initial-value problem.

$$x\frac{dy}{dx} + 2y = x^3, \qquad y(1) = -2 \tag{8.36}$$

SOLUTION   Writing the differential equation in the form (8.29), we have

$$\frac{dy}{dx} + \frac{2}{x}y = x^2 \tag{8.37}$$

and an integrating factor is

$$v = e^{\int (2/x)\,dx} = e^{2\ln x} = e^{\ln x^2} = x^2$$

Multiplying Equation (8.37) by x^2, we get

$$x^2\frac{dy}{dx} + 2xy = x^4 \quad \text{or} \quad \frac{d}{dx}(yx^2) = x^4$$

Integration with respect to x yields

$$\int \frac{d}{dx}(yx^2)\,dx = \int x^4\,dx, \qquad yx^2 = \frac{x^5}{5} + C$$

or

$$y = \tfrac{1}{5}x^3 + Cx^{-2} \tag{8.38}$$

Equation (8.38) gives the general solution of the given differential equation. Using the initial condition, we set $x = 1$ and $y = -2$ to obtain $C = -\frac{11}{5}$. Therefore, the solution of the initial-value problem is

$$y = \frac{1}{5}(x^3 - 11x^{-2})$$

The reader is encouraged to verify this result.

In Problems 1 through 18, use the method of separation of variables to find the general solution of the given differential equation.

Exercises 8.2

1. $y' = y$

2. $y' + 2y = 0$

3. $y' + \sqrt{3}y = 0$

4. $2y' + 3y = 0$

5. $\dfrac{dy}{dx} = \dfrac{4y}{x}$

6. $\dfrac{dy}{dx} = 3y$

7. $(x + 1)\dfrac{dy}{dx} = y^2$

8. $(y - 4)\dfrac{dy}{dx} = x + 3$

9. $x^2 y \dfrac{dy}{dx} = e^{-y}$

10. $e^{2x}\dfrac{dy}{dx} = 1$

11. $\dfrac{dy}{dx} = e^{x+y}$ [Hint: $e^{x+y} = e^x e^y$]

12. $2\dfrac{dy}{dx} = e^{-2x+y}$

13. $\dfrac{dy}{dx} = -3y + 2$

14. $\dfrac{dy}{dx} = \dfrac{1}{-4x + 3}$

15. $\sqrt{1 - x^2}\dfrac{dy}{dx} = -x$

16. $\dfrac{dy}{dt} = -4yt$

17. $e^x \dfrac{dy}{dx} = -xy^2$

18. $\dfrac{dy}{dx} = xe^{y-x^2}$

In Problems 19 through 26, solve the linear differential equation.

19. $y' + 2y = 3$

20. $y' - y = 10$

21. $\dfrac{dy}{dx} + 3y = x$

22. $\dfrac{dy}{dx} + \dfrac{3y}{x} = 3x^2$

23. $x\dfrac{dy}{dx} + 6y = x$

24. $y' = t^3 - 2ty$

25. $x\dfrac{dy}{dx} + (2x + 1)y = xe^{-2x}$

26. $x\dfrac{dy}{dx} = 8x^2 - 2y$

In Problems 27 through 31, solve the initial-value problem.

27. $x\dfrac{dy}{dx} - 2y = x^4,\quad y(1) = 3$

28. $y' + y = e^{-t},\quad y(0) = 2$

29. $yy' = t^3,\quad y(1) = 10$

30. $\dfrac{dy}{dx} = 5 - \dfrac{2y}{10 - x},\quad y(0) = 5$

31. $(2 + y)\dfrac{dy}{dx} = x\sqrt{y},\quad y(0) = 1$

32. Suppose that f_1 and f_2 are each solutions of the equation

$$\frac{dy}{dx} + a(x)y = 0$$

on an interval I. Show that $f_1 + f_2$ is also a solution on I.

33. Suppose that f_1 and f_2 are solutions of the equation

$$\frac{dy}{dx} + a(x)y = b(x)$$

on an interval I, where $b(x) \neq 0$ on I. Show that $f_1 + f_2$ is not a solution of the equation on I.

34. Suppose that f_1 is a solution of

$$\frac{dy}{dx} + a(x)y = b_1(x)$$

and f_2 is a solution of

$$\frac{dy}{dx} + a(x)y = b_2(x)$$

where a, b_1, and b_2 are all defined on the same interval I. Show that $f_1 + f_2$ is a solution of

$$\frac{dy}{dx} + a(x)y = b_1(x) + b_2(x)$$

on I.

35. (a) Show that there exist two values of m such that $y = e^{mx}$ is a solution of the equation

$$\frac{d^2y}{dx^2} - 4\frac{dy}{dx} + 3y = 0$$

(b) If m_1 and m_2 are the two values, show that

$$y = c_1 e^{m_1 x} + c_2 e^{m_2 x}$$

is the general solution, where c_1 and c_2 are arbitrary constants.

In Problems 36 through 40, repeat Problem 35 for the given equation.

36. $y'' + 3y' + 2y = 0$ 37. $y'' - 4y = 0$

38. $y'' - 2y = 0$ 39. $y'' - 2y' = 0$

40. $2y'' - 2y' - 3y = 0$

8.3   Some Applications

In this section we illustrate some applications of differential equations. First we return to Example 8.1.

Example 8.11   In Example 8.1, suppose that the advertising of the product reaches 5% of the population per day. Find the number of persons who would have heard of the product at time t.

SOLUTION   We are asked to solve the initial-value problem

$$\frac{dy}{dt} = k(A - y) \qquad (8.39)$$

$$y(0) = 0 \qquad (8.40)$$

Also, we are given that at the end of the first day 5% of the population is reached, that is,

$$y(1) = \tfrac{5}{100} A = \tfrac{1}{20} A \qquad (8.41)$$

Thus we seek a function $y = f(t)$ that satisfies Equations (8.39), (8.40), and (8.41). From the differential equation (8.39), we find by separation of variables

$$\int \left(\frac{1}{A - y} \frac{dy}{dt} \right) dt = \int k \, dt, \qquad -\ln |A - y| = kt + C_1$$

Note   *A is a known number and k is a constant to be determined.*

or

$$A - y = e^{-kt - C_1} = Ce^{-kt}$$

where $C = e^{-C_1}$. Hence

$$y = A - Ce^{-kt} \qquad (8.42)$$

Note   *Since $A - y > 0$, we have $|A - y| = A - y$.*

Using condition (8.40), we set $t = 0$ and $y = 0$ in (8.42) to find that $C = A$; and equation (8.42) becomes

$$y = A(1 - e^{-kt}) \qquad (8.43)$$

Using condition (8.41), we set $t = 1$ and $y = \tfrac{1}{20} A$ in (8.43) to obtain $k = -\ln \tfrac{19}{20}$. Therefore the number of persons who have been reached by advertising at any time t is given by

$$y = A(1 - e^{t \ln (19/20)})$$

Next we illustrate the use of differential equations in problems involving rates of change in the temperature of an object.

Suppose an object A with temperature T_A is immersed in a medium B with constant temperature T_B. It is found experimentally that the rate of change of temperature of A is proportional to the difference in temperature between A and B. Mathematically, this situation is expressed by

$$\frac{dT_A}{dt} = k(T_B - T_A) \qquad (8.44)$$

A decrease in temperature is indicated by a negative sign, and an increase is indicated by a positive sign.

where $k > 0$ is a constant of proportionality. Consider the following example.

Example 8.12   An object at a temperature of 180° is allowed to cool in a medium having a constant temperature of 60°. If in 1 min the temperature of the object is 120°, in how many minutes will its temperature be 90°?

SOLUTION   Here $T_B = 60°$. If T represents the temperature of the object at any time t, then

$$T(0) = 180° \tag{8.45}$$

and

$$T(1) = 120° \tag{8.46}$$

Using the differential equation (8.44), we have

$$\frac{dT}{dt} = k(60 - T) \tag{8.47}$$

Observe that the initial value of t (i.e., t = 0) is used in the lower limit of integration.

Equations (8.45), (8.46) and (8.47) form a boundary-value problem. We solve to obtain

$$\int_0^t \frac{1}{60 - T(s)} \frac{dT}{ds} \, ds = k \int_0^t ds$$

$$-\ln |60 - T(s)| \Big|_0^t = ks \Big|_0^t$$

Note 1.   Since $T \geq 60$,

$|60 - T| = T - 60$

2. $\dfrac{120}{T - 60} = e^{kt}$

$$-\ln |60 - T(t)| + \ln |60 - 180| = kt$$

$$\ln \frac{120}{T(t) - 60} = kt$$

and thus

$$T(t) = 60(1 + 2e^{-kt}) \tag{8.48}$$

In (8.48) the number k is still unknown. To evaluate k, we use (8.46) and (8.48). Thus

$$120 = 60(1 + 2e^{-k})$$

$$\tfrac{1}{2} = e^{-k}$$

and

$$\ln \tfrac{1}{2} = -k$$

Equation (8.48) becomes

$$T(t) = 60(1 + 2e^{[\ln(1/2)]t}) \tag{8.49}$$

To find t when $T = 90°$, we use Equation (8.49) and obtain

$$90 = 60(1 + 2e^{[\ln(1/2)]t})$$

$$\tfrac{1}{4} = e^{[\ln(1/2)]t}$$

and

$$\ln \tfrac{1}{4} = \left(\ln \tfrac{1}{2}\right) t$$

Hence

$$t = \frac{\ln \frac{1}{4}}{\ln \frac{1}{2}} = \frac{-\ln 4}{-\ln 2} = \frac{2 \ln 2}{\ln 2} = 2$$

that is, in 2 min the temperature of the body decreases to 90°.

Substances that decrease in quantity continuously at a rate proportional to the amount present are said to *decay* exponentially. In mathematical symbols, we write

$$\frac{dQ}{dt} = -kQ \qquad (8.50)$$

where $k > 0$ and Q is the amount of substance at time t.

Example 8.13 Suppose a radioactive substance decays according to the law in (8.50). If 20 lb are stored now, 18 lb would be left at the end of one century. Find the amount present at time t.

SOLUTION   We have the initial-value problem

$$\frac{dQ}{dt} = -kQ \qquad (8.51)$$

$$Q(0) = 20 \qquad (8.52)$$

Integration yields

$$\int_0^t \frac{1}{Q} \frac{dQ}{ds} \, ds = -\int_0^t k \, ds$$

$$\ln Q(s) \Big|_0^t = -ks \Big|_0^t$$

$$\ln Q(t) - \ln Q(0) = -kt$$

$$\ln \frac{Q(t)}{20} = -kt$$

and

$$Q(t) = 20e^{-kt} \qquad (8.53)$$

Since $Q = 18$ at $t = 1$, we have

$$18 = 20e^{-k}$$

$$0.9 = e^{-k}$$

and

$$-k = \ln 0.9$$

Therefore the desired function is defined by

$$Q(t) = 20e^{(\ln 0.9)t}$$

Carbon 14 (^{14}C) is a radioactive substance. The date at which an organism lived may be estimated by analyzing the remains of carbon 14 in the organism, since all living organisms contain carbon 12 and carbon 14. From the moment the organism dies, carbon 14 continues to decay and is no longer replaced, while carbon 12 remains unchanged. The change in the amount of carbon 14 relative to carbon 12 makes it possible to calculate the (approximate) time at which the organism lived.

Example 8.14 An archaeological expedition working in the Gobi desert came across some ancient objects. It was determined that about 30% of the original amount of ^{14}C was still present in the objects. If ^{14}C decays continuously, find the approximate age of the object (assume that the half-life of ^{14}C is 5600 yr).

*The half-life of a substance is the
time required for one-half the
amount of the substance to decay.*

SOLUTION   Let $A(t)$ be the amount of ^{14}C at time t. If $A(t)$ satisfies the differential equation (8.50), then

$$A(t) = Ce^{-kt} \qquad (8.54)$$

where C is a constant. At $t = 0$ the amount present is A_0. Hence

$$A(0) = ce^0 \Rightarrow C = A_0$$

and

$$A(t) = A_0 e^{-kt} \qquad (8.55)$$

After 5600 yr the amount of ^{14}C present is $\frac{1}{2} A_0$. Thus

$$A(5600) = A_0 e^{-5600k} = \frac{1}{2} A_0$$

$$\frac{1}{2} = e^{-5600k}$$

or

$$\ln \tfrac{1}{2} = -\ln 2$$

$$\ln \tfrac{1}{2} = \ln e^{-5600k}$$

$$k = \tfrac{1}{5600} \ln 2$$

Thus

$$A(t) = A_0 e^{(-\ln 2/5600)t}$$

In this case 30% of the ^{14}C is present. Therefore

$$\tfrac{30}{100} A_0 = A_0 e^{(-\ln 2/5600)t}$$

$$\tfrac{3}{10} = e^{(-\ln 2/5600)t}$$

and

$$\ln \tfrac{3}{10} = -\left(\frac{\ln 2}{5600} \right) t$$

Hence

$$t = -5600 \, \frac{\ln \frac{3}{10}}{\ln 2} = -5600 \, \frac{\ln 3 - \ln 10}{\ln 2}$$

$$= -5600 \, \frac{1.0986 - 2.3026}{0.6931}$$

or

$$t \approx 9700 \text{ yr}$$

In other words, the objects that were found are about 9700 years old.

1. The president of a university announces that no X-rated movies will be shown on campus. Suppose that 20% of the student body of 20,000 heard the announcement by the end of the first hour, and they transmitted this information by word of mouth. Further suppose that the rate of spread of information (per hour) is proportional to the number of students who have not heard the news. How many students would have heard the news at time t? How long will it take the news to spread through one-half the student body?

In Problems 2 through 5, the rate of change of temperature is proportional to the difference of temperatures between two mediums.

2. At a room temperature of $25°$, the temperature of a liquid is observed to be $75°$. In 15 min the temperature of the liquid is observed to be $65°$. Find the temperature of the liquid 90 min after the first observation.

3. Rosita prepares an apple pie for her friends Bob, Manuel, and Arlene. The pie is removed from the oven at $420°$ and left to cool at a room temperature of $70°$. In one-half hour the pie is at $210°$. Eager to have a piece of the pie, Bob calculates the time when the pie will be at $100°$. How long will the friends have to wait?

4. A thermometer reading $75°$ is placed outdoors. In 4 min the reading is $30°$. If the outside temperature is fixed at $20°$, find the following.

    (a) The thermometer reading 8 min after it was placed outdoors
    (b) The time it takes for the reading to drop to $21°$

5. An object is placed in a medium having a constant temperature. Initially, the temperature of the object is $40°$ higher than the medium. In 8 min the temperature of the object is $20°$ higher than the medium.

    (a) Determine k, the constant of proportionality.
    (b) In how many minutes will the difference in temperature be $10°$?

6. The population P of a country increases continuously at 4% per year. Assume $\frac{dP}{dt} = kP, k > 0$.

    (a) If the present population is 200 million, what will it be in 10 yr? 20 yr?
    (b) When will the population have doubled? tripled?

7. The birth rate of a population of ants is 3.5% per week. If there are no deaths among the ants, what is the population in 20 weeks when it is known that the population $P = 10^3$ at $t = 0$?

8. According to the 1940 census the population of Nuroville was 64,000 and in 1960 it was 80,000. If the rate of increase of population is propor-

tional to the population, find the year in which the population will be 160,000. (Use $2 \ln 2 = \ln 4 = 1.38628$, $\ln 5 = 1.60944$.)

9. A population P of mosquitoes increases by 3000 in 1 day. Given that P satisfies the initial-value problem

$$\frac{dP}{dt} = kP, \qquad P(0) = 1200$$

   (a) Find k.
   (b) Find the size of the population after 12 hr.

10. Suppose the half-life of a substance that decays continuously is 1200 yr. What is the amount of the substance left in 750 yr? 500 yr?

11. Suppose the rate of change in the amount of oil in this country is proportional to the amount present. If the constant of proportionality is 5% per year and the amount present in 1800 was 10^{12} barrels, find the amount that was present in 1900; in 1970.

12. A chemical plant discharges fumes into the atmosphere. The level h is 5 ft above the stacks at 6 A.M. and 20 ft at noon. If the rate of change in the level is proportional to h, what is the level at 5 P.M.?

13. Find the point on the curve $y = 2 \ln (x - 3)$ at which the tangent to the curve has slope $\frac{1}{2}$.

14. A particle moves according to the law

$$s = ae^{kt} + be^{-kt}$$

   where a, b, and k are constants. Show that its acceleration is proportional to its distance from the origin.

Chapter 8 Review   Part 1   *(Oral)*

Define or discuss each of the following.

1. Differential equation

2. Order of a differential equation

3. Linear differential equation

4. Nonlinear differential equation

5. Solution of a differential equation

6. General solution

7. Particular solution

8. Initial condition

9. Boundary-value problem

10. Method of separation of variables

11. Implicit solution

12. Integrating factor

13. Exponential decay

14. Half-life of a radioactive substance

Part 2 (*Written*)

1. Find the general solution of the differential equation

$$\frac{dy}{dx} = -3x^4 + 4$$

2. Find the general solution of the differential equation

$$\frac{d^2y}{dx^2} = 3 + 8x$$

3. Find a particular solution of the initial-value problem

$$\frac{dy}{dx} + 2y = 3x, \qquad y(0) = 2$$

4. Find a particular solution of the boundary-value problem

$$\frac{d^2y}{dx^2} = 2x + 1, \qquad y(0) = 2 \quad \text{and} \quad y(-2) = 3$$

5. Find a particular solution of the boundary-value problem

$$\frac{d^2y}{dx^2} + e^x = 2, \qquad y(0) = -2 \quad \text{and} \quad y(1) = \frac{3}{2}$$

6. Suppose that a rumor at your school spreads at a rate given by

$$y' = kt$$

where y is the number of people who have heard the rumor and t is the time in hours.

   (a) Find k given that $y(0) = 0$ and $y(2) = 100$.

   (b) Find y when $t = 3$, 6, and 10.

7. A radioactive substance decays at a rate given by

$$y' = -\frac{5}{100}\, y$$

where y represents the amount (in grams, g) present at time t (in mo).

   (a) Find the general solution of the equation.

   (b) Find a particular solution if $y(0) = 100$.

   (c) Find the amount of substance left 10 months later.

8. Let y denote the demand for a certain product and x denote the price per item. Suppose the rate of change of demand for the product is given by

$$\frac{dy}{dx} = -2x + 20, \qquad y(0) = 10 ,$$

Find the demand at each of the following price levels: $x = 5$, $x = 10$, $x = 15$.

9. If the marginal cost is given by

$$\frac{dC}{dx} = 40x + 1000$$

find the total cost function. If the fixed cost is 2500, find the cost of producing 100 units.

10. If the marginal profit is given by

$$\frac{dP}{dx} = -4x + 20$$

find the total profit function. If $P(0) = -20$ at what level of sales will maximum profit be realized? What is the maximum profit?

11. Let p denote the production and x the capitalization (in hundred-thousand dollars). Suppose the rate at which production changes for a unit change in capitalization, or the *marginal productivity*, for a certain company is given by

$$p'(x) = 4x^2 - 15$$

(a) If the company's present capitalization is $200,000 while producing 150 units of an item per month, find the production $p(x)$.

(b) If the company's capitalization is increased to $400,000, how much would the production increase?

In Problems 12 through 24, find the general solution of the given differential equation.

12. $y\dfrac{dy}{dx} = x^2 - 2$

13. $y\dfrac{dy}{dx} = x^2 + 2x$

14. $\dfrac{dy}{dx} + 2xy = 0$

15. $\dfrac{dy}{dx} = x^2 y$

16. $\dfrac{dy}{dx} + xy = 2x^2 y$

17. $\dfrac{dy}{dx} + 2xy = x^2$

18. $x\dfrac{dy}{dx} + 2y = x$

19. $\dfrac{dy}{dx} + y = e^{-x}$

20. $\dfrac{d^2y}{dx^2} - y = 0$

21. $\dfrac{d^2y}{dx^2} - 25y = 0$

22. $\dfrac{d^2y}{dx^2} = 100$

23. $\dfrac{d^2y}{dx^2} - 6\dfrac{dy}{dx} + 5y = 0$

24. $\dfrac{d^2y}{dx^2} - \dfrac{dy}{dx} = 0$

Numerical Applications Using Calculators

We shall give here a numerical method for solving first-order differential equations. Perhaps it should be pointed out that numerical solutions of differential equations are given in terms of the functional values for certain values of the independent variables over a fixed interval. For example, if the solution of the differential equation $y' = f(x, y)$ with initial condition x_0, y_0 is $y = F(x)$, a numerical solution to the given differential equation will be given by $y_i = F(x_i)$ for $i = 0, 1, \ldots, n$ for some n and

$$x_1 = x_0 + h, \; x_2 = x_1 + h, \; \ldots, \; x_n = x_{n-1} + h$$

with h being the constant distance between x_{i-1} and x_i.

The method we shall introduce here, which can be carried out by a pocket calculator, is called Euler's method. It is perhaps the simplest numerical method for solving a first-order differential equation $y' = f(x, y)$. Euler's method is based on the idea that the function $y = F(x)$ is assumed to be a smooth curve so that its functional values can be approximated by straight lines (see Figure 8.1), i.e.,

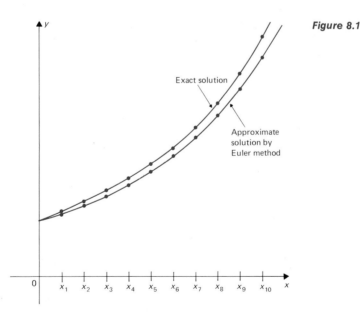

Figure 8.1

Exact solution

Approximate solution by Euler method

$$y_{i+1} \approx y_i + y_i' \cdot h \qquad \text{for} \quad i = 0, 1, \ldots$$

with initial condition x_0, y_0, $x_{i+1} = x_i + ih$, and $y_i' = f(x_i, y_i)$. We shall illustrate this method by the following example.

Let the first-order differential equation be given by $dy/dx = 2x + y$ with $x_0 = 0$, $y_0 = 1$. Euler's method can be carried out as shown in Table 8.1 ($h = 0.1$).

Table 8.1

i	$x_i = x_0 + ih$	$y_i = y_{i-1} + y'_{i-1} h$	$y'_i (x) = 2x_i + y_i$
1	0.1	1.1	1.3
2	0.2	1.23	1.63
3	0.3	1.393	1.993
4	0.4	1.5923	2.3923
5	0.5	1.83153	2.83153
6	0.6	2.114683	3.314683
7	0.7	2.4461513	3.8461513
8	0.8	2.83076643	4.43076643
9	0.9	3.273843073	5.073843073
10	1.0	3.7812273803	5.7812273803

The exact solution of $y' = 2x + y$ with initial condition $x_0 = 0$, $y_0 = 1$ can be found to be

$$y = 3e^x - 2x - 2$$

Table 8.2 is a comparison between the numerical results previously obtained by Euler's method and the exact solutions.

Table 8.2

x	y (Euler's method)	$y = 3e^x - 2x - 2$
0	1	1
0.1	1.1	1.115512754
0.2	1.23	1.264208274
0.3	1.393	1.44957642
0.4	1.5923	1.675474092
0.5	1.83153	1.946163812
0.6	2.114683	2.266356401
0.7	2.4461513	2.641258122
0.8	2.83076643	3.076622785
0.9	3.273843073	3.578809333
1.0	3.7812273803	4.154845485

Exercises

In Problems 1 through 8, apply Euler's method to solve the given initial-value problem. Compare your results with the exact solution.

1. $y' = xy^{1/3}$, $\quad x_0 = 1$, $y_0 = 1$ 2. $y' = x + y$, $\quad x_0 = 0$, $y_0 = 1$

3. $y' = -xy^2$, $\quad x_0 = 0$, $y_0 = 2$ 4. $y' = x^2 - y^2$, $\quad x_0 = -1$, $y_0 = 1$

5. $y' = -2xy$, $\quad x_0 = 0$, $y_0 = 1$ 6. $(2 + x) y' = y^3$, $\quad x_0 = -1$, $y_0 = 1$

7. $y' = ye^x$, $\quad x_0 = 0$, $y_0 = 1$ 8. $3xy' = 2y$, $\quad x_0 = 2$, $y_0 = 8$

Functions of Several Variables

9.1   Functions of Two Variables

In this chapter we shall study functions of two variables. The theory for the functions of three or more variables can be developed along similar lines. Functions of several (two or more) variables arise in many disciplines. For example, the area A of a rectangle is given by (see Figure 9.1)

$$A = xy$$

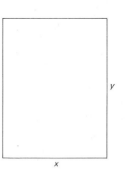

Figure 9.1

Thus the area A is a function of two independent variables. Similarly, the selling price P of an antibiotic depends upon its cost, its quality, and the selling price of the competitor. Thus P is a function of several variables.

▶ *Definition 9.1   A real-valued function of two variables is a function f that has a set of ordered pairs of real numbers as its domain, and a subset of real numbers as its range.*

In the case of functions of one variable, we write $f(x)$ to denote the value of the function at x. Similarly, if f is a function of two variables, we write $f(x, y)$ to denote the value of f at the ordered pair (x, y).

Example 9.1   Consider the relation

$$z = f(x, y) = \sqrt{1 - x^2 - y^2} \tag{9.1}$$

The relation (9.1) between x, y, and z determines a value of z or $f(x, y)$ corresponding to every ordered pair of numbers x, y such that $x^2 + y^2 \leq 1$.

For a real number b, $\sqrt{b}$ is real if and only if $b \geq 0$.

Denoting the pair of numbers (x, y) geometrically by a point on the plane, we see that the points (x, y) for which $x^2 + y^2 \leq 1$ lie on or within a circle with center at the origin and radius 1. Thus the domain of f is a region in the plane given by $\{(x, y) \mid x^2 + y^2 \leq 1\}$. The range of this function is the closed interval $[0, 1]$.

Example 9.2   Describe the domain and the range of the function f, defined by

$$z = f(x, y) = e^{x^2 + y^2}$$

SOLUTION   Domain of $f = \{(x, y) \mid x \in R, y \in R\}$, range of

$$f = \{z \mid z \geq 1\}.$$

Example 9.3   Describe the domain of the function f defined by

$$f(x, y) = \sqrt{(1 - x)(x - 2)} + \sqrt{(3 - y)(y - 5)}$$

Recall that $e^0 = 1$.

SOLUTION   Since $(1 - x)(x - 2)$ is nonnegative if $1 \leq x \leq 2$ and $(3 - y)(y - 5)$ is nonnegative if $3 \leq y \leq 5$, the domain of f is $\{(x, y) \mid 1 \leq x \leq 2; 3 \leq y \leq 5\}$. This domain is a rectangle (see Figure 9.2) in the xy-plane bounded by the lines

$$x = 1, \qquad x = 2, \qquad y = 3, \qquad y = 5$$

The y-axis and the z-axis are usually drawn in the plane of the paper (the y-axis horizontal and the z-axis vertical), and the x-axis is drawn to point toward the reader.

We have seen in Chapter 2 that the graph of a function of one variable can be represented pictorially in the plane. To graph a function of two variables, we need a three-dimensional coordinate system. We draw three mutually perpendicular coordinate axes meeting at a common point, called the *origin* (see Figure 9.3). The *coordinate lines* are the x-axis, y-axis, and z-axis. The positive directions of the axes are indicated by arrowheads. Each pair of coordinate axes determines a plane. Thus we have the xy-plane, yz-plane, and zx-plane. Any two

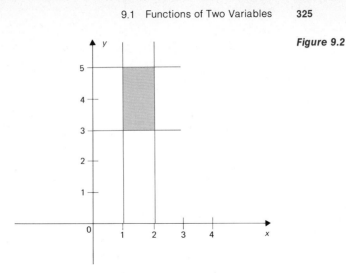

Figure 9.2

of these three planes (each unlimited in extent) divide space into four equal parts, and all three of them divide space into eight equal parts known as *octants*.

Let P be any point in space. The position of the point P is described by (a, b, c), where a, b, and c are the directed distances from the three coordinate planes, as shown in Figure 9.4. For example, the point $(3, -4, 7)$ is located 3 units in front of the yz-plane, 4 units to the left of the zx-plane, and 7 units above the xy-plane. Thus given an ordered triple of real numbers (a, b, c), we can find a point in space which represents this ordered triple. Conversely, given a point P in space, we construct the three planes through P which are parallel to the three coordinate planes. The plane parallel to the yz-plane intersects the x-axis at the *x-coordinate*; the plane parallel to the xz-plane intersects the y-axis at the *y-coordinate*; and the plane parallel to the xy-plane inter-

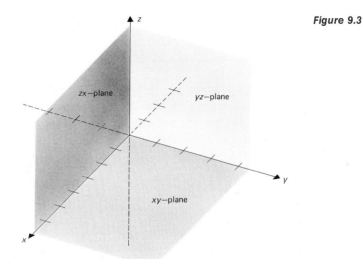

Figure 9.3

Figure 9.4

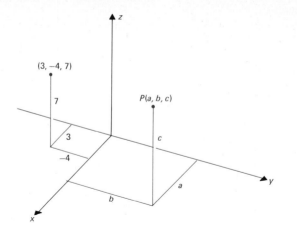

sects the z-axis at the z-*coordinate*. Thus there is a unique correspondence between points in three-dimensional space and ordered triples of real numbers.

The following information will be helpful.

We observe that all points on the x-axis have their y- and z-coordinates both zero; that is, these points have the form $(x, 0, 0)$. Similarly, the points on the y-axis and z-axis have the forms $(0, y, 0)$ and $(0, 0, z)$, respectively. We also note that all points in a plane perpendicular to the z-axis have the same value of their z-coordinates. The equation $z = c$ (where c is a real constant) is satisfied by every point (x, y, c) lying in a plane perpendicular to the z-axis and directed distance of c units from the xy-plane. Similarly,

$$y = c, \qquad x = c$$

represent planes perpendicular to the y-axis and x-axis, respectively. We now prove the following theorem.

Point	Location of the point
$(0, 0, 0)$	origin
$(a, 0, 0)$	x-axis
$(0, b, 0)$	y-axis
$(0, 0, c)$	z-axis
$(a, b, 0)$	xy-plane
$(0, b, c)$	yz-plane
$(a, 0, c)$	zx-plane

► **Theorem 9.1** *The distance between the points $P(x_1, y_1, z_1)$ and $Q(x_2, y_2, z_2)$ is*

$$d(P, Q) = \sqrt{(x_2 - x_1)^2 + (y_2 - y_1)^2 + (z_2 - z_1)^2} \qquad (9.2)$$

Note the similarity between (9.2) and the distance formula in two dimensions (Chapter 2).

PROOF   See Figure 9.5. Draw PR and QS perpendicular to the xy-plane. The coordinates of R and S are $(x_1, y_1, 0)$ and $(x_2, y_2, 0)$. Also, draw PT perpendicular to QS. Then the coordinates of T are (x_2, y_2, z_1). From the figure we see that

$$d(R, S) = \sqrt{(x_1 - x_2)^2 + (y_1 - y_2)^2}$$

In the rectangle $RPTS$, $d(P, T) = d(R, S)$. Furthermore

$$d(T, Q) = |z_2 - z_1| = \sqrt{(z_1 - z_2)^2}$$

By the Pythagorean theorem,

$$d(P, Q) = \sqrt{[d(P, T)]^2 + [d(T, Q)]^2}$$

Figure 9.5

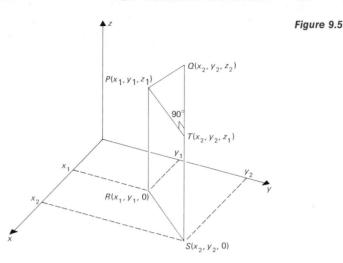

$$= \sqrt{[d(R, S)]^2 + [d(T, Q)]^2}$$
$$= \sqrt{(x_1 - x_2)^2 + (y_1 - y_2)^2 + (z_1 - z_2)^2}$$

Example 9.4   Find the distance between the points

$$P(2, 3, -1) \quad \text{and} \quad Q(-1, 0, 2)$$

SOLUTION   By direct substitution in (9.2), we obtain

$$d(P, Q) = \sqrt{(2 - (-1))^2 + (3 - 0)^2 + (-1 - 2)^2}$$
$$= \sqrt{3^2 + 3^2 + 3^2} = \sqrt{27} = 3\sqrt{3}$$

In Problems 1 through 4, find the value of the function (if defined) at the   **Exercises 9.1**
given points.

1. $f(x, y) = x^2 - xy + y$ at $(1, 0)$, $(2, -1)$ and $(-2, 3)$

2. $f(x, y) = ye^x + x^2 - 2y$ at $(0, 2)$, $(1, 0)$ and $(1, 1)$

3. $f(x, y) = 3x^2 + 4xy + y^3$ at $(0, 0)$, $(1, -1)$ and $(-1, 1)$

4. $f(x, y) = \ln xy + x \ln y$ at $(1, 1)$, $(1, e)$ and $(e, 1)$

In Problems 5 through 9, find the domain of f.

5. $f(x, y) = xy^2 + x^2 y$ 6. $f(x, y) = \sqrt{1 - \dfrac{x^2}{4} - \dfrac{y^2}{9}}$

7. $f(x, y) = \sqrt{(1 - x)(x - 2)} + \sqrt{(3 - y)(y - 4)}$

8. $f(x, y) = \dfrac{1}{\ln x + \ln y}$ 9. $f(x, y) = \dfrac{x^3 + y^3}{x^2 + y^2}$

10. Plot the following points.

   (a) $(1, 0, 0)$ (b) $(0, 2, 0)$ (c) $(0, 0, 3)$

(d) (1, 2, 0)   (e) (1, 2, 3)   (f) (2, 1, 2)
(g) (−1, 2, 0)   (h) (−1, 0, −1)   (i) (2, 2, 2)
(j) (−1, −2, −3)

11. Describe the set of all points (x, y, z) that satisfy the given conditions.

(a) $x = 0$ (b) $y = 0$ (c) $z = 3$
(d) $x = 0$ and $y = 0$ (e) $x = 1$ and $y = 0$ (f) $z = -2$
(g) $x = y = z$ (h) $x = 1, y = -1$, and $z = 2$

12. Calculate the distance between the points.

(a) $(0, 0, 0)$, (a, b, c) (b) $(1, 0, -1)$, $(0, 2, -3)$
(c) $(1, 2, 3)$, $(1, -2, 0)$ (d) $(3, 0, 1)$, $(-1, 0, -2)$
(e) $(0, 4, 5)$, $(0, -8, 0)$ (f) $(1, 2, 3)$, $(-4, -5, -6)$

9.2 Graphs

We recall from Chapter 2 that the graph of a function f of one variable is the set of all points (x, y), such that $y = f(x)$.

▶ *Definition 9.2*   *The graph of a function f of two variables is the set of all points (x, y, z), such that $z = f(x, y)$.*

For a function of one variable, the graph of $y = f(x)$ is generally a curve in the plane. However, in general, the graph of a function f given by $z = f(x, y)$ is a surface in three dimensions. Obtaining the shape, or even a rough sketch, of the graph of a function of two variables is a difficult task. Attempting to imitate the case of a function of one variable by plotting a reasonable number of points (x, y, z) in three dimensions and connecting them by a smooth surface will not succeed. Even if we are able to plot many points in space (which itself is quite difficult), we may not be able to recognize the shape of the surface. In this section we shall discuss some of the simplest surfaces.

The proof of the following theorem is beyond the scope of this book and is therefore omitted.

▶ *Theorem 9.2*   *The graph of the linear equation*

$$ax + by + cz = d \quad \text{(a, b, c not all equal to zero)} \qquad (9.3)$$

is a plane in three-dimensional space.

Example 9.5 Discuss the plane

$$2x + 3y + 4z = 12 \qquad (9.4)$$

SOLUTION   It is known that three noncollinear points in space determine a unique plane. We can find points on (9.4) by choosing two coordinates and solving for the third. If we let $x = 0$ and $y = 0$, we obtain $z = 3$. Consequently, the point $(0, 0, 3)$ is on the plane given by (9.5). Similarly, $(0, 4, 0)$ and $(6, 0, 0)$ are also points on the plane. In fact, these are the points where the plane intersects the axes. Figure 9.6 is a sketch of the plane.

Figure 9.6

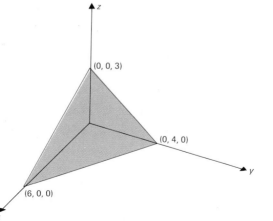

Suppose that we wish to graph the function given by

$$z = f(x) \tag{9.5}$$

in three-dimensional space. We note that one of the variables, namely y, is missing; i.e., Equation (9.5) is independent of y. Consequently, the point $(x_0, y_0, f(x_0))$ is on the graph if and only if the point $(x_0, 0, f(x_0))$ is on the graph. If follows that the graph of the surface given by (9.5) can be obtained by sliding the graph of $z = f(x)$ on the zx-plane parallel to the y-axis. Similar results hold for equations with x or z missing.

In other words, we draw the curve G given by $z = f(x)$ on the zx-plane, and the graph of $z = f(x)$ in three-dimensional space is generated by moving a horizontal line along the curve G.

Example 9.6   Sketch the graph in three-dimensional space of

$$2x + y = 4 \tag{9.6}$$

SOLUTION   Note that the variable z is missing in Equation (9.6). From the preceding discussion, we draw the graph of (9.6) in the xy-plane, which we know is a straight line. We now slide this line parallel to the z-axis, and obtain the required graph, which is a plane, as shown in Figure 9.7. In other words, we generate the graph by moving a vertical line along the straight line given by (9.6) in the xy-plane.

Figure 9.7

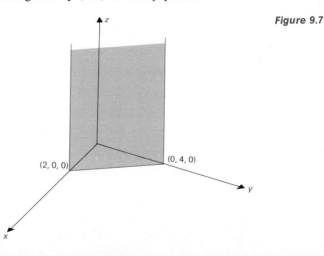

Example 9.7 Sketch the graph of

$$z = x^2 \qquad\qquad (9.7)$$

in three-dimensional space.

SOLUTION We know that the graph of $z = x^2$ in the zx-plane is a parabola. Sliding this parabola parallel to the y-axis gives the graph of (9.7) in three-dimensional space as sketched in Figure 9.8.

Figure 9.8

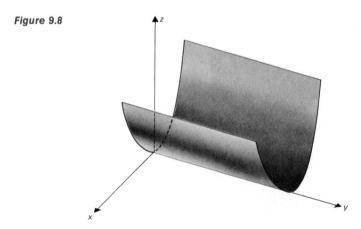

As we pointed out earlier, graphing a surface given by $z = f(x, y)$ is very difficult. However, if we hold y constant, say at $y = b$, then $z = f(x, b)$ is a function of one variable—namely, x. The graph of $z = f(x, b)$ lies on the plane $y = b$; in fact, it is the intersection of the graph of $z = f(x, y)$ with the plane $y = b$. Such an intersection is called a *cross section* of $z = f(x, y)$. Drawing several cross sections $z = f(x, b_1)$, $z = f(x, b_2)$, . . . yields a fairly good idea of the shape of the surface $z = f(x, y)$. We can find cross sections by holding x constant also.

The cross sections of a given surface resulting from the intersection of the surface with the coordinate planes are called the *traces* of the surface. For example, the trace of the surface $z = x^2 - y$ in the xy-plane is $x^2 - y = 0$. The trace in the xy-plane is obtained by letting $z = 0$. Similarly, the traces in the yz- and zx-planes are obtained by letting $x = 0$ and $y = 0$, respectively, in the equation $z = f(x, y)$.

Example 9.8 Sketch the graph of

$$x^2 + y^2 + z^2 = 9 \qquad\qquad (9.8)$$

in three-dimensional space.

SOLUTION   The trace in the *xy*-plane is given by

$$x^2 + y^2 + 0 = 9 \quad \text{or} \quad x^2 + y^2 = 9$$

which is the equation of a circle with center at $x = 0$ and $y = 0$ and radius 3. Similarly, the traces on the *yz*-plane and *xz*-plane are, respectively,

$$y^2 + z^2 = 9 \quad \text{and} \quad x^2 + z^2 = 9$$

We also note that if we take a plane parallel to any of the coordinate planes, say $z = c$ $(-3 < c < 3)$, then the cross section of this plane with (9.8) is

$$x^2 + y^2 = 9 - c^2$$

which is a circle.

These traces and cross sections suggest that the graph of (9.8) is a sphere with center at $(0, 0, 0)$ and radius 3 (see Figure 9.9).

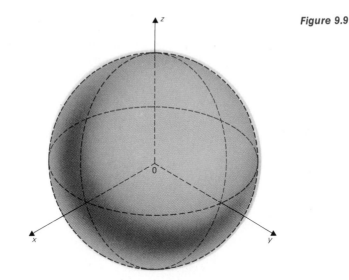

Figure 9.9

In Problems 1 through 12, sketch the graphs of the given planes.

Exercises 9.2

1. $x = 2$ 2. $y = -3$

3. $z = \frac{1}{2}$ 4. $x + y = 1$

5. $x + z = 3$ 6. $x - y = -4$

7. $2y + z = 4$ 8. $z - 3x = 6$

9. $x + y + z = 1$ 10. $x - 2y + 3z = 6$

11. $2x + y + z = 4$ 	12. $-x - y + z = 1$

In Problems 13 through 30, sketch the following graphs in three dimensions.

13. $x^2 - y = 0$ 	14. $x^2 = 4y - 2$

15. $x^2 + y^2 = 4$ 	16. $y^2 + z^2 = 1$

17. $x^2 + z^2 = 2y$ 	18. $y = x^2$

19. $x = z^2$ 	20. $x = (y - 2)^2$

21. $x^2 + y^2 + z^2 = 1$ 	22. $x^2 + y^2 + z^2 = 16$

23. $(x - 1)^2 + y^2 + z^2 = 9$ 	24. $x^2 + (y + 1)^2 + (z - 2)^2 = 4$

25. $(x - 1)^2 + (y - 2)^2 + (z - 3)^2 = 16$

26. $(x + 3)^2 + (y - 1)^2 + (z + 2)^2 = 16$

27. $z = \sqrt{4 - x^2 - y^2}$ 	28. $-x = \sqrt{9 - y^2 - z^2}$

29. $x = \sqrt{y - z^2}$ 	30. $-y = \sqrt{4z - x^2}$

9.3 Partial Differentiation

Before we discuss the notion of partial differentiation, we shall mention the ideas of limit and continuity of a function of two variables very briefly. Intuitively, we say

$$\lim_{(x, y) \to (a, b)} f(x, y) = L$$

provided that $f(x, y)$ is close to the number L whenever (x, y) is close to (a, b).

To state the idea of limit formally, we first need

▶ *Definition 9.3 Let δ be any positive real number. For a given point (a, b) in the plane, let*

$$N = \{(x, y) \mid a - \delta < x < a + \delta; b - \delta < y < b + \delta\}$$

Then the set N is called the δ-neighborhood of the point (a, b).

Note (see Figure 9.10) that the set N determines a square in the plane bounded by the lines

$$x = a - \delta, \qquad x = a + \delta, \qquad y = b - \delta, \qquad y = b + \delta$$

The center of the square is the point (a, b). The sides of the square are not included in N. To emphasize this point, some authors call it an *open neighborhood.*

We now give the formal definition of the limit.

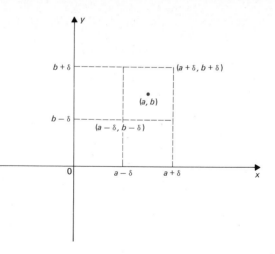

Figure 9.10
δ-neighborhood of (a, b).

Definition 9.4   If for every $\epsilon > 0$, there exists a δ-neighborhood N of (a, b), such that

◀ The quantity δ depends on ϵ.

$$|f(x, y) - L| < \epsilon$$

for every $(x, y) \in N$, then

$$\lim_{(x, y) \to (a, b)} f(x, y) = L$$

Definition 9.4 can be used to prove the usual results about limits, similar to the ones for a function of one variable. We state the following theorem without proof.

Theorem 9.3   Let   ◀

$$\lim_{(x, y) \to (a, b)} f(x, y) = L \quad and \quad \lim_{(x, y) \to (a, b)} g(x, y) = M$$

Then

(i) $\displaystyle \lim_{(x, y) \to (a, b)} [f(x, y) \pm g(x, y)] = \lim_{(x, y) \to (a, b)} f(x, y) \pm \lim_{(x, y) \to (a, b)} g(x, y)$

$$= L \pm M$$

The limit of the sum (or difference) is the sum (or difference) of the limits.

(ii) $\displaystyle \lim_{(x, y) \to (a, b)} [f(x, y) \cdot g(x, y)] = L \cdot M$

(iii) $\displaystyle \lim_{(x, y) \to (a, b)} \frac{f(x, y)}{g(x, y)} = \frac{L}{M} \qquad$ provided $M \neq 0$

The reader should write (ii) and (iii) in words.

Simple limits can be calculated using Theorem 9.3, as in the following example.

Example 9.9 Calculate

$$\lim_{(x,y)\to(1,2)} \frac{x^3+3xy^2}{x^2-y}$$

By Theorem 9.3 (iii)   **SOLUTION**

$$\lim_{(x,y)\to(1,2)} \frac{x^3+3xy^2}{x^2-y} = \frac{\lim_{(x,y)\to(1,2)}(x^3+3xy^2)}{\lim_{(x,y)\to(1,2)}(x^2-y)}$$

By Theorem 9.3 (i)

$$= \frac{\lim_{(x,y)\to(1,2)} x^3 + \lim_{(x,y)\to(1,2)}(3xy^2)}{\lim_{(x,y)\to(1,2)} x^2 - \lim_{(x,y)\to(1,2)} y}$$

$$= \frac{1^3+3(1)(2)^2}{1^2-2} = \frac{1+12}{-1} = -13$$

The definition of continuity is similar to the definition given for one variable.

► *The function f of Example 9.9 is continuous at (1, 2), since*

$$f(1,2) = \frac{1^3+3(1)(2)^2}{1^2-2} = -13$$

Definition 9.5 *Let (a, b) be a point in the domain of f. Then the function f is said to be* continuous *at (a, b) if*

$$\lim_{(x,y)\to(a,b)} f(x,y) = f(a,b)$$

A function is said to be continuous on a set S if it is continuous at every point of S.

Remark *When taking a partial derivative with respect to x (or y), we are essentially considering the behavior of the cross sections of the function on the zx- (or yz-) plane.*

Let $z = f(x,y)$ be a function of two variables. If y is held constant or fixed, then the function $z = f(x,y)$ becomes a function of x alone, and its derivative may be found by the rules discussed in Chapter 4. This derivative is called the *partial derivative of z with respect to x*, and is denoted by

$$\frac{\partial z}{\partial x}, \quad \frac{\partial f}{\partial x}, \quad \text{or} \quad f_x$$

The partial derivative with respect to y has a similar meaning. Formally, we have the following definition.

Partial derivatives ► **Definition 9.6** *Let $z = f(x,y)$ be a function of two variables. The partial derivatives of z with respect to x and with respect to y are defined by*

$$\frac{\partial z}{\partial x} = \frac{\partial f}{\partial x} = f_x = \lim_{h\to 0} \frac{f(x+h,y)-f(x,y)}{h} \tag{9.9}$$

$$\frac{\partial z}{\partial y} = \frac{\partial f}{\partial y} = f_y = \lim_{k\to 0} \frac{f(x,y+k)-f(x,y)}{k} \tag{9.10}$$

provided that the limits in (9.9) and (9.10) exist.

Notations such as $(\partial z/\partial x)_{(a,b)}$ and $f_y(a, b)$ are used to indicate the value of the partial derivative at the point (a, b).

Example 9.10 Calculate $\partial z/\partial x$ and $\partial z/\partial y$, where

$$z = x^2 y + xy^{-3}$$

Also find the value of each partial derivative at $(1, 2)$.

SOLUTION Treating y as a constant, we have

$$\frac{\partial z}{\partial x} = 2xy + y^{-3}$$

Thus

$$\left(\frac{\partial z}{\partial x}\right)_{(1, 2)} = 2(1)(2) + 2^{-3} = 4 + \tfrac{1}{8} = \tfrac{33}{8}$$

Similarly, treating x as a constant, we obtain

$$\frac{\partial z}{\partial y} = x^2 - 3xy^{-4}$$

and

$$\left(\frac{\partial z}{\partial y}\right)_{(1, 2)} = (1)^2 - 3(1)(2)^{-4} = 1 - \tfrac{3}{16} = \tfrac{13}{16}$$

The partial derivatives themselves are functions of two variables. We can, therefore, calculate the partial derivatives of $\partial z/\partial x$ and $\partial z/\partial y$, just as we calculated the partial derivatives of z. Thus we have four *second partial derivatives* (with different notations).

(i) $\dfrac{\partial}{\partial x}\left(\dfrac{\partial z}{\partial x}\right) = \dfrac{\partial^2 z}{\partial x^2} = \dfrac{\partial^2 f}{\partial x^2} = f_{xx}$

(ii) $\dfrac{\partial}{\partial y}\left(\dfrac{\partial z}{\partial x}\right) = \dfrac{\partial^2 z}{\partial y\,\partial x} = \dfrac{\partial^2 f}{\partial y\,\partial x} = f_{xy}$

(iii) $\dfrac{\partial}{\partial x}\left(\dfrac{\partial z}{\partial y}\right) = \dfrac{\partial^2 z}{\partial x\,\partial y} = \dfrac{\partial^2 f}{\partial x\,\partial y} = f_{yx}$

(iv) $\dfrac{\partial}{\partial y}\left(\dfrac{\partial z}{\partial y}\right) = \dfrac{\partial^2 z}{\partial y^2} = \dfrac{\partial^2 f}{\partial y^2} = f_{yy}$

The derivatives f_{xx} and f_{yy} are called the second partial derivatives with respect to x and y, respectively, while the derivatives f_{xy} and f_{yx} are called the *mixed partial derivatives*.

The mixed partial derivatives f_{xy} and f_{yx} are distinguished by the order in which f is successively differentiated with respect to x and y. The following theorem (we omit its proof) shows that they are equal in most cases of interest.

Definition 9.6 can be extended to functions of more than two variables. For example, if $w = f(x, y, u, v)$, then

$$\frac{\partial w}{\partial u}$$

$$= \lim_{h \to 0} \frac{f(x, y, u + h, v) - f(x, y, u, v)}{h}$$

provided that the limit exists.

Second partial derivatives

In the last equations of (ii) and (iii), note that the order of the symbols is reversed:

$$\frac{\partial^2 f}{\partial y\,\partial x} = f_{xy},$$

since $\dfrac{\partial^2 f}{\partial y\,\partial x}$ *stands for* $\dfrac{\partial}{\partial y}\left(\dfrac{\partial f}{\partial x}\right)$

and f_{xy} stands for $(f_x)_y$. Both symbols indicate that the partial derivative is first calculated with respect to x and then with respect to y.

Mixed partial derivatives

▶ **Theorem 9.4** *Let f, f_x, f_{xy}, and f_{yx} all be continuous at (a, b). Then*

$$f_{xy}(a, b) = f_{yx}(a, b)$$

Example 9.11 Calculate all the second partial derivative of

$$f(x, y) = x^3 y^4 + x^2 e^y$$

SOLUTION

$$f_x = 3x^2 y^4 + 2xe^y, \qquad f_{xx} = 6xy^4 + 2e^y, \qquad f_{xy} = 12x^2 y^3 + 2xe^y$$

$$f_y = 4x^3 y^3 + x^2 e^y, \qquad f_{yy} = 12x^3 y^2 + x^2 e^y, \qquad f_{yx} = 12x^2 y^3 + 2xe^y$$

Note that $f_{xy} = f_{yx}$, as asserted in Theorem 9.4.

Exercises 9.3

1. Evaluate the following limits and comment on the continuity of the given functions at the indicated points.

(a) $\displaystyle\lim_{(x, y) \to (2, 1)} f(x, y)$, where $f(x, y) = \dfrac{x + y}{x - y}$ if $(x \neq y)$

$$= 0 \qquad \text{if} \quad (x = y)$$

(b) $\displaystyle\lim_{(x, y) \to (1, 1)} f(x, y)$, where $f(x, y) = \dfrac{x^3 y^2 + x^2 + y}{x^2 + y^2}$ if $(x, y) \neq (0, 0)$

$$= 0 \qquad\qquad \text{if } (x, y) = (0, 0)$$

(c) $\displaystyle\lim_{(x, y) \to (2, 2)} f(x, y)$, where $f(x, y) = \dfrac{x^3 - y^3}{x^2 - y^2}$ if $(x \neq y)$

$$= 0 \qquad \text{if} \quad (x = y)$$

(d) $\displaystyle\lim_{(x, y) \to (1, 1)} f(x, y)$, where $f(x, y) = \dfrac{x - y}{\sqrt{x} - \sqrt{y}}$ if $(x \neq y)$

$$= 2 \qquad \text{if} \quad (x = y)$$

2. Calculate f_x and f_y.

(a) $f(x, y) = x^2 y - xy^2$ (b) $f(x, y) = (xy)^3$

(c) $f(x, y) = \dfrac{x}{y}$ (d) $f(x, y) = \dfrac{x + y}{x - y}$

3. Calculate f_x and f_y.

(a) $f(x, y) = xe^y$ (b) $f(x, y) = e^{x^2 + y^2}$

(c) $f(x, y) = e^{x + y} \ln (x^2 + y^2)$

(d) $f(x, y) = x^3 + y^2 \ln x + xe^{xy^2 + x^3 y}$

4. In Problem 2, calculate all second partial derivatives at $(1, 2)$.

5.   Let $z = f(x, y, u, v) = (x^2 + y^3) \cdot e^{u+v}$. Calculate

   (a) f_x, f_y, f_u, f_v (b) f_{xy}, f_{xu}, f_{uv}

   (c) $f_{xyu}, f_{uyx}, f_{uxy}$

*6.   State a theorem similar to Theorem 9.4 for functions of three variables.

9.4   Some Applications; the Total Differential

One application of partial derivatives is obvious from its counterpart in functions of a single variable. The partial derivative f_x represents the rate of change of f with respect to x, when the variable y is kept fixed. A similar comment applies to f_y.

Example 9.12   Ecology   Suppose a nuclear power plant and a chemical company are situated on a large lake. Let the pollution index P of the lake be a function of x and y, where x measures the quantity of heat entering the lake due to the power plant and y measures the amount of waste chemical pouring into the lake. We can write $P = f(x, y)$. Then $(\partial P/\partial x)_{(a, b)}$ simply means that if y is held constant at value b and x alone is allowed to vary, then the pollution index P is changing $(\partial P/\partial x)_{(a, b)}$ units for each unit change in x at the instant x takes the value a. Similar interpretation is given for $(\partial P/\partial y)_{(a, b)}$.

Example 9.13   Economics   Suppose the productivity P of a company is a function of the size of the labor force x and the amount of capital invested y: $P = f(x, y)$. Then $\partial P/\partial y$ is called the *marginal productivity of capital*, which means that for $x = a$, $y = b$, the productivity of the company increases approximately by $(\partial P/\partial y)_{(a, b)}$ (if it is positive) if the capital is increased by an additional dollar. Similarly, $\partial P/\partial x$ is called the *marginal productivity of labor*, which means that for $x = a$, $y = b$, the productivity of the company increases approximately by $(\partial P/\partial x)_{(a, b)}$ if one more person is added to the labor force.

Example 9.14   Geometry   Let $z = f(x, y)$. We mentioned earlier that geometrically this equation represents a surface. If y is held constant, say $y = b$, then we obtain a cross section of this surface, determined by the surface $z = f(x, y)$ and the plane $y = b$. Thus we have a curve $z = f(x, b)$ in the plane $y = b$. We note that z is now a function of the single variable x. The partial derivative $\partial z/\partial x$ is, therefore, the slope of the tangent to the curve of intersection of the surface and the plane $y = b$ (see Figure 9.11).

In Chapter 5, we defined the differential dy of the function $y = f(x)$ by

$$dy = f'(x) \, dx$$

Figure 9.11

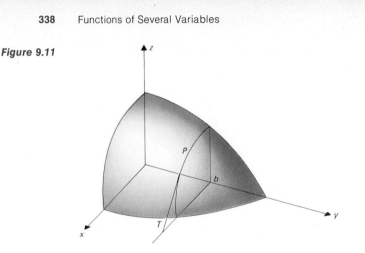

We shall define a similar term for a function of two variables.

Let $z = f(x, y)$ be a function of two variables x and y. Let (x, y) and $(x + \Delta x, y + \Delta y)$ be any two points, so that Δx and Δy are the changes in the variables x and y, respectively. Let Δz denote the consequent change in z. We have

$$z + \Delta z = f(x + \Delta x, y + \Delta y)$$

or

We have added and subtracted $f(x, y + \Delta y)$.

$$\Delta z = f(x + \Delta x, y + \Delta y) - z = f(x + \Delta x, y + \Delta y) - f(x, y) \tag{9.11}$$

$$= [f(x + \Delta x, y + \Delta y) - f(x, y + \Delta y)] + [f(x, y + \Delta y) - f(x, y)]$$

The change Δz has been expressed as the sum of two differences. Considering the first pair of functions, we note that f is evaluated at two points $(x + \Delta x, y + \Delta y)$ and $(x, y + \Delta y)$ which have the same y-value. Momentarily, we think of f as a function of x alone. Thus by the mean value theorem, we have

Of course, we are assuming that the hypotheses of the mean value theorem are satisfied.

$$f(x + \Delta x, y + \Delta y) - f(x, y + \Delta y) = f_x(c_1, y + \Delta y) \cdot \Delta x$$

where c_1 is between x and $x + \Delta x$. Similarly,

$$f(x, y + \Delta y) - f(x, y) = f_y(x, c_2) \cdot \Delta y$$

where c_2 is between y and $y + \Delta y$. Hence

$$\Delta z = f_x(c_1, y + \Delta y) \cdot \Delta x + f_y(x, c_2) \cdot \Delta y \tag{9.12}$$

Let us assume that f_x and f_y are continuous. Then when Δx and Δy are small, c_1 and c_2 are close to x and y, respectively. With these assumptions, we write

$$\Delta z \approx f_x(x, y) \, \Delta x + f_y(x, y) \, \Delta y \tag{9.13}$$

Total differential ▶ **Definition 9.7** *Let $z = f(x, y)$ be a function of two independent variables x and y, and let f, f_x, and f_y be continuous in some domain. We define*

$$dx = \Delta x, \qquad dy = \Delta y$$

$$dz = f_x \,\Delta x + f_y \,\Delta y = f_x \,dx + f_y \,dy$$

We call dz the total differential.

Thus, as in the case of a single variable, dz serves as an approximation to Δz.

Remark *The closer Δx and Δy are to zero, the closer the approximation dz is to the exact value of Δz.*

Example 9.15   Find the percentage error in the area of a rectangle, when an error of 1% is made in measuring the length and the width of the rectangle.

SOLUTION   Let x, y, and A denote the length, width, and area of the rectangle, respectively. Since $A = xy$, the partial derivatives are

$$\frac{\partial A}{\partial x} = y \quad \text{and} \quad \frac{\partial A}{\partial y} = x$$

Since we are given

$$dx = \frac{x}{100} \qquad (1\% \text{ error in length})$$

$$dy = \frac{y}{100} \qquad (1\% \text{ error in width})$$

we have

$$dA = \frac{\partial A}{\partial x}\,dx + \frac{\partial A}{\partial y}\,dy = y\,\frac{x}{100} + x\,\frac{y}{100} = \frac{2xy}{100} = \frac{2A}{100}$$

Therefore there is a 2% error in A.

1. In Example 9.12, let the pollution index P be given by

$$P = 2x^2 + 4xy + xy$$

Find $(\partial P/\partial x)_{(3,5)}$ and $(\partial P/\partial y)_{(2,3)}$. Interpret your results.

2. Repeat Problem 1 for

$$P = 3e^{x+y+xy}$$

3. In Example 9.13, suppose the productivity P of the ABC company is given by

$$P = \frac{e^{xy}}{x^2 + y^2 + y}$$

(a)   Find $\dfrac{\partial P}{\partial x}, \dfrac{\partial P}{\partial y}$.

(b)   Find $\left(\dfrac{\partial P}{\partial x}\right)_{(200, 5)}$ and $\left(\dfrac{\partial P}{\partial y}\right)_{(5, 100)}$

(c)   Interpret (b).

4.   Repeat Problem 3 for

$$P = (x^2 + e^y)^2$$

5.   A toothpaste company finds that sales per day are a function of x, the number of times that its commercial is shown on television, and y, the number of people (in millions) that see the commercial. This function is

$$f(x, y) = 50xy + 1000$$

(a)   Find f_x, f_y.
(b)   Interpret f_x when $y = 5$.
(c)   Interpret f_y when $x = 2$.

6.   Repeat Problem 5 for

$$f(x, y) = x^2 + y^2 + 20xy + 500$$

7.   Find the total differential dz if

(a) $z = xy^2 + xe^y$ (b) $z = x + y\ln x$

8.   Find the percentage error in the area of an ellipse when an error of 1% is made in the major and minor axes. [*Hint:* If $2a$ and $2b$ are the lengths of the major and minor axes, then area $= \pi ab$.]

9.   Find the percentage error in the area of an ellipse when an error of 1% is made in measuring its major axis, and an error of 2% is made in measuring its minor axis.

10.   In measuring a right circular cone, errors of 2% and 1% are made in the height and radius, respectively. Find the percentage error in the volume.

11.   In an experiment with a simple pendulum, the acceleration due to gravity g is calculated from the formula

$$T = 2\pi \sqrt{l/g}$$

where T is the period and l is the length of the pendulum. Calculate approximately the percentage error in g if l and T are measured within 1% of accuracy.

12.   Find

$$3.001 \times 8.997$$

approximately. [*Hint:* Let $f(x, y) = xy$. Then $f(3, 9) = 27$, $dx = 0.001$, and $dy = -0.003$.]

13.   Suppose that the cost C of producing two items x and y is given by $C = f(x, y)$. Then $\partial C/\partial x$ and $\partial C/\partial y$ are called the *partial marginal costs* with respect to x and y, respectively. Find the partial marginal costs with respect to x and with respect to y for the following cost functions.

(a) $C = 4x^2 + xy - 13y^2$ (b) $C = y \ln x + 3x^2 y$

(c) $C = xe^y + y$ (d) $C = x \ln (16x + e^y)$

9.5   Maxima and Minima

The work in Chapter 5 regarding the maximum and minimum values of a function of one variable can be generalized to functions of two variables. We begin with the following definition.

Definition 9.8   Let f be a function of two variables with domain D. We say that f has a relative maximum value *at a point P(a, b) if there exists a δ-neighborhood N ⊂ D of P, such that for every (x, y) ∈ N,*

$$f(a, b) \geq f(x, y)$$

◀ *In the case of two variables, δ-neighborhoods are used instead of open intervals.*

Relative minimum value is defined analogously. We call f(a, b) an extreme value *if it is a maximum or a minimum value.*

We recall that if a function of one variable has a relative extremum at a point and is differentiable there, then its derivative at that point is zero. A similar result for a function of two variables is given as follows.

Theorem 9.5   Let f(a, b) be a relative extreme value of a function f. If both f_x and f_y exist at (a, b), then   ◀

$$f_x(a, b) = 0 \quad \text{and} \quad f_y(a, b) = 0$$

PROOF   If $f(a, b)$ is a relative extreme value of the function f of two variables x and y, then it must also be a relative extreme value of the function $f(x, b)$ of one variable x for $x = a$. Consequently (from Chapter 5), its derivative, $f_x(a, b)$, for $x = a$ must necessarily be zero. Similarly, we can show that

$$f_y(a, b) = 0$$

We remark that the aforementioned conditions are necessary, but not sufficient; that is, the converse of Theorem 9.5 is not true. (See Example 9.17.)

The following theorem states a second-derivative test that is used to decide whether or not the function has a relative maximum or a relative minimum.

Theorem 9.6   Let a function f of two variables have partial derivatives at all points of a δ-neighborhood of a point (a, b). Suppose that   ◀

$$f_x(a, b) = 0 \quad \text{and} \quad f_y(a, b) = 0$$

and let

$$A = f_{xx}(a, b), \quad B = f_{xy}(a, b), \quad \text{and} \quad C = f_{yy}(a, b)$$

Then we have the following observations.

(i) *f(a, b) is a relative maximum value if $AC - B^2 > 0$ and $A < 0$.*
(ii) *f(a, b) is a relative minimum value if $AC - B^2 > 0$ and $A > 0$.*
(iii) *f has a* saddle point *at (a, b) if $AC - B^2 < 0$.*
(iv) *The case $AC - B^2 = 0$ is doubtful and needs further considerations.*

Figure 9.12
Saddle point.

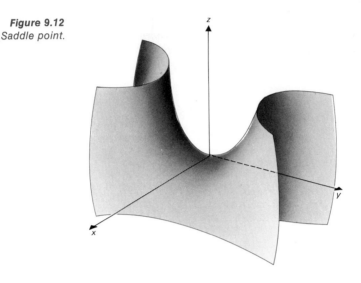

The proof of this theorem is beyond the scope of this book and is therefore omitted.

Working Rules

Step 1 Find f_x and f_y.

Step 2 Solve $f_x = 0$ and $f_y = 0$ simultaneously.

Step 3 Find $A = f_{xx}$, $B = f_{xy}$, $C = f_{yy}$, and evaluate at the solution set of Step 2.

Step 4 Check $AC - B^2$ for its sign. Check A for its sign. Find the part of Theorem 9.6 that is applicable.

Example 9.16 Find the extreme values of

$$f(x, y) = x^2 + y^2 - 4x - 6y + 1$$

SOLUTION We have

$$f_x = 2x - 4 \quad \text{and} \quad f_y = 2y - 6$$

Setting $f_x = 0$ and $f_y = 0$, we get $x = 2$ and $y = 3$. Thus $(2, 3)$ is the only possible location for a relative extremum. Applying the second-derivative test to the given function at $(2, 3)$, we have

$$A = f_{xx} = 2, \quad B = f_{xy} = 0, \quad \text{and} \quad C = f_{yy} = 2$$

Therefore

$$AC - B^2 = 2 \cdot 2 - 0^2 = 4 > 0$$

Since $A = 2 > 0$, then part (ii) of Theorem 9.6 applies and hence the function f has a relative minimum value at $(2, 3)$.

Example 9.17 Find the extreme values of

$$f(x, y) = xy(3 - x - y)$$

SOLUTION We have

$$f_x = 3y - 2xy - y^2 = y(3 - 2x - y)$$
$$f_y = 3x - x^2 - 2xy = x(3 - 2y - x)$$
$$A = f_{xx} = -2y$$
$$B = f_{xy} = 3 - 2x - 2y$$
$$C = f_{yy} = -2x$$

We now solve the equations

$$\begin{cases} f_x = 0 \\ f_y = 0 \end{cases}$$

Or

$$\begin{cases} y(3 - 2x - y) = 0 \\ x(3 - 2y - x) = 0 \end{cases}$$

from which we obtain four pairs of equations to consider:

(i) $y = 0$, $x = 0$

(ii) $3 - 2x - y = 0$, $x = 0$

(iii) $y = 0$, $3 - 2y - x = 0$

(iv) $3 - 2x - y = 0$, $3 - 2y - x = 0$

Recall that $ab = 0 \Rightarrow a = 0$ or $b = 0$.

Solving these equations, we find the following possible locations of the extreme values:

$$(0, 0), \quad (0, 3), \quad (3, 0), \quad (1, 1)$$

For $(0, 0)$, we have $A = 0$, $B = 3$, and $C = 0$, so that $AC - B^2 < 0$. Hence f has a saddle point at $(0, 0)$. For $(0, 3)$, we have $A = -6$, $B = -3$, and $C = 0$, so that $AC - B^2 < 0$. Hence $(0, 3)$ is another saddle point of f. Similarly, we can show that $(3, 0)$ is also a saddle point of f. For $(1, 1)$, we have $A = -2$, $B = -1$, and $C = -2$, so that $AC - B^2 > 0$. Since $A = -2 < 0$, part (i) of Theorem 9.6 applies. Consequently, $f(1, 1)$ is a relative maximum value.

In Problems 1 through 8, find all relative extrema and saddle points.

Exercises 9.5

1. $f(x, y) = x^2 - y^2$

2. $f(x, y) = x^2 + y^2 - 1$

3. $f(x, y) = x^2 - y^2 + 2x - 4y + 3$

4. $f(x, y) = x^2 - xy$

5. $f(x, y) = x^2 - xy + y^2$

6. $f(x, y) = x^3 - 4y^2$

7. $f(x, y) = y^3 + x^2 - 3x$

8. $f(x, y) = xy(x - y)$

9. Find the points on the surface

$$z = \sqrt{xy + 1}$$

whose distance from the origin is minimum. [*Hint:* Let

$$u = d[(0, 0, 0), (x, y, z)]^2 = x^2 + y^2 + z^2$$

Find the minimum of u, when $z^2 = xy + 1$.]

10. A rectangular tank is open at the top and holds 32 cu ft. Find its dimensions so that the surface area of the tank is a minimum.

11. Find the dimensions of a rectangular box of maximum volume, if the box has no top and has a surface area of 108 sq in.

12. Consider all triangles with a given perimeter, $2s$. Show that the one with the largest area is an equilateral triangle. [*Hint:* If a, b, and c are the sides of a triangle, then $a + b + c = 2s$, and the area A is given by

$$A = \sqrt{s(s - a)(s - b)(s - c)}\,]$$

13. The sum of three positive numbers is 30. Show that their product is maximum when they are equal.

14. Suppose a rectangular box of volume 64 cu ft is to be constructed from three different materials. The cost of the bottom is 50¢/sq ft; of the sides, is 30¢/sq ft; and of the top, is 40¢/sq ft. Find the dimensions of the most economical box.

15. Do Problem 14 if the bottom costs 50¢/sq ft, and the sides and top cost 20¢/sq ft.

16. The cost of a day's production at ABC Company is given by

$$C = x^2 + y^2 + xy - 20x - 25y + 1500$$

where x is the labor force and y (in thousands of tons) is the amount of raw material. Find the x and y that minimize the cost C.

17. Do Problem 16 if

$$C = 3x^2 + 2y^2 + 3xy - 66x - 58y + 1600$$

18. A company produces x units of product A at a selling price of $500 - 2x$ and y units of product B at a selling price of $400 - 3y$. Given that the cost of producing both products is

$$C(x, y) = x^2 + 20x + 40y + y^2$$

(a) Show that the profit function P is given by

$$P(x, y) = 480x - 3x^2 + 360y - 4y^2$$

(b) How many items of each product should be produced to maximize profits?

9.6   Extremal Problems with Constraints

In many maxima and minima problems, the functions defined are subject to certain restrictions (or conditions) called constraints. For example, suppose that we wish to find the minimum value of the function

$$z = f(x, y) = x^2 + y^2 \qquad (9.14)$$

The solution is obvious, since $x^2 + y^2 \geq 0$ for all $x, y \in R$ and $x^2 + y^2 = 0$ if and only if $x = 0$ and $y = 0$. Consequently, the minimum value of f is 0, and it is attained at $(0, 0)$. However, we may be asked to find the minimum value of the function f defined by (9.14), subject to the constraint

$$x + y - 1 = 0 \qquad (9.15)$$

Obviously, f cannot have minimum value at $(0, 0)$ since $(0, 0)$ does not satisfy (9.15). To solve the problem in this case, we solve Equation (9.15) for y:

$$y = 1 - x \qquad (9.16)$$

Substituting this value of y in (9.14), we have

$$z = x^2 + y^2 = x^2 + (1 - x)^2 \qquad (9.17)$$

$$= x^2 + 1 - 2x + x^2 = 2x^2 - 2x + 1$$

Since z is a function of one variable x, we can use the method of Chapter 5. We find $\dfrac{dz}{dx}$ and set it equal to zero:

$$\frac{dz}{dx} = 4x - 2 = 0 \Rightarrow x = \tfrac{1}{2}$$

Also, using the second-derivative test we find

$$\frac{d^2z}{dx^2} = 4 > 0$$

Therefore $x = \tfrac{1}{2}$ gives the minimum value of z. Substituting $x = \tfrac{1}{2}$ in (9.16), we get $y = \tfrac{1}{2}$. Thus the minimum value of f is attained at $\left(\tfrac{1}{2}, \tfrac{1}{2}\right)$, and

$$f\left(\tfrac{1}{2}, \tfrac{1}{2}\right) = \left(\tfrac{1}{2}\right)^2 + \left(\tfrac{1}{2}\right)^2 = \tfrac{1}{4} + \tfrac{1}{4} = \tfrac{1}{2}$$

The following example illustrates how to use a similar method when three variables are involved.

Example 9.18   Find the minimum value of

$$u = x^2 + y^2 + z^2 \qquad (9.18)$$

subject to the constraint

$$x + y + z = 30 \qquad (9.19)$$

SOLUTION   Solving (9.19) for z, we obtain

$$z = 30 - x - y \qquad (9.20)$$

We substitute this value of z in (9.18):

$$u = x^2 + y^2 + (30 - x - y)^2 \qquad (9.21)$$

Then

$$u_x = 2x + 2(30 - x - y)(-1) = 4x + 2y - 60$$

$$u_y = 2y + 2(30 - x - y)(-1) = 2x + 4y - 60$$

$$u_{xx} = 4; \qquad u_{xy} = 2; \qquad u_{yy} = 4$$

Solving

$$\begin{cases} u_x = 0 \\ u_y = 0 \end{cases}$$

we find that $x = 10$, $y = 10$. Also, $u_{xx} \cdot u_{yy} > (u_{xy})^2$, and $u_{xx} > 0$. Hence u is minimum for $x = 10$, $y = 10$, and

$$z = 30 - x - y = 30 - 10 - 10 = 10$$

and this minimum value is

$$u = 10^2 + 10^2 + 10^2 = 300$$

The previous method works, since the constraint equation (9.19) could be solved for one of the variables in terms of the other variables. Suppose that Equation (9.19) was replaced by

$$x^3 y^2 + 2xy^2 + y^5 z^3 + 7xy + 8yz = \sqrt{71} \qquad (9.22)$$

It would indeed be difficult to solve (9.22) for any one of the three variables. There is an alternative method developed by Lagrange. We state this method in the following theorem.

Lagrange's method of undetermined multipliers ▶ **Theorem 9.7** *Let* f, g_1, g_2, $\ldots$, g_m *be differentiable functions of the variables* x_1, x_2, $\ldots$, x_n. *Let* f *be subject to the following constraints.*

$$g_1(x_1, x_2, \ldots, x_n) = 0$$

$$g_2(x_1, x_2, \ldots, x_n) = 0$$

$$\vdots \qquad (9.23)$$

$$g_m(x_1, x_2, \ldots, x_n) = 0$$

Then the points at which f *attains an extreme value are found among the points* $(x_1, x_2, \ldots, x_n)$ *for which there exist constants* λ_1, λ_2, $\ldots$, λ_m *(each* λ_i *is called a Lagrange multiplier), such that*

$$\begin{cases} \dfrac{\partial f}{\partial x_1} - \lambda_1 \dfrac{\partial g_1}{\partial x_1} - \lambda_2 \dfrac{\partial g_2}{\partial x_1} - \cdots - \lambda_m \dfrac{\partial g_m}{\partial x_1} = 0 \\[2mm] \dfrac{\partial f}{\partial x_2} - \lambda_1 \dfrac{\partial g_1}{\partial x_2} - \lambda_2 \dfrac{\partial g_2}{\partial x_2} - \cdots - \lambda_m \dfrac{\partial g_m}{\partial x_1} = 0 \\[2mm] \vdots \\ \dfrac{\partial f}{\partial x_n} - \lambda_1 \dfrac{\partial g_1}{\partial x_n} - \lambda_2 \dfrac{\partial g_2}{\partial x_n} - \cdots - \lambda_m \dfrac{\partial g_m}{\partial x_n} = 0 \\[2mm] g_1(x_1, x_2, \ldots, x_n) = 0 \\ g_2(x_1, x_2, \ldots, x_n) = 0 \\ \vdots \\ g_m(x_1, x_2, \ldots, x_n) = 0 \end{cases} \tag{9.24}$$

The proof of this theorem is omitted. Suppose we consider a special case of this theorem, when the function f of two variables x and y is subject to a single constraint $g(x, y) = 0$. Then the theorem asserts that the points (x, y) at which the function f has an extreme value are among the points (x, y) for which there exists a real number λ, such that

$$\begin{cases} \dfrac{\partial f}{\partial x} - \lambda \dfrac{\partial g}{\partial x} = 0 \\[2mm] \dfrac{\partial f}{\partial y} - \lambda \dfrac{\partial g}{\partial y} = 0 \\[2mm] g(x, y) = 0 \end{cases} \tag{9.25}$$

Example 9.19 Find the extreme values of

$$f(x, y) = xy \tag{9.26}$$

subject to the constraint

$$g(x, y) = x^2 + y^2 - 1 = 0 \tag{9.27}$$

SOLUTION Since

$$\frac{\partial f}{\partial x} = y, \qquad \frac{\partial f}{\partial y} = x, \qquad \frac{\partial g}{\partial x} = 2x, \qquad \frac{\partial g}{\partial y} = 2y$$

Equations (9.25) become

$$\begin{cases} y - \lambda(2x) = 0 \\ x - \lambda(2y) = 0 \\ x^2 + y^2 - 1 = 0 \end{cases} \tag{9.28}$$

We solve the first equation in (9.28) for y and substitute in the second equation to obtain $\qquad\qquad y = \lambda(2x)$ $\qquad\qquad$ (9.29)

and $\qquad\qquad\qquad\qquad x(1 - 4\lambda^2) = 0$ $\qquad\qquad$ (9.30)

From (9.30), we note that either $x = 0$ or $1 - 4\lambda^2 = 0$. If $x = 0$, then from (9.29) we must have $y = 0$. But $x = 0$, $y = 0$ does not satisfy (9.27). Hence $1 - 4\lambda^2 = 0$ or

$$\lambda = \pm\tfrac{1}{2} \qquad\qquad (9.31)$$

Substituting these values of λ in (9.29), we get

$$y = x \quad \text{or} \quad y = -x \qquad\qquad (9.32)$$

For $y = \pm x$, we have from the third equation of (9.28)

$$x^2 + y^2 = 1$$
$$x^2 + (\pm x)^2 = 1$$
$$x^2 + x^2 = 1$$
$$2x^2 = 1$$

Then $x = \pm 1/\sqrt{2}$ and $y = \pm x = \pm 1/\sqrt{2}$. Thus the extreme values of f occur among

$$\left(\frac{1}{\sqrt{2}}, \frac{1}{\sqrt{2}}\right), \quad \left(\frac{1}{\sqrt{2}}, -\frac{1}{\sqrt{2}}\right), \quad \left(-\frac{1}{\sqrt{2}}, \frac{1}{\sqrt{2}}\right), \quad \left(-\frac{1}{\sqrt{2}}, -\frac{1}{\sqrt{2}}\right)$$

We now compute the values of f at these points.

$$f\left(\frac{1}{\sqrt{2}}, \frac{1}{\sqrt{2}}\right) = \left(\frac{1}{\sqrt{2}}\right)\left(\frac{1}{\sqrt{2}}\right) = \frac{1}{2}$$

$$f\left(\frac{1}{\sqrt{2}}, -\frac{1}{\sqrt{2}}\right) = \left(\frac{1}{\sqrt{2}}\right)\left(-\frac{1}{\sqrt{2}}\right) = -\frac{1}{2}$$

$$f\left(-\frac{1}{\sqrt{2}}, \frac{1}{\sqrt{2}}\right) = \left(-\frac{1}{\sqrt{2}}\right)\left(\frac{1}{\sqrt{2}}\right) = -\frac{1}{2}$$

$$f\left(-\frac{1}{\sqrt{2}}, -\frac{1}{\sqrt{2}}\right) = \left(-\frac{1}{\sqrt{2}}\right)\left(-\frac{1}{\sqrt{2}}\right) = \frac{1}{2}$$

Thus the maximum value of f is $\tfrac{1}{2}$, which is attained at the points $(1/\sqrt{2}, 1/\sqrt{2})$ and $(-1/\sqrt{2}, -1/\sqrt{2})$. The minimum value of f is $-\tfrac{1}{2}$, attained at the points $(1/\sqrt{2}, -1/\sqrt{2})$ and $(-1\sqrt{2}, 1/\sqrt{2})$.

Exercises 9.6

1. State Theorem 9.7 for a function f of three variables x, y, and z, subject to the constraint $g(x, y, z) = 0$.

2. State Theorem 9.7 for a function f of three variables, x, y, and z, subject to two constraints $g_1(x, y, z) = 0$ and $g_2(x, y, z) = 0$.

3. Find the maximum and minimum values of $f(x, y) = x + y$ subject to $x^2 + y^2 = 1$.

4. Find the maximum and minimum values of $f(x, y) = x^2 - y^2$ subject to $x^2 + y^2 = 1$.

5. By Lagrange's method, find the maximum and minimum values of $f(x, y) = x^2 + y^2$ subject to $x + y = 1$.

6. Do Example 9.18, using Lagrange's method. [*Hint:* Use Problem 1.]

In Problems 7 through 11, use Lagrange's method.

7. Do Problem 10 of Exercises 9.5.

8. Do Problem 11 of Exercises 9.5.

9. Do Problem 13 of Exercises 9.5.

10. Do Problem 14 of Exercises 9.5.

11. Find the minimum value of

$$f(x, y, z) = x^2 + y^2 + z^2$$

subject to

$$x + 2y + z = 1 \quad \text{and} \quad 2x - y - 3z = 4$$

9.7 Double Integrals

In Chapter 7 we saw that if $y = f(x)$ is a continuous function and non-negative [$f(x) \geq 0$] on $[a, b]$, then $\int_a^b f(x)\, dx$ represents the area of the region bounded by the curve $y = f(x)$, the x-axis, and the lines $x = a$ and $x = b$. Now we look for the extension of the integral that will represent volume. It turns out that finding the volume of a solid involves performing integration twice, so the process is called the *double integral*. The rigorous definition of the double integral is beyond the scope of this book. However, we shall give an intuitive description for the fact that we need to integrate twice while considering the volume of a solid.

Let D be a region in the xy-plane which is bounded by two functions $y = g(x)$ and $y = h(x)$, and let $z = f(x, y)$ be the surface that lies above the region D. We are interested in finding the volume under $z = f(x, y)$ bounded by D (see Figure 9.13).

We imagine a slice cut from the solid by a plane perpendicular to the x-axis at $x = c$. This cross-sectional area is given by the integral

$$\int_{h(c)}^{g(c)} f(c, y)\, dy \tag{9.33}$$

If $x = a$ and $x = b$ are the lines that bound the region D, then we can find the cross-sectional area given by (9.33) for every $x \in [a, b]$. Thus we obtain the area function

$$A(x) = \int_{h(x)}^{g(x)} f(x, y)\, dy$$

where x is held fixed and the limits of integration depend upon where the cutting plane is taken. In other words, the y-limits are functions of

Double integrals appeared in the first part of the eighteenth century. They were used to determine the gravitational attraction exerted by a lamina on particles. The procedure of evaluating double integrals by repeated integration was given by Euler. In their works on gravitational attraction, Lagrange and Laplace used multiple integration. Important advances in the theory of multiple integrals were made by Lebesgue and Fubini.

Figure 9.13

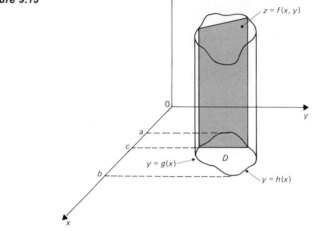

x, the functions that represent the boundary curves for the region D. Finally, we see that the volume V of the solid under $z = f(x, y)$ bounded by D is the sum of all the areas $A(x)$. This sum leads to another integration. Therefore we have

$$V = \int_a^b A(x) \, dx = \int_a^b \left[\int_{h(x)}^{g(x)} f(x, y) \, dy \right] dx \qquad (9.34)$$

The expression on the right-hand side of (9.34) is called the *iterated integral*. We usually delete the brackets in the expression for the iterated integral, and write

$$\int_a^b \left[\int_{h(x)}^{g(x)} f(x, y) \, dy \right] dx = \int_a^b \int_{h(x)}^{g(x)} f(x, y) \, dy \, dx \qquad (9.35)$$

To evaluate a given iterated (double) integral, we proceed as follows:

(a)   Integrate $\int f(x, y) \, dy$ with respect to y with x held fixed, and evaluate the resulting integral between the limits $y = h(x)$ and $y = g(x)$.

(b)   Then integrate the result obtained in (a) with respect to x between the limits $x = a$ and $x = b$.

Example 9.20   Evaluate the iterated integral

$$I = \int_2^4 \int_1^{2x+1} (3x + y) \, dy \, dx$$

SOLUTION

$$I = \int_2^4 \left[3xy + \frac{y^2}{2} \right]_1^{2x+1} dx$$

$$= \int_2^4 \left\{ \left[3x(2x + 1) + \frac{(2x + 1)^2}{2} \right] - \left[3x + \frac{1}{2} \right] \right\} dx$$

$$= \int_2^4 (8x^2 + 2x)\, dx = \left[\frac{8x^3}{3} + x^2 \right]_2^4$$

$$= \left[\frac{8(4)^3}{3} + (4)^2 \right] - \left[\frac{8(2)^3}{3} + (2)^2 \right] = \frac{484}{3}$$

To evaluate the iterated integral $\int_c^d \int_{g(y)}^{h(y)} f(x, y)\, dx\, dy$, we merely interchange the role of x and y. In other words, we first integrate with respect to x and then with respect to y.

Example 9.21   Evaluate the iterated integral

$$I = \int_0^2 \int_0^y (x + e^{y^2})\, dx\, dy$$

SOLUTION

$$I = \int_0^2 \left(\left[\frac{x^2}{2} + xe^{y^2} \right]_0^y \right) dy = \int_0^2 \left(\frac{y^2}{2} + ye^{y^2} \right) dy$$

$$= \left[\frac{y^3}{6} + \frac{e^{y^2}}{2} \right]_0^2 = \left(\frac{8}{6} + \frac{e^4}{2} \right) - \left(0 + \frac{1}{2} \right) = \frac{5}{6} + \frac{e^4}{2}$$

Example 9.22   Find the volume of the solid whose base is the triangle in the xy-plane bounded by the x-axis, the line $y = x$, and the line $x = 1$, and whose top is in the plane $z = 2x + y + 1$.

SOLUTION   For any x between 0 and 1, y varies from $y = 0$ to $y = x$ (see Figure 9.14). Hence the volume V of the solid is given by

$$V = \int_0^1 \int_0^x (2x + y + 1)\, dy\, dx = \int_0^1 \left\{ \left[2xy + \frac{y^2}{2} + y \right]_0^x \right\} dx$$

Note   *Integrating with respect to y, we treat x as a constant.*

Figure 9.14

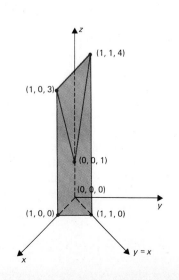

$$= \int_0^1 \left(2x^2 + \frac{x^2}{2} + x \right) dx = \int_0^1 \left(\frac{5}{2} x^2 + x \right) dx$$

$$= \left[\frac{5}{6} x^3 + \frac{x^2}{2} \right]_0^1 = \frac{5}{6} + \frac{1}{2} = \frac{4}{3}$$

Example 9.23 Find the volume of the solid whose base D in the xy-plane is bounded by the curves $y = x^2$ and $y = x$, and whose top is the surface $z = x^2 y^3$.

SOLUTION We first sketch the region D in the xy-plane (see Figure 9.15).

Figure 9.15

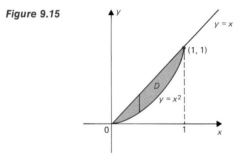

Now suppose x between 0 and 1 is held fixed. Then y varies from $y = x^2$ to $y = x$ in the region D. Consequently, the limits for the inner integral are x^2 and x, while the limits for the outer integral are 0 and 1. Hence the required volume V is given by

$$V = \int_0^1 \int_{x^2}^x x^2 y^3 \, dy \, dx = \int_0^1 \left\{ \left[x^2 \frac{y^4}{4} \right]_{x^2}^x \right\} dx$$

$$= \int_0^1 \left(\frac{x^6}{4} - \frac{x^{10}}{4} \right) dx = \frac{1}{4} \int_0^1 (x^6 - x^{10}) \, dx$$

$$= \frac{1}{4} \left(\frac{x^7}{7} - \frac{x^{11}}{11} \right) \Big]_0^1 = \frac{1}{4} \left(\frac{1}{7} - \frac{1}{11} \right) = \frac{1}{77}$$

We could also compute the required volume by first integrating with respect to x and then with respect to y. For this method, we observe that if y is held fixed between 0 and 1, then x varies from $x = y$ to $x = y^{1/2}$ (see Figure 9.16). Hence

$$V = \int_0^1 \int_y^{y^{1/2}} x^2 y^3 \, dx \, dy = \frac{1}{3} \int_0^1 \left[y^3 x^3 \right]_y^{y^{1/2}} dy$$

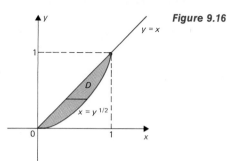

Figure 9.16

$$= \frac{1}{3} \int_0^1 (y^{9/2} - y^6)\, dy = \frac{1}{3} \left[\frac{y^{11/2}}{11/2} - \frac{y^7}{7} \right]_0^1$$

$$= \frac{1}{3} \left[\frac{2}{11} - \frac{1}{7} \right] = \frac{1}{77}$$

In Problems 1 through 24, evaluate the double iterated integrals.   Exercises 9.7

1. $\displaystyle\int_0^2 \int_0^1 (1 + x)\, dx\, dy$

2. $\displaystyle\int_1^3 \int_0^2 3xy^2\, dx\, dy$

3. $\displaystyle\int_0^2 \int_1^3 3xy^2\, dy\, dx$

4. $\displaystyle\int_0^3 \int_0^3 (x + y^2)\, dx\, dy$

5. $\displaystyle\int_1^2 \int_0^1 (y + x^2)\, dx\, dy$

6. $\displaystyle\int_0^3 \int_3^4 yx^2\, dx\, dy$

7. $\displaystyle\int_0^1 \int_0^1 (x^2 + y)\, dy\, dx$

8. $\displaystyle\int_0^1 \int_0^1 (x^2 + y)\, dx\, dy$

9. $\displaystyle\int_0^1 \int_2^4 (4 + y)\, dx\, dy$

10. $\displaystyle\int_2^4 \int_0^1 (4 + y)\, dy\, dx$

11. $\displaystyle\int_0^1 \int_0^1 e^{2x + y}\, dx\, dy$

12. $\displaystyle\int_0^1 \int_0^1 e^{2x + y}\, dy\, dx$

13. $\displaystyle\int_0^2 \int_1^x (2x + y + 1)\, dy\, dx$

14. $\displaystyle\int_0^1 \int_0^{x^2} (xy + e^{x^3})\, dy\, dx$

15. $\displaystyle\int_{-1}^1 \int_y^{y^2} (y^2 + 2xy - x^2)\, dx\, dy$

16. $\displaystyle\int_0^2 \int_0^{1 - x^2} y\, dy\, dx$

17. $\displaystyle\int_1^4 \int_0^x \sqrt{x - y}\, dy\, dx$

18. $\displaystyle\int_0^2 \int_0^x e^{x^2}\, dy\, dx$

19. $\displaystyle\int_0^1 \int_0^{8x} \sqrt{x + y}\, dy\, dx$

20. $\displaystyle\int_1^{\ln 8} \int_0^{\ln y} e^{x + y}\, dx\, dy$

21. $\displaystyle\int_0^2 \int_{\sqrt{y}}^1 dx\, dy$

22. $\displaystyle\int_{-1}^2 \int_{x^2 + 1}^{3x + 4} dy\, dx$

23. $\displaystyle\int_0^1 \int_{y^2}^y \sqrt{\frac{x}{y}}\, dx\, dy$

24. $\displaystyle\int_0^\pi \int_0^{x^2} \frac{y}{x}\, dy\, dx$

25. Find the volume of the solid whose base is the triangle in the xy-plane bounded by the x-axis, the line $y = x$, and the line $x = 1$, and whose top is the plane $z = x + y + 2$.

26. Find the volume of the solid whose base is the region in the xy-plane that is bounded by $x = 0$, $y = 0$, and $y = 1 - x$, and whose top is bounded by the surface $z = x^2 + y^2$.

27. Find the volume of the solid under the surface $z = xy$ bounded by $y = 3x$ and $y = 4 - x^2$ and the plane $z = 0$.

28. Find the volume of the solid bounded by the surfaces $z = 0$, $z = x$, and $y^2 + x = 2$.

29. Find the volume of the solid bounded by the planes $y = x$, $y = 2 - x$, $y = 0$, $z = 0$, and $x + y + z = 3$.

30. Find the volume of the solid bounded by the surfaces $z = 0$, $x = 0$, $y^2 + x = 4$, and $z - y = 2$.

Chapter 9 Review

Part 1 (*Oral*)

Define or discuss the following.

1. A real-valued function of two variables

2. Three-dimensional coordinate system

3. Graph of $z = f(x, y)$

4. Neighborhood of a point in the plane

5. $f(x, y) \to L$ as $(x, y) \to (a, b)$.

6. f is continuous at (a, b)

7. $\dfrac{\partial f}{\partial x}, \dfrac{\partial f}{\partial y}, \dfrac{\partial^2 f}{\partial y\, \partial x}$

8. Total differential

9. Relative maximum and relative minimum

10. Lagrange's method of undetermined multipliers

Part 2 (*Written*)

1. Plot the following points.

   (a) (1, 0, 1)   (b) (0, 0, 2)   (c) (−1, 2, 3)

2. Calculate the distance between the points.

   (a) (1, 2, 3), (−1, 0, 1)   (b) (−2, 1, −3), (3, −1, −4)

3. (a) What is the equation of the xy-plane?
   (b) What is the equation of the yz-plane?
   (c) What is the equation of the xz-plane?

(d) What is the equation of the plane containing the points
(0, 0, 1), (1, 0, 0), and (0, 2, 0)?

4. Sketch the graph of the plane $2x + 3y + 4z = 12$.

5. Sketch the graph of $x^2 + y^2 + z^2 = 16$ by drawing several cross
sections.

In Problems 6 through 15, find f_x, f_y, f_{xx}, f_{yy}, and f_{yx}.

6. $f(x, y) = 2x - 5y^2$ 7. $f(x, y) = xy^2 + 7$

8. $f(x, y) = 8x - 5x^2y + 7xy^3$ 9. $f(x, y) = \sqrt{x^2 + y^2}$

10. $f(x, y) = \dfrac{x - y}{x + y}$ 11. $f(x, y) = e^x e^y - xe^{y^2 + 2y}$

12. $f(x, y) = \ln x^2 y^3$ 13. $f(x, y) = \ln (x^2 + y^3)$

14. $f(x, y) = xye^{x + y}$ 15. $f(x, y) = x^2 y^3 \ln xy$

In Problems 16 through 19, find $f_x(1, 2)$, $f_y(-1, 3)$, and $f_{xy}(2, 3)$.

16. $f(x, y) = 3x - 4y^2$ 17. $f(x, y) = 4xy$

18. $f(x, y) = \dfrac{x + y}{x - y}$ 19. $f(x, y) = xe^y$

20. The manufacturing cost of a widget is given by

$$M(x, y) = 4xy^2 + 10x^2 + 5y^2 + 5$$

where x is the cost of labor and y is the cost of the material.
Find and interpret the following.

(a) $\left(\dfrac{\partial M}{\partial x}\right)_{(3,\,2)}$ (b) $\left(\dfrac{\partial M}{\partial y}\right)_{(4,\,5)}$

21. The period T of a pendulum of length l is given by

$$T = 2\pi \sqrt{\dfrac{l}{g}}$$

Find the error in g due to errors of 3% in l and 1% in T.

In Problems 22 through 25, find any relative maxima or relative minima
for the given function:

22. $f(x, y) = x^2 - xy + y^2 + 2x + 2y + 7$

23. $f(x, y) = 8x + 8y - 10x^2 - 4y^2 + 4xy + 40$

24. $f(x, y) = x^2 + y^2 + 2x + 2y - xy - 6$

25. $f(x, y) = 4x + 4y + 2xy - 5x^2 - 2y^2 + 4$

26. If $u = x^2 + y^2 + z^2$ and $x + y + z = 30$, find the values of x, y, and
z for which u is a minimum.

27. Find the points on the surface $z^2 = xy + 1$ whose distance from the origin is minimum.

28. Find the dimensions of a rectangular tank (without the top) having a volume of 32 cu units and minimum surface area.

29. Find the dimensions of a rectangular box (without the top) having maximum volume and a surface area of 108 sq in.

30. Find the minimum value of

$$D = \sqrt{x^2 + y^2 + z^2}$$

subject to the constraint

$$x + y + z = 30$$

31. Find the maximum and minimum distances from the origin to the curve

$$5x^2 + 6xy + 5y^2 = 8$$

32. Discuss the maxima and minima of

$$f(x, y) = x^2 + y^2 + z^2$$

subject to the constraint

$$\begin{cases} x + y - z = 1 \\ 2x - 3y + 4z = 2 \end{cases}$$

Numerical Applications Using Calculators

Calculators can be used to approximate total differentials. Consider the following example:

Example 9.24 Use total differentials to approximate

$$\frac{\sqrt[3]{27.5}}{3.01}$$

SOLUTION Define $z = f(x, y) = \dfrac{\sqrt[3]{x}}{y}$.

If we let $x = 27$ and $y = 3$, then $z = 1$. The error in x is $dx = 0.5$ and the error in y is $dy = 0.01$. Therefore the corresponding error in z is

$$dz = f_x \, dx + f_y \, dy$$

$$= \frac{1}{3y \sqrt[3]{x^2}} \, dx - \frac{\sqrt[3]{x}}{y^2} \, dy$$

$$= \frac{0.5}{3(3)(9)} - \frac{3(0.01)}{3^2}$$

$$= 0.0061728395 - 0.0033333333$$

$$= 0.0028395056$$

Thus $\sqrt[3]{27.5}/3.01 \approx z + dz = 1.0028395056$. We note that direct computation yields

$$\frac{\sqrt[3]{27.5}}{3.01} \approx 1.00279248$$

Use a pocket calculator to approximate each of the following by total differentials and check the accuracy by direct computation.

Exercises

1. $\sqrt{(3.0001)(2.997)}$ [*Hint:* Let $f(x, y) = \sqrt{xy}$, $f(3, 3) = 3$, $dx = 0.0001$.]

2. $\dfrac{\sqrt{36.01}}{3.94}$

3. $\dfrac{\sqrt[3]{8.002}}{5.01}$

4. 5.9998×1.000001

5. $\sqrt[5]{(31.0001)(242.998)}$

CHAPTER 10

The Trigonometric Functions

Trigonometry was originally invented by the ancient Greeks to measure triangles. The word *trigonometry* means *triangle measurement*, and has been used extensively in several fields, especially in surveying, astronomy, and navigation. Trigonometry has come to identify the branch of mathematics concerned with the properties of *circular* or *trigonometric functions*. These functions are used to describe cyclical phenomena.

10.1 The Trigonometric Functions

Since angles play the central role in trigonometry, we shall discuss them first. We say that an angle is generated when a half-line (or a ray) rotates in a plane about its endpoint. The endpoint is called the *vertex* of the angle, and we refer to the initial position of the half-line as the *initial side* and its final position as the *terminal side* (see Figure 10.1).

For a given half-line in the plane, there are two possible directions of rotation from the initial position—clockwise rotation and counter-clockwise rotation. We shall follow the usual convention that a clockwise rotation produces a negative angle, while counterclockwise rotation generates a positive angle. An angle is said to be in *standard position* if a rectangular coordinate system is associated with it in such a way that the vertex of the angle is the origin and the initial side of the angle is the positive *x*-axis.

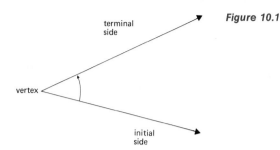

Figure 10.1

terminal side

vertex

initial side

To measure an angle, we require a basic unit of measure. Since we are interested in measuring the amount of rotation by a half-line, the most natural unit to employ is a complete rotation. Commonly, two kinds of units are used to express the measure of an angle: the *degree* and the *radian*.

Measure of an angle

Definition 10.1 An angle is said to have a measure of one degree if it is generated by a half-line rotating $\frac{1}{360}$ of a complete rotation about its endpoint. Thus one complete rotation measures 360 degrees.

◀ *The degree unit was introduced by the early Babylonians; it is believed that the number 360 was chosen to suit astronomical calculations. One solar year (the period during which the Earth completes one revolution around the sun) is $365\frac{1}{4}$ days, while one lunar year (the period during which the moon completes twelve revolutions around the Earth) is 354 days. Perhaps the number 360 was found to be a suitable approximation to these numbers! An advantage of the number 360 is that it is divisible by "most" of the small integers.*

To denote the degree measure symbolically, we use the symbol ° written at the top and to the right of the measure of an angle. For example, an angle with a measure of 45 degrees is denoted by 45°. Thus we have

$$360° = 1 \text{ complete rotation}$$

$$180° = \tfrac{1}{2} \text{ of a complete rotation}$$

$$90° = \tfrac{1}{4} \text{ of a complete rotation} = 1 \text{ right angle}$$

One *minute* (written 1′) has $\frac{1}{60}$ the measure of a degree, and one *second* (written 1″) has $\frac{1}{60}$ the measure of a minute. In summary,

$$1 \text{ complete rotation} = 4 \text{ right angles} = 360°$$

$$1 \text{ right angle} = 90°$$

$$1° = 60′$$

$$1′ = 60″$$

An angle whose measure is 27 degrees 47 minutes and 33 seconds is symbolically denoted by an angle 27°47′33″.

Although the degree-minute-second system is adequate for all practical purposes of angle measurement, the system best suited for the operations of calculus is the radian measure.

Definition 10.2 Let the vertex of the angle to be measured be placed at the center of a circle of radius r and let the length of the arc AB of

◀

the circle subtended by the two sides of the angle be s. Then the radian *measure θ of the angle is given by θ = s/r (see Figure 10.2).*

Figure 10.2

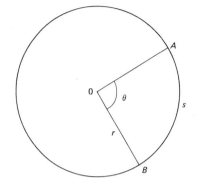

In Definition 10.2, it is easy to see that if $s = r$, then $\theta = 1$ radian (rad). Consequently, 1 rad is the measure of the angle subtended by an arc whose length is equal to the radius of the circle.

It is easy to correlate degree measure with radian measure. We recall from geometry that the complete circle (of radius r) has arc length $2\pi r$. Thus the radian measure of the complete rotation at the center of the circle is $2\pi r/r = 2\pi$ rad. In the degree system this angle measures 360°. Consequently

Note π (read "pie") is an irrational number whose approximate value is 3.14.

$$2\pi \text{ rad} = 360°$$

The following table gives the correspondence between degree and radian measures of some of the special angles.

Degrees	0°	15°	30°	45°	60°	90°	120°	150°	180°	270°	360°
Radians	0	$\dfrac{\pi}{12}$	$\dfrac{\pi}{6}$	$\dfrac{\pi}{4}$	$\dfrac{\pi}{3}$	$\dfrac{\pi}{2}$	$\dfrac{2\pi}{3}$	$\dfrac{5\pi}{6}$	π	$\dfrac{3\pi}{2}$	2π

The classical definition of the trigonometric functions is given for acute angles θ by the ratios of the lengths of the sides in a right triangle as follows (see Figure 10.3).

The angle θ is said to be acute if $0 < \theta < \pi/2$.

Figure 10.3

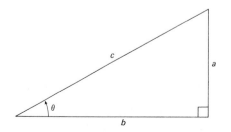

$$\text{sine of } \theta = \sin \theta = \frac{a}{c}$$

$$\text{cosine of } \theta = \cos \theta = \frac{b}{c}$$

$$\text{tangent of } \theta = \tan \theta = \frac{a}{b}$$

$$\text{cosecant of } \theta = \csc \theta = \frac{c}{a}$$

$$\text{secant of } \theta = \sec \theta = \frac{c}{b}$$

$$\text{cotangent of } \theta = \cot \theta = \frac{b}{a},$$

where a is the length of the side opposite angle θ, b is the length of the side adjacent to angle θ, and c is the length of the hypotenuse.

   Remark   Since similar triangles have proportional sides, the values of the six trigonometric functions of an angle θ depend only on the measure of the angle θ and not on the particular right triangle used for the computation of the trigonometric functions.

   The approach just described defines the trigonometric functions for acute angles only. The approach cannot be used to compute trigonometric functions for obtuse angles (angles larger than 90° or $\pi/2$ rad) since a right triangle cannot have an obtuse angle (Why?). To extend some of the results of calculus to the trigonometric functions, we employ a different approach based on analytic geometry.

Definition 10.3   *Let u be a real number and let an angle whose radian measure is u be placed in standard position (see Figure 10.4). Let*   ◄

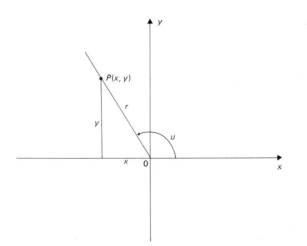

Figure 10.4

P(x, y) be a point different from the origin on the terminal side of this angle and let $r = \sqrt{x^2 + y^2}$. Then the six trigonometric functions of u are defined by

$$\sin u = \frac{y}{r}, \qquad \csc u = \frac{r}{y}$$

$$\cos u = \frac{x}{r}, \qquad \sec u = \frac{r}{x}$$

$$\tan u = \frac{y}{x}, \qquad \cot u = \frac{x}{y}$$

It is easy to see that this definition of the trigonometric functions agrees with the right-triangle definition for acute angles. By measuring the angle clockwise if $u < 0$ and counterclockwise if $u \geq 0$, we can measure an angle of u radians for any real number u.

Domain of sine and cosine

Since $r = \sqrt{x^2 + y^2}$, r is always positive. Consequently, the cosine and sine functions are defined for all real numbers u. In other words, the domain of these two functions is the set R of real numbers.

Domain of tangent and secant

It is also clear from the definition that the tangent and secant functions are not defined on the set $A = \{u \,|\, u = (2k + 1)\,\pi/2, k \in Z\}$ since for any $u \in A$, the related angle has its terminal side on the y-axis, and hence $x = 0$. Thus the domain of the tangent and secant functions is the set $R - A$.

Domain of cotangent and cosecant

For a similar reason, the contangent and cosecant functions are not defined on the set $B = \{u \,|\, u = k\pi, k \in Z\}$. Consequently, the domain of the cotangent and cosecant functions is the set $R - B$.

Range of trigonometric functions

Since $|x| \leq \sqrt{x^2 + y^2} = r$ and $|y| \leq \sqrt{x^2 + y^2} = r$, we have

$$\left|\frac{x}{r}\right| \leq 1 \quad \text{and} \quad \left|\frac{y}{r}\right| \leq 1$$

Thus

$$-1 \leq \frac{x}{r} \leq 1, \qquad -1 \leq \frac{y}{r} \leq 1$$

The range of the sine and cosine functions is the set $I = \{t \,|\, -1 \leq t \leq 1\}$. The range of the tangent and cotangent functions is R, and the range of the secant and cosecant functions is $R - C$, where C is the open interval $(-1, 1)$.

We observe from Definition 10.3 that we need only define the sine and cosine functions explicitly. The other four trigonometric functions can be defined in terms of these two functions as follows.

$$\tan u = \frac{y}{x} = \frac{y/r}{x/r} = \frac{\sin u}{\cos u}, \qquad \cos u \neq 0$$

$$\cot u = \frac{x}{y} = \frac{x/r}{y/r} = \frac{\cos u}{\sin u}, \qquad \sin u \neq 0$$

$$\sec u = \frac{r}{x} = \frac{r/r}{x/r} = \frac{1}{\cos u}, \qquad \cos u \neq 0$$

$$\csc u = \frac{r}{y} = \frac{r/r}{y/r} = \frac{1}{\sin u}, \qquad \sin u \neq 0$$

Signs of the Trigonometric Functions

Since r is always positive, it is clear from the definitions of the trigonometric functions that their signs depend upon x and y. We know that the signs of x and y depend upon the quadrant in which the terminal side of the angle belongs. For example, if the terminal side of the angle is in the second quadrant, then x is negative and y is positive. Therefore the sine and cosecant are positive, while the remaining functions are all negative. For convenience, we write the signs of the trigonometric functions in tabular form. In Table 10.1 we write the signs of the sine, cosine, and tangent functions only. Since the functions cosecant, secant, and cotangent are reciprocals of the functions sine, cosine, and tangent, respectively, their signs can be easily determined from the table or Figure 10.5.

Note $\cot u = \dfrac{\cos u}{\sin u} = \dfrac{1}{\tan u}$

Figure 10.5

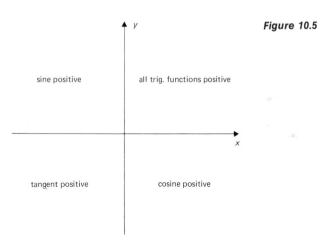

In Table 10.1 note that we have not given the sign of the trigonometric functions for the angles whose terminal sides lie along one of the axes, i.e., for the angles $u = 0$, $u = \pi/2$, $u = \pi$, $u = 3\pi/2$, $u = 5\pi/2$, etc.

Table 10.1

Quadrant	I	II	III	IV
sine	+	+	−	−
cosine	+	−	−	+
tangent	+	−	+	−

Example 10.1   Let the positive angle θ be in standard position. Evaluate the six trigonometric functions of θ if its terminal side passes through the point $(12, -5)$.

SOLUTION   From Figure 10.6 we see that

Figure 10.6

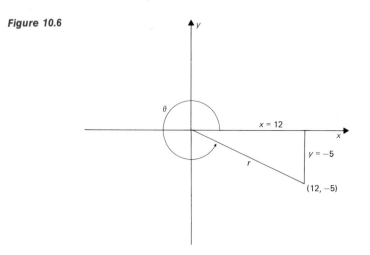

$$r = \sqrt{(12)^2 + (-5)^2} = \sqrt{169} = 13$$

Thus

$$\sin \theta = \frac{y}{r} = \frac{-5}{13}, \qquad \csc \theta = -\frac{13}{5}$$

$$\cos \theta = \frac{x}{r} = \frac{12}{13}, \qquad \sec \theta = \frac{13}{12}$$

$$\tan \theta = \frac{y}{x} = \frac{-5}{12}, \qquad \cot \theta = -\frac{12}{5}$$

Example 10.2   Find the value of the six trigonometric functions of $\pi/2$.

SOLUTION   For the angle $\theta = \pi/2$, we have

$$x = 0, \qquad y = r$$

Therefore $\tan \pi/2$ and $\sec \pi/2$ are undefined, and

$$\sin \frac{\pi}{2} = \frac{y}{r} = 1, \qquad\qquad \cos \frac{\pi}{2} = \frac{x}{r} = \frac{0}{r} = 0$$

$$\csc \frac{\pi}{2} = \frac{r}{y} = \frac{r}{r} = 1, \qquad \cot \frac{\pi}{2} = \frac{x}{y} = \frac{0}{r} = 0$$

Example 10.3 Suppose the terminal side of θ in standard position lies in the third quadrant and $\cos \theta = -\frac{3}{5}$. Find the values of the other trigonometric functions of θ.

SOLUTION In Figure 10.7, we note that

$$r^2 = x^2 + y^2$$

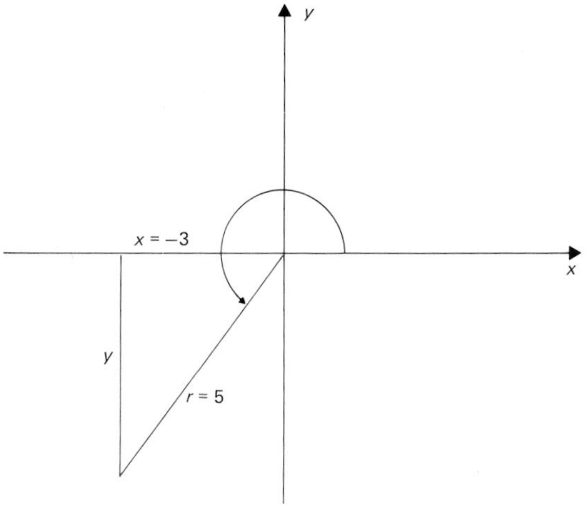

Figure 10.7

Thus $y^2 = r^2 - x^2 = (5)^2 - (-3)^2 = 25 - 9 = 16$ and $y = \pm 4$. Since y is negative in the third quadrant, we have $y = -4$. Hence the values of the remaining trigonometric functions are

$$\sin \theta = -\tfrac{4}{5}, \qquad \tan \theta = \tfrac{4}{3}, \qquad \cot \theta = \tfrac{3}{4}$$

$$\sec \theta = -\tfrac{5}{3}, \qquad \csc \theta = -\tfrac{5}{4}$$

Throughout these problems, assume that the given angle is in standard position.

Exercises 10.1

1. Identify the quadrant in which the terminal side of θ lies in the following cases.
    (a) $\sin \theta < 0, \quad \cos \theta > 0$ (b) $\cot \theta < 0, \quad \csc \theta > 0$
    (c) $\tan \theta > 0, \quad \sec \theta > 0$ (d) $\sec \theta < 0, \quad \tan \theta > 0$

2. In which quadrants may the terminal side of θ lie in the following cases?
    (a) $\cot \theta > 0$ (b) $\csc \theta > 0$ (c) $\cos \theta > 0$

3. What are the values of $\sin \theta$, $\cos \theta$, and $\tan \theta$ if the terminal side of θ passes through the point P?
    (a) $P(-7, 24)$ (b) $P(-7, -12)$ (c) $P(-3, 4)$

4. Evaluate $\sin \theta$, $\cot \theta$, and $\sec \theta$ in the following cases.
    (a) $\cos \theta = \frac{10}{13}$ (b) $\csc \theta = -\frac{4}{3}$

5. Evaluate the six trigonometric functions of θ for the following.

    (a) $\theta = 0$ (b) $\theta = \pi$ (c) $\theta = \dfrac{3\pi}{2}$

6. If the terminal side of θ belongs to the fourth quadrant and $\sec \theta = \sqrt{2}$, show that

$$\frac{1 + \tan \theta + \csc \theta}{1 + \cot \theta - \csc \theta} = -1$$

10.2   Graphs of Trigonometric Functions

In Section 10.1, we indicated that the sine and cosine functions are defined for all real numbers x, and each function has $I = \{y | -1 \le y \le 1\}$ as its range. It will be shown (in Section 10.5) that these two functions are differentiable everywhere (and hence are continuous). Consequently, to sketch the graphs of these functions, we need to know their values at several points. First we find the values of these functions at the so-called *special angles*, namely, the angles 0, $\pi/6$, $\pi/4$, $\pi/3$, and $\pi/2$. Then we use the results to compute their values at several other angles.

From Section 10.1 we know that

$$\sin 0 = 0, \qquad \sin \frac{\pi}{2} = 1 \qquad \cos 0 = 1, \qquad \cos \frac{\pi}{2} = 0$$

Now let us calculate the values of the trigonometric functions at $\pi/4$. Since $\pi/4 = 45°$, the right triangle in Figure 10.8 is an isosceles tri-

Figure 10.8

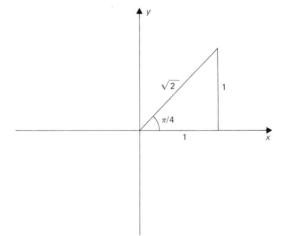

angle. Thus if we choose $x = 1$, then $y = 1$, and $r = \sqrt{1^2 + 1^2} = \sqrt{2}$. Consequently,

$$\sin \frac{\pi}{4} = \frac{1}{\sqrt{2}} \qquad \cos \frac{\pi}{4} = \frac{1}{\sqrt{2}}$$

Knowing the sine and cosine of $\pi/4$, we can also calculate the values of the other trigonometric functions at $\pi/4$.

$$\tan \frac{\pi}{4} = \frac{\sin \frac{\pi}{4}}{\cos \frac{\pi}{4}} = 1 \qquad \cot \frac{\pi}{4} = \frac{1}{\tan \frac{\pi}{4}} = 1$$

$$\csc \frac{\pi}{4} = \frac{1}{\sin \frac{\pi}{4}} = \sqrt{2} \qquad \sec \frac{\pi}{4} = \frac{1}{\cos \frac{\pi}{4}} = \sqrt{2}$$

We now wish to calculate the values of the trigonometric functions at $\pi/3$. Before we do so, let us consider the equilateral triangle in Figure 10.9. Each side of this triangle is 2 units long and each angle is

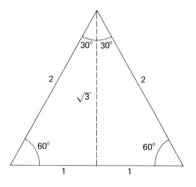

Figure 10.9

$60° = \pi/3$ rad. If we divide this triangle into two right triangles (by dropping a perpendicular from the vertex to the base) as shown in Figure 10.9, then by the Pythagorean theorem, we find the altitude of this triangle to be $\sqrt{3}$.

Thus placing the angle $\pi/3$ in standard position as in Figure 10.10 and choosing $r = 2$, we have $x = 1$, $y = \sqrt{3}$. Therefore

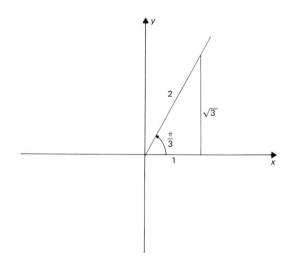

Figure 10.10

$$\sin \frac{\pi}{3} = \frac{\sqrt{3}}{2} \qquad \csc \frac{\pi}{3} = \frac{2}{\sqrt{3}}$$

$$\cos \frac{\pi}{3} = \frac{1}{2} \qquad \sec \frac{\pi}{3} = 2$$

$$\tan \frac{\pi}{3} = \sqrt{3} \qquad \cot \frac{\pi}{3} = \frac{1}{\sqrt{3}}$$

From the preceding argument, we can also deduce the values of the trigonometric functions at $\pi/6 = 30°$ (see Figure 10.11).

Figure 10.11

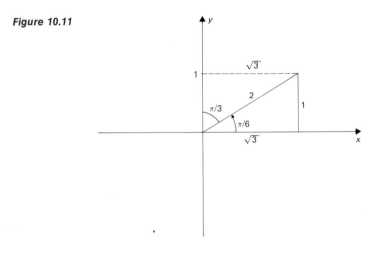

$$\sin \frac{\pi}{6} = \frac{1}{2} \qquad \csc \frac{\pi}{6} = 2$$

$$\cos \frac{\pi}{6} = \frac{\sqrt{3}}{2} \qquad \sec \frac{\pi}{6} = \frac{2}{\sqrt{3}}$$

$$\tan \frac{\pi}{6} = \frac{1}{\sqrt{3}} \qquad \cot \frac{\pi}{6} = \sqrt{3}$$

The technique followed in Table 10.2 should be useful in memorizing the values of the sine and cosine of 0, $\pi/6$, $\pi/4$, $\pi/3$, $\pi/2$.

Table 10.2

angle	0	$\pi/6$	$\pi/4$	$\pi/3$	$\pi/2$
	0°	30°	45°	60°	90°
sine	$\sqrt{0/4}$	$\sqrt{1/4}$	$\sqrt{2/4}$	$\sqrt{3/4}$	$\sqrt{4/4}$
cosine	$\sqrt{4/4}$	$\sqrt{3/4}$	$\sqrt{2/4}$	$\sqrt{1/4}$	$\sqrt{0/4}$

We observe that the value of r is not a factor in the final values of the trigonometric functions. Therefore we can use any value of r that we wish. Thus we shall use $r = 1$ from now on.

We shall presently see that it is sufficient to know the values of the trigonometric functions only for the acute angles (angles in the first quadrant). From these values we can compute the values for angles in other quadrants. We now state an important theorem, whose proof is omitted.

Theorem 10.1 *For θ, a real number,* ◀

$$\sin(-\theta) = -\sin\theta \qquad (10.1)$$
$$\cos(-\theta) = \cos\theta \qquad (10.2)$$

$$\sin\left(\frac{\pi}{2} - \theta\right) = \cos\theta \qquad (10.3)$$
$$\cos\left(\frac{\pi}{2} - \theta\right) = \sin\theta \qquad (10.4)$$

$$\sin\left(\frac{\pi}{2} + \theta\right) = \cos\theta \qquad (10.5)$$
$$\cos\left(\frac{\pi}{2} + \theta\right) = -\sin\theta \qquad (10.6)$$

$$\sin(\pi + \theta) = -\sin\theta \qquad (10.7)$$
$$\cos(\pi + \theta) = -\cos\theta \qquad (10.8)$$

$$\sin(\pi - \theta) = \sin\theta \qquad (10.9)$$
$$\cos(\pi - \theta) = -\cos\theta \qquad (10.10)$$

$$\sin\left(\frac{3\pi}{2} + \theta\right) = -\cos\theta \qquad (10.11)$$
$$\cos\left(\frac{3\pi}{2} + \theta\right) = \sin\theta \qquad (10.12)$$

$$\sin\left(\frac{3\pi}{2} - \theta\right) = -\cos\theta \qquad (10.13)$$
$$\cos\left(\frac{3\pi}{2} - \theta\right) = -\sin\theta \qquad (10.14)$$

$$\sin(2\pi - \theta) = -\sin\theta \qquad (10.15)$$
$$\cos(2\pi - \theta) = \cos\theta \qquad (10.16)$$

$$\sin(2\pi + \theta) = \sin\theta \qquad (10.17)$$
$$\cos(2\pi + \theta) = \cos\theta \qquad (10.18)$$

A useful scheme for remembering these results is the following: Suppose that we pair the six trigonometric functions as follows.

$$\left\{ \begin{matrix} \sin\theta \\ \cos\theta \end{matrix} \right. \qquad \left\{ \begin{matrix} \tan\theta \\ \cot\theta \end{matrix} \right. \qquad \left\{ \begin{matrix} \sec\theta \\ \csc\theta \end{matrix} \right.$$

These pairings are called the *cofunctional pairs*. The second of each pair is called the cofunction of the first, and vice versa. With this nomenclature in mind, and denoting by $f(\theta)$ any one of the six trigonometric functions and $\text{cof}(\theta)$ its cofunction, we can state that

$$f\left(\theta \pm n\frac{\pi}{2}\right) = \begin{cases} \pm f(\theta), & n \text{ even} \\ \pm \text{cof}(\theta), & n \text{ odd} \end{cases} \qquad (10.19)$$

The sign $\pm$ in the second member must be chosen to correspond to the sign of $f(\theta \pm n\pi/2)$. For example, if θ is an acute angle, then

$$\sec\left(\theta + \frac{3\pi}{2}\right) = \pm\csc\theta$$

since $n = 3$ is odd. Because $\theta + 3\pi/2$ has its terminal side in the fourth quadrant, the cosine and the secant are positive; thus

$$\sec\left(\theta + \frac{3\pi}{2}\right) = \csc\theta$$

Example 10.4 Show that

$$\frac{4}{3}\cot^2\frac{\pi}{6} + 3\,\sin^2\frac{\pi}{3} - 2\,\csc^2\frac{\pi}{3} - \frac{3}{4}\tan^2\frac{\pi}{6} = \frac{10}{3}$$

SOLUTION By direct substitution, the left-hand side (LHS) of the preceding equation is

$$\frac{4}{3}(\sqrt{3})^2 + 3\left(\frac{\sqrt{3}}{2}\right)^2 - 2\left(\frac{2}{\sqrt{3}}\right)^2 - \frac{3}{4}\left(\frac{1}{\sqrt{3}}\right)^2$$

$$= \frac{4}{3}\cdot 3 + 3\cdot\frac{3}{4} - 2\cdot\frac{4}{3} - \frac{3}{4}\cdot\frac{1}{3} = 4 + \frac{9}{4} - \frac{8}{3} - \frac{1}{4} = \frac{10}{3}$$

Example 10.5 Evaluate

(a) $\tan\left(-\frac{11\pi}{3}\right)$ (b) $\sec\left(\frac{11\pi}{4}\right)$

SOLUTION

(a) $\tan\left(-\frac{11\pi}{3}\right) = -\tan\frac{11\pi}{3} = -\tan\left(3\pi + \frac{2\pi}{3}\right) = -\tan\frac{2\pi}{3}$

$$= -\tan\left(\frac{\pi}{2} + \frac{\pi}{6}\right) = \cot\frac{\pi}{6} = \frac{\cos\dfrac{\pi}{6}}{\sin\dfrac{\pi}{6}} = \sqrt{3}$$

(b) $\sec\dfrac{11\pi}{4} = \dfrac{1}{\cos\dfrac{11\pi}{4}} = \dfrac{1}{\cos\left(2\pi + \dfrac{3\pi}{4}\right)} = \dfrac{1}{\cos\dfrac{3\pi}{4}}$

$$= \frac{1}{\cos\left(\pi - \dfrac{\pi}{4}\right)} = \frac{1}{-\cos\dfrac{\pi}{4}} = -\sqrt{2}$$

The trigonometric functions belong to a large class of functions called *periodic* functions, in which there is a regular repetition of the values of the function over a certain interval. For example, we observe that as θ increases from 0 to 2π, the point $P(x, y)$, such that $x^2 + y^2 = 1$, completes one revolution around the unit circle. Then, as θ continues to increase, the point $P(x, y)$ begins its second trip around the unit circle and the trigonometric functions begin to repeat their earlier behavior.

Periodicity ▶ *Definition 10.4   A function f is called* periodic *if there exists a nonzero fixed number p, such that f(x + p) is defined and*

$$f(x + p) = f(x) \qquad\qquad (10.20)$$

for all x in the domain of f. If p is the smallest positive number for which (10.20) *holds, then p is called the* period *of the function f.*

We can see that if P (see Figure 10.12) represents the real number

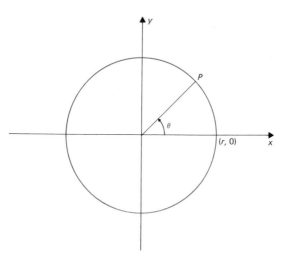

Figure 10.12

θ $(0 \le \theta < 2\pi)$ on the circumference of a circle of radius r, then P also represents all real numbers which are elements of the set

$$S = \{\alpha \mid \alpha = 2\pi n + \theta;\ n \in Z\}$$

From the definition of the sine and cosine functions of a real number θ, we find that these functions depend on the coordinates (x, y) of the point P corresponding to the number θ. Since all real numbers of the set S are represented by the same point P, it follows that the values of the sine and cosine functions are equal for every element of S.

Consequently, for a real number θ and $n \in Z$,

$$\sin(\theta + 2\pi n) = \sin\theta \quad \text{and} \quad \cos(\theta + 2\pi n) = \cos\theta$$

If $p = 2\pi n$, then

$$\sin(\theta + p) = \sin\theta \quad \text{and} \quad \cos(\theta + p) = \cos\theta$$

It can be shown that the smallest value of $|p|\ (p \ne 0)$ with the property just mentioned is 2π. Hence the sine and cosine functions are periodic with period 2π.

Since $\sec\theta = 1/\cos\theta$ and $\csc\theta = 1/\sin\theta$, the secant and cosecant functions are also periodic functions with period 2π. Since

$$\tan(\theta + \pi) = \frac{\sin(\theta + \pi)}{\cos(\theta + \pi)} = \frac{-\sin\theta}{-\cos\theta} = \tan\theta$$

the tangent function and the cotangent function (cot $\theta = 1/\tan \theta$) are periodic functions with period π.

Graph of $y = \sin \theta$

Since the sine function is periodic with period 2π, the graph of this function repeats the basic shape obtained on the interval $[0, 2\pi]$. Table 10.3 summarizes the variation of $\sin \theta$ as θ varies from 0 to 2π.

Table 10.3

θ varies	$\sin \theta$ varies
0 to $\dfrac{\pi}{2}$	0 to 1
$\dfrac{\pi}{2}$ to π	1 to 0
π to $\dfrac{3\pi}{2}$	0 to -1
$\dfrac{3\pi}{2}$ to 2π	-1 to 0

Having noted a broad pattern of the variation of the values of $y = \sin \theta$, we can obtain a good approximation of the graph by plotting a number of points $(\theta, \sin \theta)$ as θ takes values from 0 to 2π.

Table 10.4 gives the corresponding values of θ and $\sin \theta$, and Figure 10.13 shows the rough sketch of the graph of $y = \sin \theta$.

Table 10.4

θ	$\sin \theta$	θ	$\sin \theta$
0	0	π	0
$\dfrac{\pi}{6}$	$\dfrac{1}{2}$	$\dfrac{7\pi}{6}$	$-\dfrac{1}{2}$
$\dfrac{\pi}{4}$	$\dfrac{1}{\sqrt{2}}$	$\dfrac{5\pi}{4}$	$-\dfrac{1}{\sqrt{2}}$
$\dfrac{\pi}{3}$	$\dfrac{\sqrt{3}}{2}$	$\dfrac{4\pi}{3}$	$-\dfrac{\sqrt{3}}{2}$
$\dfrac{\pi}{2}$	1	$\dfrac{3\pi}{2}$	-1
$\dfrac{2\pi}{3}$	$\dfrac{\sqrt{3}}{2}$	$\dfrac{5\pi}{3}$	$-\dfrac{\sqrt{3}}{2}$
$\dfrac{3\pi}{4}$	$\dfrac{1}{\sqrt{2}}$	$\dfrac{7\pi}{4}$	$-\dfrac{1}{\sqrt{2}}$
$\dfrac{5\pi}{6}$	$\dfrac{1}{2}$	$\dfrac{11\pi}{6}$	$-\dfrac{1}{2}$
		2π	0

Figure 10.13

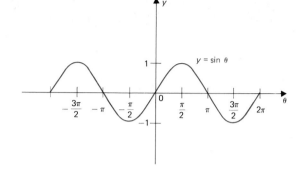

The tangent function is periodic with period π. Thus the graph of $y = \tan \theta$ repeats the basic shape obtained on the interval $[0, \pi]$. Table 10.5 gives the variation of $y = \tan \theta$ as θ varies from 0 to π. We note that $\tan \pi/2$ is undefined. Therefore the graph has a *discontinuity* or break at $\theta = \pi/2$.

Graph of $y = \tan \theta$

θ varies	$\tan \theta$ varies
0 to $\dfrac{\pi}{2}$	0 to $+\infty$
$\dfrac{\pi}{2}$ to π	$-\infty$ to 0

Table 10.5

Table 10.6 gives the corresponding values of θ and $\tan \theta$ as θ varies from 0 to π, and Figure 10.14 shows the rough sketch of the graph of $y = \tan \theta$.

θ	$\tan \theta$	θ	$\tan \theta$
0	0	$\left(\dfrac{\pi}{2}\right)+$	$-\infty$
$\dfrac{\pi}{6}$	$\dfrac{1}{\sqrt{3}} \approx 0.58$	$\dfrac{2\pi}{3}$	$-\sqrt{3}$
$\dfrac{\pi}{4}$	1	$\dfrac{3\pi}{4}$	-1
$\dfrac{\pi}{3}$	$\sqrt{3} \approx 1.73$	$\dfrac{5\pi}{6}$	$-\dfrac{1}{\sqrt{3}}$
$\left(\dfrac{\pi}{2}\right)-$	$+\infty$	π	0

Table 10.6

$\left(\dfrac{\pi}{2}\right)-$ *means that* θ *approaches* $\pi/2$ *through values less than* $\pi/2$, *and* $\left(\dfrac{\pi}{2}\right)+$ *means that* θ *approaches* $\pi/2$ *through values greater than* $\pi/2$.

Figure 10.14

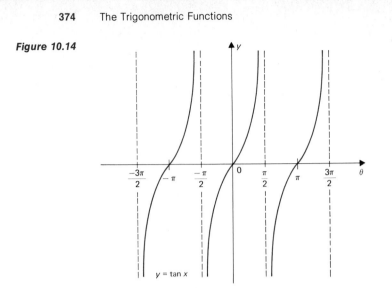

$y = \tan x$

As indicated earlier, the secant function is periodic with period 2π, and it suffices to sketch the graph of $y = \sec\ \theta = 1/\cos\ \theta$ over $[0, 2\pi]$. Table 10.7 presents the values of θ and $\sec\ \theta$ as θ varies from 0 to 2π, and Figure 10.15 shows the sketch of the graph.

Graph of $y = \sec\ \theta$

Figure 10.15

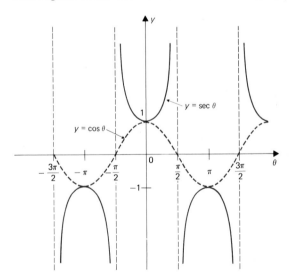

$y = \sec\ \theta$

$y = \cos\ \theta$

Exercises 10.2

1. Find the sines and cosines of the following real numbers.

   (a) $\dfrac{3\pi}{4}$ (b) $\dfrac{16\pi}{3}$ (c) $-\dfrac{47\pi}{6}$ (d) $\dfrac{73\pi}{6}$

2. Express the following in terms of trigonometric functions of positive real numbers less than $\pi/2$.

   (a) $\sin \dfrac{7\pi}{13}$ (b) $\csc \dfrac{19\pi}{7}$ (c) $\sec \dfrac{101\pi}{3}$

   (d) $\cot \left(-\dfrac{13\pi}{4}\right)$ (e) $\tan \left(-\dfrac{5\pi}{3}\right)$ (f) $\cos \left(-\dfrac{2\pi}{3}\right)$

Table 10.7

θ	sec θ	θ	sec θ
0	1	π	-1
$\dfrac{\pi}{6}$	$\dfrac{2}{\sqrt{3}}$	$\dfrac{7\pi}{6}$	$-\dfrac{2}{\sqrt{3}}$
$\dfrac{\pi}{4}$	$\sqrt{2}$	$\dfrac{5\pi}{4}$	$-\sqrt{2}$
$\dfrac{\pi}{3}$	2	$\dfrac{4\pi}{3}$	-2
$\left(\dfrac{\pi}{2}\right)-$	$+\infty$	$\left(\dfrac{3\pi}{2}\right)-$	$-\infty$
$\left(\dfrac{\pi}{2}\right)+$	$-\infty$	$\left(\dfrac{3\pi}{2}\right)+$	$+\infty$
$\dfrac{2\pi}{3}$	-2	$\dfrac{5\pi}{3}$	2
$\dfrac{3\pi}{4}$	$-\sqrt{2}$	$\dfrac{7\pi}{4}$	$\sqrt{2}$
$\dfrac{5\pi}{6}$	$-\dfrac{2}{\sqrt{3}}$	$\dfrac{11\pi}{6}$	$\dfrac{2}{\sqrt{3}}$
		2π	1

In Problems 3 through 7, show that the given equations are true.

3. $\sin\dfrac{\pi}{3}\cos\dfrac{\pi}{6}+\cos\dfrac{\pi}{3}\sin\dfrac{\pi}{6}=1$

4. $\cos^2\dfrac{\pi}{4}-\sin^2\dfrac{\pi}{4}=\cos\dfrac{\pi}{2}$

5. $\tan^2\dfrac{\pi}{3}+4\cos^2\dfrac{\pi}{4}+3\cos^2\dfrac{\pi}{3}=\dfrac{23}{4}$

6. $\dfrac{1}{2}\sin^2 30°+\dfrac{1}{4}\tan^2 45°+\dfrac{5}{4}\sec^3 60°=\dfrac{41}{4}$

7. $\tan^2\dfrac{4\pi}{3}+\sin^2\dfrac{2\pi}{3}+\cos^2\dfrac{4\pi}{3}=\sec^2\dfrac{\pi}{3}$

8. Sketch the graph of the function $f(x)=\cos x$.

9. Sketch the graph of the function $f(x)=\cot x$.

10. Sketch the graph of the function $f(x)=\csc x$.

10.3 Some Important Formulas

In Section 10.1, we came across some of the basic trigonometric identities. For example,

$$\tan \theta = \frac{\sin \theta}{\cos \theta} \quad \text{and} \quad \sec \theta = \frac{1}{\cos \theta}$$

In this section, we shall first discuss some of the consequences of one of the most basic identities, the Pythagorean identity. We state and prove this in the next theorem.

▶   *Theorem 10.2*   *For any real number* θ,

$$\sin^2 \theta + \cos^2 \theta = 1 \tag{10.21}$$

PROOF   Let θ be any angle in standard position, and let $P(x, y)$ be a point on the terminal side of θ, such that $d(0, P) = r$. Then by the Pythagorean theorem,

$$x^2 + y^2 = r^2 \tag{10.22}$$

Dividing both sides of (10.22) by r^2 ($r^2 \neq 0$), we obtain

$$\left(\frac{x}{r}\right)^2 + \left(\frac{y}{r}\right)^2 = 1 \tag{10.23}$$

But from the definition of the trigonometric functions, we have

$$\cos \theta = \frac{x}{r}, \quad \sin \theta = \frac{y}{r}$$

Therefore from (10.23) we get

$$\cos^2 \theta + \sin^2 \theta = 1$$

which proves Theorem 10.2.

Example 10.6   For any real number θ, prove the following.

Why do we require $\sin \theta \neq 0$?

$$1 + \cot^2 \theta = \csc^2 \theta, \quad \sin \theta \neq 0 \tag{10.24}$$

$$1 + \tan^2 \theta = \sec^2 \theta, \quad \cos \theta \neq 0 \tag{10.25}$$

SOLUTION   Dividing both sides of (10.21) by $\sin^2 \theta$, we get

$$\left(\frac{\sin \theta}{\sin \theta}\right)^2 + \left(\frac{\cos \theta}{\sin \theta}\right)^2 = \left(\frac{1}{\sin \theta}\right)^2$$

or

$$1 + \cot^2 \theta = \csc^2 \theta$$

Similarly, dividing both sides of (10.21) by $\cos^2 \theta$, we get

$$\left(\frac{\sin \theta}{\cos \theta}\right)^2 + \left(\frac{\cos \theta}{\cos \theta}\right)^2 = \left(\frac{1}{\cos \theta}\right)^2$$

or

$$\tan^2 \theta + 1 = \sec^2 \theta$$

Example 10.7 If $\cot \theta = \frac{12}{5}$, $\pi < \theta < 3\pi/2$, find the values of $\sec \theta$ and $\sin \theta$.

SOLUTION From (10.24) we obtain

$$\csc^2 \theta = 1 + \cot^2 \theta = 1 + \left(\frac{12}{5}\right)^2 = 1 + \frac{144}{25} = \frac{169}{25}$$

Thus we have

$$\csc \theta = \pm \frac{13}{5}$$

Since $\pi < \theta < 3\pi/2$, $\csc \theta$ is negative. Therefore,

$$\csc \theta = -\frac{13}{5}$$

and hence

$$\sin \theta = \frac{1}{\csc \theta} = -\frac{5}{13}$$

We find $\sec \theta$ through a series of identities:

$$\sec \theta = \frac{1}{\cos \theta} = \frac{\sin \theta}{\cos \theta} \cdot \frac{1}{\sin \theta} = \tan \theta \cdot \csc \theta = \left(\frac{5}{12}\right)\left(-\frac{13}{5}\right) = -\frac{13}{12}$$

Example 10.8 Prove that

$$\frac{\sin \theta}{1 + \cos \theta} + \frac{1 + \cos \theta}{\sin \theta} = 2 \csc \theta, \qquad \sin \theta \neq 0$$

SOLUTION $\displaystyle \frac{\sin \theta}{1 + \cos \theta} + \frac{1 + \cos \theta}{\sin \theta} = \frac{\sin^2 \theta + (1 + \cos \theta)^2}{\sin \theta \, (1 + \cos \theta)}$

$$= \frac{\sin^2 \theta + 1 + 2 \cos \theta + \cos^2 \theta}{\sin \theta \, (1 + \cos \theta)}$$

$$= \frac{2 + 2 \cos \theta}{\sin \theta \, (1 + \cos \theta)} \qquad \sin^2 \theta + \cos^2 \theta = 1;$$

$$\qquad\qquad\qquad\qquad \sin \theta \neq 0 \Rightarrow \cos \theta \neq -1$$

$$= \frac{2(1 + \cos \theta)}{\sin \theta \, (1 + \cos \theta)}$$

$$= \frac{2}{\sin \theta} = 2 \csc \theta$$

Let us consider a function f with domain D. Let x_1 and x_2 be any two elements of D, such that $x_1 + x_2 \in D$. It is often of interest and importance to know how $f(x_1 + x_2)$ is related to $f(x_1)$ and $f(x_2)$. A formula that connects these quantities in some way may be called an *addition formula*. We see immediately that a linear function f, defined by *Addition formula* $f(x) = mx$ (for some real number m), satisfies the relation

$$f(x_1 + x_2) = f(x_1) + f(x_2) \qquad (10.26)$$

It is easy to see that the cosine function does not satisfy (10.26). For example, let $x_1 = \pi/6$ and $x_2 = \pi/6$. Then

$$\cos\left(\frac{\pi}{6} + \frac{\pi}{6}\right) = \cos\left(\frac{\pi}{3}\right) = \frac{1}{2},$$

whereas

$$\cos\left(\frac{\pi}{6}\right) + \cos\left(\frac{\pi}{6}\right) = \frac{\sqrt{3}}{2} + \frac{\sqrt{3}}{2} = \sqrt{3},$$

and

$$\sqrt{3} \neq \frac{1}{2}$$

Thus, in general,

$$\cos(x_1 + x_2) \neq \cos x_1 + \cos x_2$$

Now we state (without proof) the addition formulas for trigonometric functions.

▶ *Theorem 10.3* *For any real numbers α and β,*

$$\cos(\alpha - \beta) = \cos\alpha \cos\beta + \sin\alpha \sin\beta \qquad (10.27)$$

$$\cos(\alpha + \beta) = \cos\alpha \cos\beta - \sin\alpha \sin\beta \qquad (10.28)$$

$$\sin(\alpha - \beta) = \sin\alpha \cos\beta - \cos\alpha \sin\beta \qquad (10.29)$$

$$\sin(\alpha + \beta) = \sin\alpha \cos\beta + \cos\alpha \sin\beta \qquad (10.30)$$

$$\tan(\alpha - \beta) = \frac{\tan\alpha - \tan\beta}{1 + \tan\alpha \tan\beta} \qquad (10.31)$$

$$\tan(\alpha + \beta) = \frac{\tan\alpha + \tan\beta}{1 - \tan\alpha \tan\beta} \qquad (10.32)$$

Remark *All the formulas (10.27) through (10.32) are valid when α and β are multiples of certain other numbers. For example,*

(i) If $\alpha = 2x$ and $\beta = 3y$, then

$\cos(2x + 3y) =$
$\quad \cos 2x \cos 3y - \sin 2x \sin 3y$

(ii) If $\alpha = x/2$ and $\beta = y/5$, then

$\sin\left(\dfrac{x}{2} - \dfrac{y}{5}\right) =$

$\sin\dfrac{x}{2}\cos\dfrac{y}{5} - \cos\dfrac{x}{2}\sin\dfrac{y}{5}$

From Theorem 10.3 we can also establish the following identities.

$$\cos 2\theta = \cos^2\theta - \sin^2\theta \qquad (10.33)$$

$$\cos 2\theta = 2\cos^2\theta - 1 \qquad (10.34)$$

$$\cos 2\theta = 1 - 2\sin^2\theta \qquad (10.35)$$

$$\cos 2\theta = \frac{1 - \tan^2\theta}{1 + \tan^2\theta} \qquad (10.36)$$

$$\cos\theta = \cos^2\frac{\theta}{2} - \sin^2\frac{\theta}{2} \qquad (10.37)$$

$$\cos\theta = 2\cos^2\frac{\theta}{2} - 1 \qquad (10.38)$$

$$\cos \theta = 1 - 2 \sin^2 \frac{\theta}{2} \tag{10.39}$$

$$\cos \theta = \frac{1 - \tan^2 \frac{\theta}{2}}{1 + \tan^2 \frac{\theta}{2}} \tag{10.40}$$

$$\sin 2\theta = 2 \sin \theta \cos \theta \tag{10.41}$$

$$\sin 2\theta = \frac{2 \tan \theta}{1 + \tan^2 \theta} \tag{10.42}$$

$$\sin \theta = 2 \sin \frac{\theta}{2} \cos \frac{\theta}{2} \tag{10.43}$$

$$\sin \theta = \frac{2 \tan \frac{\theta}{2}}{1 + \tan^2 \frac{\theta}{2}} \tag{10.44}$$

$$\tan 2\theta = \frac{2 \tan \theta}{1 - \tan^2 \theta} \tag{10.45}$$

$$\tan \theta = \frac{2 \tan \frac{\theta}{2}}{1 - \tan^2 \frac{\theta}{2}} \tag{10.46}$$

In Problems 1 through 9, prove the given identities.   Exercises 10.3

1. $\sin^4 \theta - \cos^4 \theta = \sin^2 \theta - \cos^2 \theta$

2. $\sin^6 \theta + \cos^6 \theta = 1 - 3 \sin^2 \theta \cos^2 \theta$

3. $\cot^4 \theta + \cot^2 \theta = \csc^4 \theta - \csc^2 \theta$

4. $\csc^6 \theta - \cot^6 \theta = 1 + 3 \csc^2 \theta \cot^2 \theta$

5. $\dfrac{1}{1 + \tan \theta} = \dfrac{\cot \theta}{1 + \cot \theta}$

6. $\dfrac{1}{1 + \cos \theta} + \dfrac{1}{1 - \cos \theta} = 2 \csc^2 \theta$

7. $(\sin \theta - \cos \theta)(\tan \theta + \cot \theta) = \sec \theta - \csc \theta$

8. $(\csc \theta - \sin \theta)(\sec \theta - \cos \theta) = \dfrac{1}{\tan \theta + \cot \theta}$

9. $\dfrac{1 - \tan^2 \theta}{1 + \tan^2 \theta} = 2 \cos^2 \theta - 1$

10. Find $\sin 15°$, $\cos 15°$. [*Hint*: $15° = 45° - 30° = [(\pi/4) - (\pi/6)]$ rad.]

11. Find $\sin 75°$, $\cos 75°$.

12. Suppose α and β are acute angles, $\sin \alpha = \frac{8}{17}$ and $\sin \beta = \frac{5}{13}$. Prove the following.

(a) $\sin (\alpha - \beta) = \frac{21}{221}$ (b) $\cos (\alpha + \beta) = \frac{140}{221}$

13. Prove that

$$\cot (\alpha - \beta) = \frac{\cot \alpha \cot \beta + 1}{\cot \beta - \cot \alpha}$$

14. Prove that

$$\tan \left(\frac{\pi}{4} + \theta\right) \tan \left(\frac{\pi}{4} - \theta\right) = 1$$

10.4   Inverse Trigonometric Functions

In Chapter 2 we saw that a function f does not necessarily have an inverse function f^{-1}. We recall that f^{-1} exists if f is one-to-one. In Section 10.2 we noted that all the trigonometric functions are periodic. The period for the sine and cosine functions is 2π, whereas the period for the tangent function is π. Thus, for example, if

$$\sin x = \frac{1}{2}$$

then,

$$x = \frac{\pi}{6} \pm 2n\pi \quad \text{or} \quad \frac{5\pi}{6} \pm 2n\pi, \quad n \in I$$

Therefore the sine function is not one-to-one, and hence its inverse function does not exist.

However, we can restrict the domain (that is, choose a subset of the domain) over which the sine function is defined so that the restricted function is one-to-one. We see from the graph of the sine function that there are many alternative ways of choosing an interval over which the sine function has an inverse. For example, some of these subintervals are

To determine from the graph of a function $y = f(x)$ whether or not the function f has an inverse, we draw horizontal lines through each point in the range of f on the y-axis. If each such line cuts the graph of f at exactly one point, then f has an inverse.

$$I_1 = \left\{x \,\middle|\, -\frac{\pi}{2} \le x \le \frac{\pi}{2}\right\}$$

$$I_2 = \left\{x \,\middle|\, -\frac{3\pi}{2} \le x \le -\frac{\pi}{2}\right\}$$

$$I_3 = \left\{x \,\middle|\, \frac{\pi}{2} \le x \le \frac{3\pi}{2}\right\}$$

In such cases as these, it is customary to choose that interval for which the following conditions are satisfied.

(a) $|x|$ is the smallest value corresponding to a given value of the function.

(b) The full range is maintained for the values of the function.

With (a) and (b) in mind, such an interval for the sine function is the set

$$\left\{x \;\middle|\; -\frac{\pi}{2} \le x \le \frac{\pi}{2}\right\}$$

Hence the restricted sine function defined over the interval $\{x \mid -\pi/2 \le x \le \pi/2\}$ has an inverse function. Further, the domain of the inverse function is the set $\{y \mid -1 \le y \le 1\}$, and the range is the set $\{x \mid -\pi/2 \le x \le \pi/2\}$. Formally, we have the following definition.

Definition 10.5 *The restricted sine function with the domain* $[-\pi/2,\ \pi/2]$ *has an inverse function called the* arcsine function, *defined by the relation* ◀

$$y = \arcsin x \qquad \text{if and only if} \qquad x = \sin y,\ -\frac{\pi}{2} \le y \le \frac{\pi}{2}$$

When considering the relation $y = \arcsin x$, it is helpful to think "y is the angle whose sine is x."

Let $y = f(x) = \arcsin x$. The domain and the range of the arcsine function are $-1 \le x \le 1$ and $-\pi/2 \le y \le \pi/2$, respectively. We can construct the following table to give us the values for the graph of the arcsine function (see Figure 10.16).

Table 10.8

x	$y = \arcsin x$
-1	$-\pi/2$
$-\sqrt{3}/2$	$-\pi/3$
$-1/\sqrt{2}$	$-\pi/4$
$-1/2$	$-\pi/6$
0	0
$1/2$	$\pi/6$
$1/\sqrt{2}$	$\pi/4$
$\sqrt{3}/2$	$\pi/3$
1	$\pi/2$

Figure 10.16

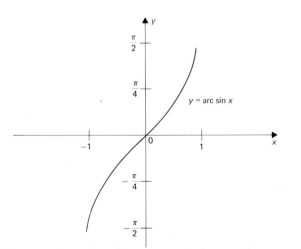

The graph of the arcsine function can be constructed in an indirect fashion as well. In Chapter 2 we indicated that the graph of f^{-1} can be constructed from the graph of f by reflecting the graph of f through the line $y = x$. Consequently, we draw the graph of $y = \sin x$, $-\pi/2 \le x \le \pi/2$, and from this obtain the graph of $y = \arcsin x$, $-1 \le x \le 1$, as shown in Figure 10.17. Since $\csc x = 1/\sin x$, $\sin x \ne 0$, both of these functions have the same period. By a similar argument we define the inverse cosecant function.

Figure 10.17

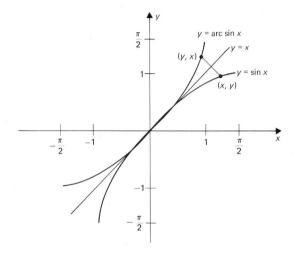

Inverse cosecant function ▶

Definition 10.6 *The* inverse cosecant function *is defined by the correspondence*

$$y = \operatorname{arccsc} x \quad \textit{if and only if } \csc y = x, \; -\frac{\pi}{2} \le y \le \frac{\pi}{2}, \, y \ne 0$$

The domain of the arccsc *function is* $\{x \, | \, 1 \le |x|\}$ *and its range is* $\{y \, | -\pi/2 \le y \le \pi/2, \; y \ne 0\}$.

The graph of the arccsc function can be obtained either by reflecting the graph of $y = \csc x$, $-\pi/2 \le x \le \pi/2$, $x \ne 0$, through the line $y = x$; or by constructing a table of its coordinates. Figure 10.18 is a rough sketch of the graph of $y = \operatorname{arccsc} x$, $1 \le |x|$.

We now look for the definition of the *inverse tangent function*. The tangent function $y = \tan x$ is defined with the domain $\{x \, | \, x \in R; \; x \ne (2n + 1)\pi/2, \, n \in I\}$, and range $\{y \, | \, y \in R\}$. From the graph of the tangent function, we can see that this function is not one-to-one and thus cannot have an inverse. However, if the domain is restricted to $\{x \, | \, -\pi/2 < x < \pi/2\}$, then the restricted tangent function is one-to-one. Consequently, we have

Figure 10.18

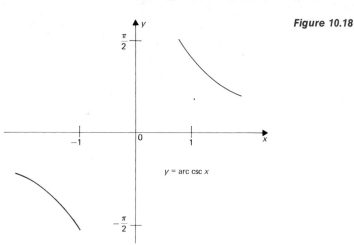

y = arc csc x

Definition 10.7 The restricted tangent function with the domain $\{x|-\pi/2 < x < \pi/2\}$ *has an inverse function called the* arctan function, *defined by the relation*   ◄   arctan function

$$y = \arctan x \quad \textit{if and only if} \quad x = \tan y, \quad -\frac{\pi}{2} < y < \frac{\pi}{2}$$

The domain of the arctan function is $\{x|x \in R\}$ *and its range is* $\{y|-\pi/2 < y < \pi/2\}$.

The graph of the arctangent function (see Figure 10.19) can be ob-

Figure 10.19

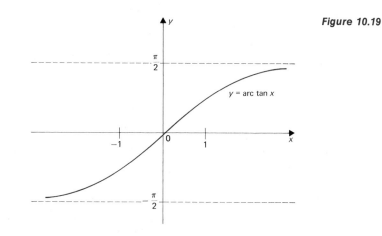

y = arc tan x

tained either by reflecting the graph of $y = \tan x$, $-\pi/2 < x < \pi/2$, through the line $y = x$; or by constructing a table of its coordinates.

The definitions and graphs of the arccosine, arcsecant, and arccotangent functions are left for the students. (See Problem 1, Exercises 10.4).

Example 10.9   Evaluate the following.

(a) $\arcsin \dfrac{\sqrt{3}}{2}$ (b) $\cos \left(\arcsin \dfrac{3}{5}\right)$ (c) $\sin \left(\arctan \dfrac{3}{4}\right)$

SOLUTION

(a)   Let $\theta = \arcsin(\sqrt{3}/2)$. Then by definition, we have

$$\sin \theta = \frac{\sqrt{3}}{2}, \qquad -\frac{\pi}{2} \le \theta \le \frac{\pi}{2}$$

Thus $\theta = \pi/3$.

(b)   Let $\theta = \arcsin \frac{3}{5}$, where $-\pi/2 \le \theta \le \pi/2$. Then

$$\sin \theta = \tfrac{3}{5}$$

Since $\sin \theta$ is positive, θ is a positive acute angle. Therefore

$$\cos \theta = \sqrt{1 - \sin^2 \theta} = \sqrt{1 - \tfrac{9}{25}} = \sqrt{\tfrac{16}{25}} = \tfrac{4}{5}$$

that is

$$\cos \left(\arcsin \tfrac{3}{5}\right) = \tfrac{4}{5}$$

(c)   Let $\theta = \arctan \frac{3}{4}$, where $-\pi/2 < \theta < \pi/2$. Then

$$\tan \theta = \tfrac{3}{4}$$

Since $\tan \theta$ is positive, θ is a positive acute angle. If we construct the right triangle with θ in standard position as shown in Figure 10.20, we find by the Pythagorean theorem that the hypotenuse is 5. Consequently,

$$\sin \theta = \tfrac{3}{5}$$

that is, $\sin \left(\arctan \tfrac{3}{4}\right) = \tfrac{3}{5}$

Figure 10.20

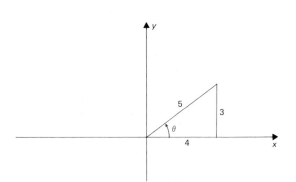

1. State why the following definitions are valid.   **Exercises 10.4**

   (a) arccos $x = y$ if and only if $x = \cos y$, $0 \le y \le \pi$

   (b) arcsec $x = y$ if and only if $x = \sec y$, $0 \le y \le \pi$, $y \ne \dfrac{\pi}{2}$

   (c) arccot $x = y$ if and only if $x = \cot y$, $0 < y < \pi$

2. Sketch the graph of the following using the definitions in Problem 1.
   (a) arccos x (b) arcsec x (c) arccot x

3. Calculate the following.
   (a) $\arccos\left(\frac{1}{2}\sqrt{2}\right)$ (b) $\arctan(\sqrt{3})$ (c) arccot (0)
   (d) $\arcsin(-1)$ (e) arcsec (-2) (f) arccsc $\sqrt{2}$

4. Calculate the following.

   (a) $\arcsin\left(\sin\dfrac{3\pi}{2}\right)$ (b) $\cos\left(\arcsin\dfrac{4}{5}\right)$

   (c) $\sin(\arctan 1)$ (d) $\sin\left(\text{arccot } \dfrac{3}{4}\right)$

   (e) $\arcsin\left(\sin\dfrac{5\pi}{4}\right)$ (f) $\arccos\left(\cos\dfrac{9\pi}{4}\right)$

   (g) $\sin\left(\arctan\dfrac{5}{12}\right)$ (h) $\arctan\left[\tan\left(-\dfrac{19\pi}{7}\right)\right]$

10.5 An Important Limit

To differentiate the trigonometric function, we need to evaluate certain limits:

$$\lim_{h \to 0} \frac{\sin h}{h}, \qquad \lim_{h \to 0} \frac{\cos h - 1}{h}$$

In this section, we shall prove that

$$\lim_{h \to 0} \frac{\sin h}{h} = 1$$

and use this result to evaluate several other limits.

Theorem 10.4   ◀

$$\lim_{h \to 0} \frac{\sin h}{h} = 1 \qquad\qquad (10.47)$$

Note h is a real number, and $\sin h$ is the sine of angle h measured in radians.

PROOF

Case 1   Let h approach zero through positive values. Since we are interested in small positive values of h, it is sufficient to consider h such that

$$0 < h < \frac{\pi}{2}$$

We place angle h in standard position and draw a unit circle (circle with radius 1) with center at the origin. We also construct the two right triangles POM and TOA as shown in Figure 10.21. From Figure 10.21 we can observe that

Figure 10.21

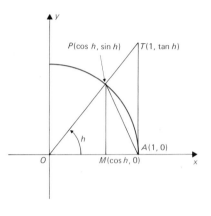

area of ΔPOA < area of sector POA < area of ΔTOA (10.48)

Further, we see that

$$d(O, A) = d(O, P) = 1$$

$$d(P, M) = \sin h$$

$$d(T, A) = \tan h = \frac{\sin h}{\cos h}$$

Area of sector POA =

$$\pi[d(O, A)]^2 \frac{h}{2\pi}$$

Consequently, (10.48) is equivalent to

$$\tfrac{1}{2} \, d(O, A) \cdot d(P, M) < \tfrac{1}{2} \, [d(O, A)]^2 \, h < \tfrac{1}{2} \, d(O, A) \cdot d(T, A)$$

or

$$\tfrac{1}{2} \cdot 1 \cdot \sin h < \tfrac{1}{2} \cdot 1^2 \cdot h < \tfrac{1}{2} \cdot 1 \cdot \frac{\sin h}{\cos h}   \quad (10.49)$$

Multiplying inequality (10.49) by $2/\sin h$ $(\sin h \neq 0)$, we obtain

$$1 < \frac{h}{\sin h} < \frac{1}{\cos h}   \quad (10.50)$$

Taking the reciprocals, we have

$$1 > \frac{\sin h}{h} > \cos h   \quad (10.51)$$

As $h \to 0^+$, $\cos h$ approaches 1 $(\cos 0 = 1)$. Thus $(\sin h)/h$ is "squeezed" between two functions, both of which approach 1 as $h \to 0^+$. Therefore $(\sin h)/h$ also approaches 1 as $h \to 0^+$. This proves Case 1.

Case 2 Let h approach zero through negative values, that is, $h \to 0^-$. If h is a negative number, we write $h = -k$, where k is positive. Then we have

$$\lim_{h \to 0^-} \frac{\sin h}{h} = \lim_{-k \to 0^-} \frac{\sin (-k)}{-k} = \lim_{k \to 0^+} \frac{-\sin k}{-k}$$

$$= \lim_{k \to 0^+} \frac{\sin k}{k} = 1 \qquad\qquad \text{Case 1}$$

Therefore

$$\lim_{h \to 0^+} \frac{\sin h}{h} = \lim_{h \to 0^-} \frac{\sin h}{h} = 1$$

Hence

$$\lim_{h \to 0} \frac{\sin h}{h} = 1$$

Theorem 10.5   ◄

$$\lim_{h \to 0} \frac{\cos h - 1}{h} = 0$$

PROOF   We shall prove this theorem by using Theorem 10.4.

$$\frac{\cos h - 1}{h} = \frac{\cos h - 1}{h} \cdot \frac{\cos h + 1}{\cos h + 1}, \qquad \cos h \neq -1$$

$$= \frac{-\sin^2 h}{h(\cos h + 1)} = (-1) \frac{\sin h}{h} \cdot \frac{\sin h}{\cos h + 1}$$

Thus

$$\lim_{h \to 0} \frac{\cos h - 1}{h} = \lim_{h \to 0} (-1) \frac{\sin h}{h} \cdot \frac{\sin h}{\cos h + 1}$$

$$= (-1) \lim_{h \to 0} \frac{\sin h}{h} \cdot \lim_{h \to 0} \frac{\sin h}{\cos h + 1}$$

$$= (-1) \cdot 1 \cdot 0 = 0$$

$$\lim_{h \to 0} \frac{\sin h}{\cos h + 1} = $$
$$\frac{\lim_{h \to 0} \sin h}{\lim_{h \to 0} (\cos h + 1)} = \frac{0}{2} = 0$$

Example 10.10   Evaluate the following limits.

(a) $\displaystyle\lim_{x \to 0} \frac{\sin 5x}{x}$
(b) $\displaystyle\lim_{x \to 0} \frac{\sin mx}{\sin nx}$
(c) $\displaystyle\lim_{\theta \to 0} \frac{\sin^2 \dfrac{\theta}{2}}{\theta^2}$

SOLUTION

(a) Let $h = 5x$. Then

$$\lim_{x \to 0} \frac{\sin 5x}{x} = \lim_{x \to 0} \frac{\sin 5x}{x} \cdot \frac{5}{5} = \lim_{x \to 0} \frac{5 \sin 5x}{5x}$$

$$= \lim_{h \to 0} \frac{5 \sin h}{h} = 5 \lim_{h \to 0} \frac{\sin h}{h} = 5$$

(b)

We assume $m \cdot n \neq 0$.

$$\lim_{x \to 0} \frac{\sin mx}{\sin nx} = \lim_{x \to 0} \frac{\dfrac{\sin mx}{mx} \cdot mx}{\dfrac{\sin nx}{nx} \cdot nx} = \lim_{x \to 0} \frac{m \cdot \dfrac{\sin mx}{mx}}{n \cdot \dfrac{\sin nx}{nx}}$$

$$= \frac{m \left(\displaystyle\lim_{mx \to 0} \frac{\sin mx}{mx} \right)}{n \left(\displaystyle\lim_{nx \to 0} \frac{\sin nx}{nx} \right)} = \frac{m \cdot 1}{n \cdot 1} = \frac{m}{n}$$

(c)

$$\lim_{\theta \to 0} \frac{\sin^2 \dfrac{\theta}{2}}{\theta^2} = \lim_{\theta \to 0} \frac{\sin^2 \dfrac{\theta}{2}}{\left(\dfrac{\theta}{2}\right)^2 \cdot 4} = \lim_{\theta \to 0} \frac{1}{4} \cdot \left(\frac{\sin \dfrac{\theta}{2}}{\dfrac{\theta}{2}} \right)^2 = \frac{1}{4} \lim_{\theta/2 \to 0} \left(\frac{\sin \dfrac{\theta}{2}}{\dfrac{\theta}{2}} \right)^2$$

$$= \frac{1}{4} (1)^2 = \frac{1}{4}$$

Example 10.11 Evaluate the following limits.

(a) $\displaystyle\lim_{\theta \to 0} \frac{1 - \cos \theta}{\theta^2}$ (b) $\displaystyle\lim_{\theta \to 0} \frac{\tan \theta - \sin \theta}{\sin^3 \theta}$

An alternative solution to (a):

$$\lim_{\theta \to 0} \frac{1 - \cos \theta}{\theta^2} = \lim_{\theta \to 0} \frac{2 \sin^2 \dfrac{\theta}{2}}{\theta^2}$$

$$= \lim_{\theta \to 0} \frac{2 \sin^2 \dfrac{\theta}{2}}{\left(\dfrac{\theta}{2}\right)^2 \cdot 4} = \frac{2}{4} \lim_{\theta \to 0} \left(\frac{\sin \dfrac{\theta}{2}}{\dfrac{\theta}{2}} \right)^2$$

$$= \frac{1}{2}$$

SOLUTION

(a) $\displaystyle\lim_{\theta \to 0} \frac{1 - \cos \theta}{\theta^2} = \lim_{\theta \to 0} \frac{(1 - \cos \theta)}{\theta^2} \cdot \frac{1 + \cos \theta}{1 + \cos \theta}$

$$= \lim_{\theta \to 0} \frac{1 - \cos^2 \theta}{\theta^2 (1 + \cos \theta)}$$

$$= \lim_{\theta \to 0} \frac{\sin^2 \theta}{\theta^2 (1 + \cos \theta)}$$

$$= \lim_{\theta \to 0} \frac{1}{1 + \cos \theta} \cdot \lim_{\theta \to 0} \left(\frac{\sin \theta}{\theta} \right)^2$$

$$= \frac{1}{2} \cdot 1 = \frac{1}{2}$$

(b) $\displaystyle \frac{\tan \theta - \sin \theta}{\sin^3 \theta} = \frac{\dfrac{\sin \theta}{\cos \theta} - \sin \theta}{\sin^3 \theta} = \frac{\sin \theta \left(\dfrac{1}{\cos \theta} - 1 \right)}{\sin^3 \theta}$

$$= \frac{\sin \theta}{\sin^3 \theta} \cdot \frac{1 - \cos \theta}{\cos \theta} = \frac{1 - \cos \theta}{\cos \theta \sin^2 \theta}$$

$$= \frac{1 - \cos \theta}{\cos \theta \sin^2 \theta} \cdot \frac{1 + \cos \theta}{1 + \cos \theta}$$

$$= \frac{1 - \cos^2 \theta}{\cos \theta \sin^2 \theta (1 + \cos \theta)}$$

$$= \frac{\sin^2 \theta}{\cos \theta \sin^2 \theta (1 + \cos \theta)} = \frac{1}{\cos \theta(1 + \cos \theta)}$$

Therefore

$$\lim_{\theta \to 0} \frac{\tan \theta - \sin \theta}{\sin^3 \theta} = \lim_{\theta \to 0} \frac{1}{\cos \theta(1 + \cos \theta)} = \frac{1}{\lim_{\theta \to 0} \cos \theta(1 + \cos \theta)}$$

$$= \frac{1}{\lim_{\theta \to 0} \cos \theta \lim_{\theta \to 0} (1 + \cos \theta)} = \frac{1}{1 \cdot 2} = \frac{1}{2}$$

In Problems 1 through 5, calculate the value (if any) of each of the limits.   Exercises 10.5

1. $\lim_{x \to 0} \dfrac{\sin 2x}{x}$

2. $\lim_{x \to 0} \dfrac{\sin^2 x}{x^2}$

3. $\lim_{x \to 0^+} \dfrac{\sin x}{\sqrt{x}} \quad \left[Hint: \quad \dfrac{\sin x}{\sqrt{x}} = \sqrt{x} \, \dfrac{\sin x}{x} \right]$

4. $\lim_{x \to 0} \dfrac{\tan x}{x}$

5. $\lim_{x \to 0} \dfrac{\tan^2 x - \sin^2 x}{x^4}$

$$\left[Hint: \quad \frac{\tan^2 x - \sin^2 x}{x^4} = \frac{\tan x - \sin x}{x^3} \cdot \left(\frac{\tan x}{x} + \frac{\sin x}{x} \right) \right]$$

10.6   Derivatives of the Trigonometric Functions

We now use Theorems 10.4 and 10.5 to calculate the derivative of $\sin x$.

Theorem 10.6   *If $f(x) = \sin x$, then*

$$f'(x) = \cos x \qquad (10.52)$$

◀ *Theorem 10.6 may be written in other notations:*

$$y = \sin x$$
$$\frac{dy}{dx} = \cos x$$
$$\frac{d}{dx}(\sin x) = \cos x$$
$$D(\sin x) = \cos x$$
$$(\sin x)' = \cos x$$

PROOF   Setting up the difference quotient, we have

$$\frac{f(x+h) - f(x)}{h} = \frac{\sin (x+h) - \sin x}{h}$$

$$= \frac{\sin x \cos h + \cos x \sin h - \sin x}{h}$$

$$= \sin x \left(\frac{\cos h - 1}{h} \right) + \cos x \left(\frac{\sin h}{h} \right)$$

Thus

$$f'(x) = \lim_{h \to 0} \frac{f(x+h) - f(x)}{h} = \sin x \lim_{h \to 0} \frac{\cos h - 1}{h} + \cos x \lim_{h \to 0} \frac{\sin h}{h}$$

Theorems 10.4 and 10.5

$$= (\sin x) \cdot 0 + (\cos x) \cdot 1 = \cos x$$

Therefore the derivative of sin x is cos x.

By applying the chain rule, we find a more general formula:

$$\frac{d}{dx}(\sin u) = \cos u \cdot \frac{du}{dx} \tag{10.53}$$

where u is a differentiable function of x.

Example 10.12 Differentiate the following with respect to x.

(a) $\sin(3x^2 + 2x + 1)$ (b) $\sin(e^{x^2})$

Note $u = 3x^2 + 2x + 1,$

$\dfrac{du}{dx} = 6x + 2$

SOLUTION

(a)   Using (10.53), with $u = 3x^2 + 2x + 1$, we have

$$\frac{d}{dx}\sin(3x^2 + 2x + 1) = \frac{d}{dx}(\sin u) = \cos u \cdot \frac{du}{dx}$$
$$= [\cos(3x^2 + 2x + 1)](6x + 2)$$
$$= 2(3x + 1)\cos(3x^2 + 2x + 1)$$

Note If $u = e^{x^2}$, then

$\dfrac{du}{dx} = \dfrac{d}{dx}(e^{x^2})$

$= e^{x^2}\dfrac{d}{dx}(x^2) = e^{x^2}(2x)$

(b)   Again using (10.53), with $u = e^{x^2}$, we have

$$\frac{d}{dx}[\sin(e^{x^2})] = \frac{d}{dx}\sin u = \cos u \cdot \frac{du}{dx}$$
$$= \cos(e^{x^2}) \cdot e^{x^2} \cdot 2x = 2xe^{x^2}\cos(e^{x^2})$$

▶   **Theorem 10.7** *If $f(x) = \cos x$, then*
$$f'(x) = -\sin x \tag{10.54}$$

Recall that

$\sin\left(\dfrac{\pi}{2} - x\right) = \cos x$ *and*

$\cos\left(\dfrac{\pi}{2} - x\right) = \sin x$

PROOF Since

$$f(x) = \cos x = \sin\left(\frac{\pi}{2} - x\right)$$

we have

$$\frac{d}{dx}\cos x = \frac{d}{dx}\sin\left(\frac{\pi}{2} - x\right) = \cos\left(\frac{\pi}{2} - x\right)\frac{d}{dx}\left(\frac{\pi}{2} - x\right)$$

Apply (10.53)

$u = \dfrac{\pi}{2} - x \Rightarrow \dfrac{du}{dx} = -1$

$$= \left[\cos\left(\frac{\pi}{2} - x\right)\right] \cdot (-1) = -\cos\left(\frac{\pi}{2} - x\right) = -\sin x$$

Again, by the chain rule, we have

$$\frac{d}{dx}(\cos u) = -\sin u \cdot \frac{du}{dx} \qquad (10.55)$$

where u is a differentiable function of x.

Since the other trigonometric functions can be expressed in terms of the sine and cosine functions, we can find their derivatives by using the rules of differentiation established in Chapter 4. We have

Theorem 10.8   ◀

(i) $\dfrac{d}{dx}\tan x = \sec^2 x$

(ii) $\dfrac{d}{dx}\cot x = -\csc^2 x$

(iii) $\dfrac{d}{dx}\sec x = \sec x \tan x$

(iv) $\dfrac{d}{dx}\csc x = -\csc x \cot x$

PROOF   We shall prove only (i) and (iii).

(i) $\dfrac{d}{dx}\tan x = \dfrac{d}{dx}\left(\dfrac{\sin x}{\cos x}\right)$

$$= \frac{\cos x \dfrac{d}{dx}(\sin x) - \sin x \dfrac{d}{dx}(\cos x)}{\cos^2 x}$$

$$= \frac{(\cos x)(\cos x) - (\sin x)(-\sin x)}{\cos^2 x}$$

$$= \frac{\cos^2 x + \sin^2 x}{\cos^2 x} = \frac{1}{\cos^2 x} = \sec^2 x$$

Recall that

$$\frac{d}{dx}\left(\frac{u}{v}\right) = \frac{v \cdot u' - u \cdot v'}{v^2}$$

(iii) $\dfrac{d}{dx}\sec x = \dfrac{d}{dx}\left(\dfrac{1}{\cos x}\right)$

$$= \frac{(\cos x)\dfrac{d}{dx}(1) - 1\dfrac{d}{dx}(\cos x)}{\cos^2 x}$$

The derivative of a constant is zero.

$$= \frac{(\cos x)\cdot 0 - (1)(-\sin x)}{\cos^2 x}$$

$$= \frac{\sin x}{\cos^2 x} = \frac{1}{\cos x}\cdot\frac{\sin x}{\cos x} = \sec x \tan x$$

Applying the chain rule, we may write

$$\frac{d}{dx}\tan u = \sec^2 u \cdot \frac{du}{dx} \qquad (10.56)$$

$$\frac{d}{dx} \cot u = -\csc^2 u \cdot \frac{du}{dx} \qquad\qquad (10.57)$$

$$\frac{d}{dx} \sec u = \sec u \tan u \cdot \frac{du}{dx} \qquad\qquad (10.58)$$

$$\frac{d}{dx} \csc u = -\csc u \cot u \cdot \frac{du}{dx} \qquad\qquad (10.59)$$

Example 10.13   If $y = \tan^3 (2x + 5)$, find dy/dx.

SOLUTION   Let $u = \tan (2x + 5)$ and $v = 2x + 5$. Then

$$y = u^3, \qquad u = \tan v, \qquad v = 2x + 5$$

and

$$\frac{dy}{du} = 3u^2, \qquad \frac{du}{dv} = \sec^2 v, \qquad \frac{dv}{dx} = 2$$

By the chain rule,

$$\frac{dy}{dx} = \frac{dy}{du} \cdot \frac{du}{dv} \cdot \frac{dv}{dx} = 3u^2 \cdot \sec^2 v \cdot 2 = 6u^2 \sec^2 v$$

$$= 6 \tan^2 (2x + 5) \sec^2 (2x + 5)$$

Knowing the derivatives of the trigonometric functions, we are able to calculate the derivatives of their inverse functions.

▶   **Theorem 10.9**   *If* $y = \arcsin x \ (-\pi/2 < y < \pi/2)$, *then*

$$\frac{dy}{dx} = \frac{1}{\sqrt{1 - x^2}} \qquad\qquad (10.60)$$

PROOF   For $y = \arcsin x$, we have

$$x = \sin y \qquad \left(-\frac{\pi}{2} < y < \frac{\pi}{2}\right)$$

We calculate the derivative on both sides with respect to x, treating y as a function of x. Thus

$$\frac{d}{dx} (x) = \frac{d}{dx} (\sin y)$$

Using implicit differentiation

$$1 = \cos y \cdot \frac{dy}{dx}$$

$$\frac{dy}{dx} = \frac{1}{\cos y} \qquad (\cos y \neq 0)$$

Since $-\pi/2 < y < \pi/2$, $\cos y > 0$. Thus $\cos y = \sqrt{1 - \sin^2 y}$. Therefore

$$\frac{dy}{dx} = \frac{1}{\cos y} = \frac{1}{\sqrt{1 - \sin^2 y}} = \frac{1}{\sqrt{1 - x^2}}$$

Similar calculations yield the following theorem.

Theorem 10.10   ◄

If $y = \arccos x$ $(0 < y < \pi)$, then $\dfrac{dy}{dx} = -\dfrac{1}{\sqrt{1 - x^2}}$ (10.61)

If $y = \arctan x$ $\left(-\dfrac{\pi}{2} < y < \dfrac{\pi}{2}\right)$, then $\dfrac{dy}{dx} = \dfrac{1}{1 + x^2}$ (10.62)

If $y = \text{arccot } x$ $(0 < y < \pi)$, then $\dfrac{dy}{dx} = -\dfrac{1}{1 + x^2}$ (10.63)

The proof of Theorem 10.10 is left as an exercise for the students.

Example 10.14   Find the derivative of

$$\arcsin\left(\frac{2x}{1 + x^2}\right), \qquad (|x| < 1)$$

SOLUTION   Let

$$y = \arcsin\left(\frac{2x}{1 + x^2}\right) \quad \text{and} \quad u = \frac{2x}{1 + x^2}$$

Then

$$y = \arcsin u, \qquad u = \frac{2x}{1 + x^2} \;;\; \frac{dy}{du} = \frac{1}{\sqrt{1 - u^2}}$$

and

$$\frac{du}{dx} = \frac{(1 + x^2)\,2 - 2x(2x)}{(1 + x^2)^2} = \frac{2 + 2x^2 - 4x^2}{(1 + x^2)^2} = \frac{2(1 - x^2)}{(1 + x^2)^2}$$

By the chain rule,

$$\frac{dy}{dx} = \frac{dy}{du} \cdot \frac{du}{dx} = \frac{1}{\sqrt{1 - u^2}} \cdot \frac{2(1 - x^2)}{(1 + x^2)^2}$$

$$= \frac{1}{\sqrt{1 - \dfrac{4x^2}{(1 + x^2)^2}}} \cdot \frac{2(1 - x^2)}{(1 + x^2)^2}$$

$$= \frac{(1 + x^2)}{\sqrt{(1 + x^2)^2 - 4x^2}} \cdot \frac{2(1 - x^2)}{(1 + x^2)^2}$$

$$= \frac{(1 + x^2)}{1 - x^2} \cdot \frac{2(1 - x^2)}{(1 + x^2)^2} = \frac{2}{1 + x^2}$$

Alternative method: Let $x = \tan \theta$. Then

$$y = \arcsin\left(\frac{2 \tan \theta}{1 + \tan^2 \theta}\right)$$
$$= \arcsin(\sin 2\theta) = 2\theta$$

Thus

$$y = 2\theta \quad \text{and} \quad \theta = \arctan x$$

and

$$\frac{dy}{d\theta} = 2 \quad \text{and} \quad \frac{d\theta}{dx} = \frac{1}{1 + x^2}$$

By the chain rule,

$$\frac{dy}{dx} = \frac{dy}{d\theta} \cdot \frac{d\theta}{dx} = 2 \cdot \frac{1}{1 + x^2} = \frac{2}{1 + x^2}$$

Example 10.15   Find the maximum and minimum values of

$$4 \sin \theta + 3 \cos \theta \qquad (0 \le \theta \le 2\pi)$$

SOLUTION   Let $f(\theta) = 4 \sin \theta + 3 \cos \theta$. Then

$$f'(\theta) = 4 \cos \theta - 3 \sin \theta$$

Thus $f'(\theta) = 0$, if

$$4 \cos \theta - 3 \sin \theta = 0$$

$$4 \cos \theta = 3 \sin \theta$$

$$\tan \theta = \tfrac{4}{3}$$

In the interval $[0, 2\pi]$, there are two values, α and β, in the first and third quadrants such that

$$\tan \alpha = \tfrac{4}{3} \quad \text{and} \quad \tan \beta = \tfrac{4}{3}$$

If α is in the first quadrant, then

$$\sin \alpha = \frac{4}{\sqrt{4^2 + 3^2}} = \frac{4}{5} \quad \text{and} \quad \cos \alpha = \frac{3}{\sqrt{4^2 + 3^2}} = \frac{3}{5}$$

Since β is in the third quadrant, $\sin \beta$ and $\cos \beta$ are negative. Thus we have

$$\sin \beta = -\tfrac{4}{5} \quad \text{and} \quad \cos \beta = -\tfrac{3}{5}$$

We take the second derivative and evaluate it at α and β:

$$f''(\theta) = -4 \sin \theta - 3 \cos \theta$$

$$f''(\alpha) = -4 \cdot \tfrac{4}{5} - 3 \cdot \tfrac{3}{5} = -5 \qquad \text{(a negative number)}$$

$$f''(\beta) = -4\left(-\tfrac{4}{5}\right) - 3\left(-\tfrac{3}{5}\right) = 5 \qquad \text{(a positive number)}$$

Therefore $f(\alpha)$ is a maximum value and $f(\beta)$ is a minimum value of the function.

By substitution, we find that $f(\alpha) = 5$ and $f(\beta) = -5$, which are the maximum and the minimum values respectively, of f.

Exercises 10.6

In Problems 1 through 14, differentiate with respect to x.

1. $x \sin x$

2. $\dfrac{4x + 1}{\cos x}$

3. $\dfrac{\sin x + \cos x}{\sin x - \cos x}$

4. $\dfrac{\sin (3x + 1)}{\cos (2x + 3)}$

5. $\tan^4 (7x^2 + 3x + 9)$

6. $\sin^2 (2x) \cos^3 (2x)$

7. $e^{\sin x - \tan x}$

8. $\ln (x^2 \sin x)$

9. $e^x \sin^2 (\cos x)$

10. $\arcsin (2x)$

11. $(x^2 + 1) \arctan (x)$ 12. $\arcsin \sqrt{1 - x^2}$

13. $e^{\sin x + \arctan x}$ 14. $\ln (x^2 + \tan x + \text{arccsc } x)$

15. Find $f'(\pi/4)$ if $f(x) = (1 + x) \tan x$.

16. A particle moves in a straight line with the equation of motion $x = a \cos (bt + c)$, where a, b, and c are constants and x is the distance measured from a point 0 on the straight line. Show that the acceleration at any time t is proportional to the distance from 0.

17. Water is being poured into a conical vessel (which is held with its vertex down) at the rate of 2 cu ft/min. The semivertical angle of the conical vessel is arctan $\frac{1}{2}$ (see Figure 10.22). Find the rate at which the water is rising in the vessel when the depth of the water is 2 ft.

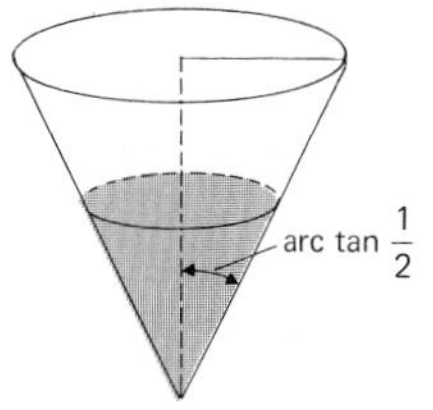

Figure 10.22

arc tan $\dfrac{1}{2}$

18. A bell-tent consists of a cylindrical portion resting on the ground and a conical portion on top of the cylinder (see Figure 10.23). Prove that if

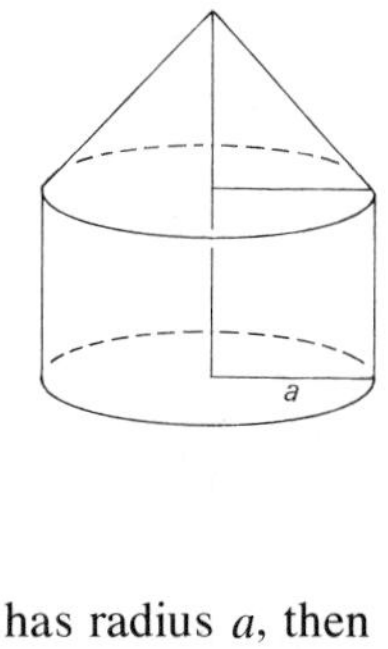

Figure 10.23

the volume of the bell-tent is V and the circular base has radius a, then the amount of canvas used is a minimum when the semivertical angle of the cone is arccos $\left(\frac{2}{3}\right)$. *Hint:* A cone with radius a and height h has volume $= \frac{1}{3} \pi a^2 h$, and

$$\text{surface area} = \pi a \text{ length of the slant height}$$

$$= \pi a \cdot \sqrt{a^2 + h^2}$$

*19. A flagpole 45 ft high stands at the top of a building 85 ft high (see Figure 10.24). If an observer's eye is 5 ft above the ground, how far should the observer stand from the foot of the building in order that the vertical angle θ subtended by the flagpole at the observer's eye be maximum?

Figure 10.24

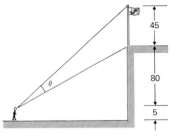

10.7 Integration of the Trigonometric Functions

From the fundamental theorem of calculus, it is clear that there is an integration formula to correspond to each differentiation formula. The integrals corresponding to the derivatives in the previous section are as follows.

$$\int \cos x \, dx = \sin x + C \tag{10.64}$$

$$\int \sin x \, dx = -\cos x + C \tag{10.65}$$

$$\int \sec^2 x \, dx = \tan x + C \tag{10.66}$$

$$\int \csc^2 x \, dx = -\cot x + C \tag{10.67}$$

$$\int \sec x \tan x \, dx = \sec x + C \tag{10.68}$$

$$\int \csc x \cot x \, dx = -\csc x + C \tag{10.69}$$

$$\int \frac{1}{\sqrt{1-x^2}} \, dx = \arcsin x + C, \quad -1 < x < 1 \tag{10.70}$$

$$\int \frac{-1}{\sqrt{1-x^2}} \, dx = \arccos x + C, \quad -1 < x < 1 \tag{10.71}$$

$$\int \frac{1}{1+x^2} \, dx = \arctan x + C \tag{10.72}$$

$$\int \frac{-1}{1+x^2} \, dx = \operatorname{arccot} x + C \tag{10.73}$$

Recall that

$$\int \frac{1}{x} \, dx = \ln |x| + C.$$

$u = \cos x; \quad du = -\sin x \, dx$

The integrals of $\tan x$, $\cot x$, $\sec x$, and $\csc x$ can be obtained using the techniques of Chapter 8.

$$\int \tan x \, dx = \int \frac{\sin x}{\cos x} \, dx = -\int \frac{1}{u} \, du \tag{10.74}$$

$$= -\ln |u| + C = -\ln |\cos x| + C$$

$$= \ln \left| \frac{1}{\cos x} \right| + C = \ln |\sec x| + C$$

$$\int \cot x \, dx = \int \frac{\cos x}{\sin x} \, dx = \int \frac{1}{u} \, du \qquad (10.75) \qquad u = \sin x; \quad du = \cos x \, dx$$

$$= \ln |u| + C = \ln |\sin x| + C$$

$$\int \sec x \, dx = \int \frac{\sec x(\sec x + \tan x)}{\sec x + \tan x} \, dx = \int \frac{1}{u} \, du \qquad (10.76)$$

$u = \sec x + \tan x;$
$du = (\sec^2 x + \sec x \tan x)dx$
$\qquad = \sec x(\sec x + \tan x) \, dx$

$$= \ln |u| + C = \ln |\sec x + \tan x| + C$$

$$\int \csc x \, dx = -\int \frac{-\csc x(\csc x + \cot x)}{\csc x + \cot x} \, dx = -\int \frac{1}{u} \, du \qquad (10.77)$$

$u = \csc x + \cot x;$
$du = (-\csc^2 x - \csc x \cot x)dx$
$\qquad = -\csc x(\csc x + \cot x) \, dx$

$$= -\ln |u| + C = -\ln |\csc x + \cot x| + C$$

We shall now give some examples to illustrate how to solve simple integration problems involving trigonometric functions.

Example 10.16 Integrate $\int \sin^5 x \cos x \, dx$.

SOLUTION Let $u = \sin x$, then $du = \cos x \, dx$. Thus

$$\int \sin^5 x \cos x \, dx = \int u^5 \, du = \frac{u^6}{6} + C = \frac{\sin^6 x}{6} + C$$

To integrate $\int f(\sin x) \cos x \, dx$, we use the substitution $u = \sin x$; similarly, for $\int f(\cos x) \sin x \, dx$, we use $u = \cos x$.

Example 10.17 Integrate

(a) $\int \cos^5 x \, dx$ (b) $\int \cos^4 x \, dx$

SOLUTION

(a) $\displaystyle\int \cos^5 x \, dx = \int \cos^4 x \cos x \, dx = \int (1 - \sin^2 x)^2 \cos x \, dx$

$\qquad = \displaystyle\int (1 - u^2)^2 \, du = \int (1 - 2u^2 + u^4) \, du$

$\qquad = u - \dfrac{2u^3}{3} + \dfrac{u^5}{5} + C$

$\qquad = \sin x - \dfrac{2}{3} \sin^3 x + \dfrac{1}{5} \sin^5 x + C$

$u = \sin x$, $du = \cos x \, dx$. This technique is used to integrate $\sin^n x$ or $\cos^n x$ when n is odd.

(b) $\displaystyle\int \cos^4 x \, dx = \int (\cos^2 x)^2 \, dx = \int \left(\frac{1 + \cos 2x}{2}\right)^2 dx$

$\qquad = \dfrac{1}{4} \displaystyle\int (1 + 2 \cos 2x + \cos^2 2x) \, dx$

$\qquad = \dfrac{1}{4} \displaystyle\int dx + \dfrac{2}{4} \int \cos 2x \, dx + \dfrac{1}{4} \int \cos^2 2x \, dx$

$\qquad = \dfrac{x}{4} + \dfrac{2}{4} \cdot \dfrac{\sin 2x}{2} + \dfrac{1}{4} \displaystyle\int \left(\frac{1 + \cos 4x}{2}\right) dx$

$\qquad = \dfrac{x}{4} + \dfrac{1}{4} \sin 2x + \dfrac{1}{8} \displaystyle\int (1 + \cos 4x) \, dx$

*$\cos 2x = 2 \cos^2 x - 1$
$\qquad = 1 - 2 \sin^2 x$*

$\int \cos ax \, dx = \left(\dfrac{\sin ax}{a}\right) + C$, which can be seen by letting $u = ax$, $du = a \, dx$. This technique is used to evaluate $\sin^n x$ or $\cos^n x$ when n is even.

$$= \frac{x}{4} + \frac{1}{4} \sin 2x + \frac{1}{8} \left[x + \frac{\sin 4x}{4} \right] + C$$

$$= \frac{x}{4} + \frac{x}{8} + \frac{1}{4} \sin 2x + \frac{1}{32} \sin 4x + C$$

$$= \frac{3}{8} x + \frac{1}{4} \sin 2x + \frac{1}{32} \sin 4x + C$$

Note *The following important substitutions should be remembered.*

(i) *If the integrand has $\sqrt{a^2 - x^2}$ as a factor,
let $x = a \sin \theta$ or $x = a \cos \theta$.*

(ii) *If the integrand has $\sqrt{a^2 + x^2}$ as a factor,
let $x = a \tan \theta$.*

(iii) *If the integrand has $\sqrt{x^2 - a^2}$ as a factor,
let $x = a \sec \theta$.*

These substitutions are sometimes useful even if the above factors are in the denominator.

Example 10.18 Integrate $\int x \sqrt{1 - x^2}\, dx$.

SOLUTION Let $x = \sin \theta$; then $dx = \cos \theta\, d\theta$.

$$\int x \sqrt{1 - x^2}\, dx = \int \sin \theta \sqrt{1 - \sin^2 \theta}\, \cos \theta\, d\theta$$

$$= \int \sin \theta \cos \theta \cos \theta\, d\theta = \int \sin \theta \cos^2 \theta\, d\theta$$

$$= - \int \cos^2 \theta (-\sin \theta)\, d\theta = - \int u^2\, du = -\frac{u^3}{3} + C$$

$$= -\frac{\cos^3 \theta}{3} + C = -\frac{1}{3} \cos^3 (\arcsin x) + C$$

Example 10.19 Integrate

$$\int \frac{dx}{\sqrt{7 + 6x - x^2}}$$

SOLUTION We have $7 + 6x - x^2 = 16 - (x - 3)^2 = (4)^2 - (x - 3)^2$. Thus

Let $x - 3 = 4 \sin \theta$ and $dx = 4 \cos \theta\, d\theta$.

$$\int \frac{dx}{\sqrt{7 + 6x - x^2}} = \int \frac{dx}{\sqrt{(4)^2 - (x - 3)^2}} = \int \frac{4 \cos \theta\, d\theta}{\sqrt{(4)^2 - (4 \sin \theta)^2}}$$

$$= \int \frac{4 \cos \theta\, d\theta}{4 \sqrt{1 - \sin^2 \theta}} = \int \frac{\cos \theta\, d\theta}{\cos \theta} = \int d\theta = \theta + C$$

Now

$$x - 3 = 4 \sin \theta \Rightarrow \frac{x - 3}{4} = \sin \theta$$

or

$$\theta = \arcsin \frac{x - 3}{4}$$

Therefore the given integral $= \theta + C = \arcsin \frac{x - 3}{4} + C$.

Example 10.20 Integrate $\int \frac{dx}{e^x + e^{-x}}$

SOLUTION   We let $u = e^x$, then $du = e^x \, dx$.

$$\int \frac{1}{e^x + e^{-x}} \, dx = \int \frac{e^x}{e^{2x} + 1} \, dx = \int \frac{du}{u^2 + 1}$$

$$= \arctan u + C = \arctan e^x + C$$

Integrate in Problems 1 through 24.

1. $\displaystyle\int \frac{\sec^2 x}{1 + \tan x} \, dx$

2. $\displaystyle\int \frac{\sin x}{1 + \cos x} \, dx$

3. $\displaystyle\int \cos x \, e^{\sin x} \, dx$

4. $\displaystyle\int \sec^2 x \tan^3 x \, dx$

5. $\displaystyle\int x \cos^2(x^2 + 1) \, dx$

6. $\displaystyle\int e^{\sin x \cos x} \cos 2x \, dx$

7. $\displaystyle\int \frac{\sin x}{\cos^2 x} \, dx$

8. $\displaystyle\int \frac{1}{\cos^2 x (\tan x + 5)} \, dx$

9. $\displaystyle\int \sin^2 x \cos^2 x \, dx$

10. $\displaystyle\int x \cos x \, dx$

11. $\displaystyle\int e^{\tan x} \sec^2 x \, dx$

12. $\displaystyle\int x^2 \sin x \, dx$

13. $\displaystyle\int x \tan^2 x \, dx$

14. $\displaystyle\int_0^{\pi/2} \sin^2 x \, dx$

15. $\displaystyle\int_0^{\pi/2} \sin x \cos^2 x \, dx$

16. $\displaystyle\int_0^{\pi/4} \tan^5 x \, dx$

17. $\displaystyle\int x \sin x^2 \cos x^2 \, dx$

18. $\displaystyle\int \sqrt{a^2 - x^2} \, dx$

19. $\displaystyle\int \frac{dx}{x^2 + 2x + 10}$

20. $\displaystyle\int x \sqrt{9 - x^2} \, dx$

21. $\displaystyle\int (x + 1) \tan (x^2 + 2x) \, dx$

22. $\displaystyle\int \frac{x \sec^2 x^2}{\tan x^2} \, dx$

23. $\displaystyle\int \frac{e^{\sin x} \cos x \, dx}{1 + e^{\sin x}}$

24. $\displaystyle\int \frac{dx}{x^2 + a^2}$

10.8 Applications

Business

In many manufacturing situations, it is found that the same basic process with slight adjustments produces two different products. If Company C produces the two products A and B, then both these products compete for the resources of Company C. Let us assume that Company C has a fixed amount of money to produce both these products. Thus if the company wishes to increase the amount of production of product A, it will have to decrease the amount of production of product B. The amounts which can be produced (with the fixed resources of C) will satisfy a certain relation which usually can be placed in the form of an equation.

Example 10.21   The United Grain Company produces two brands of cereals. The possible amounts x and y (in millions of cases) per year of brands A and B, respectively, that can be produced by the company are related by

$$x^2 + y^2 + 2x + 4y = 31 \qquad (10.78)$$

The company can sell brand A for \$2.00 per case, and brand B for \$1.50 per case. Find x and y so as to maximize the revenue of the company.

SOLUTION   We can write (10.78) as

$$(x^2 + 2x + 1) + (y^2 + 4y + 4) = 31 + 1 + 4$$

or

$$(x + 1)^2 + (y + 2)^2 = 6^2 \qquad (10.79)$$

We can see that (10.79) represents a circle (see Figure 10.25) with center at $A(-1, -2)$ and radius 6. If $P(x, y)$ is a point on the circle (10.79), then from Figure 10.25, we note that

Figure 10.25

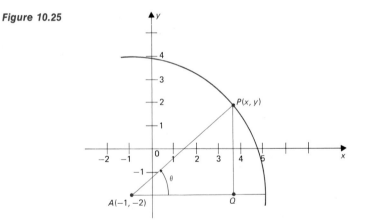

$$\sin \theta = \frac{d(P, Q)}{d(A, P)} = \frac{y + 2}{6} \qquad (10.80)$$

and

$$\cos \theta = \frac{d(A, Q)}{d(A, P)} = \frac{x + 1}{6} \qquad (10.81)$$

From (10.80) and (10.81), we get

$$\begin{cases} x = 6 \cos \theta - 1 \\ y = 6 \sin \theta - 2 \end{cases} \qquad (10.82)$$

Since brand A sells for \$2 per case and brand B sells for \$3/2 per case, the revenue R is given by

$$R = 2x + \tfrac{3}{2} y \qquad (10.83)$$

Substituting from (10.82), we obtain

$$R = 2(6 \cos \theta - 1) + \tfrac{3}{2} (6 \sin \theta - 2) \qquad (10.84)$$

or

$$R = 12 \cos \theta + 9 \sin \theta - 5 \qquad (10.85)$$

To maximize R, we differentiate R with respect to θ,

$$R' = -12 \sin \theta + 9 \cos \theta \qquad (10.86)$$

and set $R' = 0$:

$$12 \sin \theta = 9 \cos \theta$$

or

$$\tan \theta = \tfrac{3}{4} \qquad (10.87)$$

From (10.87), we get

$$\sec^2 \theta = 1 + \tan^2 \theta = 1 + \tfrac{9}{16} = \tfrac{25}{16}$$

or

$$\sec \theta = \tfrac{5}{4} \Rightarrow \cos \theta = \tfrac{4}{5}$$

We also have

$$\sin \theta = \sqrt{1 - \cos^2 \theta} = \sqrt{1 - \tfrac{16}{25}} = \sqrt{\tfrac{9}{25}} = \tfrac{3}{5}$$

If we substitute these values of $\sin \theta$ and $\cos \theta$ in

$$R'' = -12 \cos \theta - 9 \sin \theta \qquad (10.88)$$

we see that R'' is negative. Therefore $\sin \theta = \tfrac{3}{5}$ and $\cos \theta = \tfrac{4}{5}$ gives the maximum revenue. Hence

$$x = 6 \cos \theta - 1 = 6 \cdot \left(\tfrac{4}{5}\right) - 1 = \tfrac{24}{5} - 1 = \tfrac{19}{5}$$

and

$$y = 6 \sin \theta - 2 = 6 \cdot \left(\tfrac{3}{5}\right) - 2 = \tfrac{18}{5} - 2 = \tfrac{8}{5}$$

give the maximum revenue.

Physics

Recall that if a particle moves a distance $x = f(t)$ in time t, then dx/dt represents its instantaneous velocity and d^2x/dt^2 represents its instantaneous acceleration.

For a particle moving along a plane curve, we study its motion by resolving it along two perpendicular directions. For example, if the particle travels with a velocity u along a line which makes an angle α

Figure 10.26

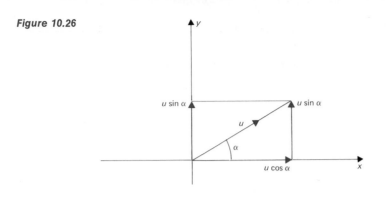

with the positive direction of the x-axis (see Figure 10.26), then we consider u as being composed of two velocities—one along the x-axis, $u \cos \alpha$; the other along the y-axis, $u \sin \alpha$. This is exactly like the relationship between the lengths of the hypotenuse and legs of a right triangle.

Example 10.22   Projectile Problem A projectile is fired from a cannon with a velocity u ft/sec making an angle θ $(0 < \theta < \pi/2)$ with the horizontal. Assuming that gravity is the only force on the projectile, find the equation of the path of the projectile.

   SOLUTION   Let us take the cannon as the origin of the coordinate plane. After t sec, if $P(x, y)$ is the position of the projectile, then x denotes the horizontal displacement of the projectile and y denotes its height above the cannon (see Figure 10.27). As discussed earlier, the initial velocities (at $t = 0$) along the x-axis and y-axis are

Figure 10.27

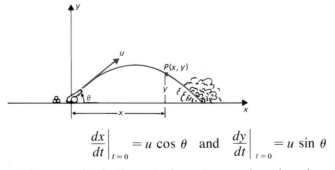

$$\frac{dx}{dt}\bigg|_{t=0} = u \cos \theta \quad \text{and} \quad \frac{dy}{dt}\bigg|_{t=0} = u \sin \theta$$

Since gravity is the only force (we are ignoring air resistance, etc.) pulling on the projectile, there is no acceleration in the horizontal direction, while the acceleration in the vertical direction is g. Therefore

Acceleration due to gravity is denoted by g and g is approximately 32 ft/sec².

In (10.89) we take −g, since the accelerations d²y/dt² and g are in opposite directions.

$$\frac{d^2x}{dt^2} = 0 \quad \text{and} \quad \frac{d^2y}{dt^2} = -g \tag{10.89}$$

Integrating (10.89), we get

$$\frac{dx}{dt} = c_1 \quad \text{and} \quad \frac{dy}{dt} = -gt + c_2 \qquad (10.90)$$

Initially, $x = y = 0$ when $t = 0$, and $dx/dt = u \cos \theta$, $dy/dt = u \sin \theta$. Therefore

$$u \cos \theta = c_1 \quad \text{and} \quad u \sin \theta = c_2$$

Thus (10.90) becomes

$$\frac{dx}{dt} = u \cos \theta \quad \text{and} \quad \frac{dy}{dt} = -gt + u \sin \theta \qquad (10.91)$$

Integrating (10.91), we get

$$\begin{cases} x = (u \cos \theta)\, t + c_3 \\ y = -\frac{1}{2} gt^2 + (u \sin \theta)\, t + c_4 \end{cases} \qquad (10.92)$$

Again we apply the initial conditions that $t = 0$ when $x = 0$ and $y = 0$:

$$c_3 = 0 = c_4$$

Thus

$$\begin{cases} x = (u \cos \theta)\, t \\ y = -\frac{1}{2} gt^2 + (u \sin \theta)\, t \end{cases} \qquad (10.93)$$

We eliminate t between the two equations of (10.93):

$$y = x \tan \theta - \frac{1}{2} \frac{gx^2}{u^2} \sec^2 \theta \qquad (10.94)$$

Substitute $t = \dfrac{x}{u \cos \theta}$ in the second equation of (10.93).

which represents a parabola. Thus the path of a projectile moving under gravity is a parabola (see Figure 10.27).

Example 10.23 Simple Harmonic Motion If a particle moves in a straight line with an acceleration that is proportional to the distance of the particle from a fixed point and always directed toward that point, then the particle is executing *simple harmonic motion* (SHM). Write the differential equation of a particle in SHM and solve it.

SHM is the simplest type of oscillatory motion about a position of equilibrium. It occurs in the swing of a simple pendulum, the rocking of a ship, the vibration of a tuning fork.

   SOLUTION   Let us choose the origin to be the fixed point and the x-axis to be the straight line (see Figure 10.28). Then the differential

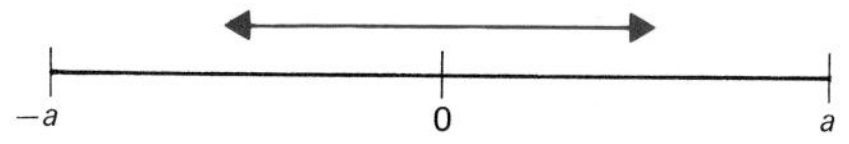

Figure 10.28

equation of the particle in SHM is given by

$$\frac{d^2x}{dt^2} = -\omega^2 x, \qquad (10.95)$$

The negative sign is taken since d^2x/dt^2 and $\omega^2 x$ are in opposite directions.

where ω^2 is the constant of proportionality. Multiplying both sides of (10.95) by $2(dx/dt)$ and integrating, we obtain

$$\left(\frac{dx}{dt}\right)^2 = -\omega^2 x^2 + c \tag{10.96}$$

To verify that integrating (10.95) gives (10.96), we differentiate both sides of (10.96) with respect to t:

$$2\left(\frac{dx}{dt}\right)\left(\frac{d^2x}{dt^2}\right) = -\omega^2 2x \frac{dx}{dt}$$

which is equivalent to (10.95).

Since the particle is undergoing negative acceleration, it has to momentarily rest at some point, say $x = a$. Then $dx/dt = 0$ at $x = a$. Consequently, from (10.96), we get

$$c = \omega^2 a^2 \tag{10.97}$$

Substituting this value of c in (10.96), we get

$$\frac{dx}{dt} = \pm\, \omega \sqrt{a^2 - x^2} \tag{10.98}$$

Solving the differential equation (10.98), we obtain

$$\arcsin\left(\frac{x}{a}\right) = \omega t + c_1 \quad \text{or} \quad \arccos\left(\frac{x}{a}\right) = \omega t + c_2$$

where c_1 and c_2 are the constants of integration. The solution, therefore, is

$$x = a \sin\left(\omega t + c_1\right) \quad \text{or} \quad x = a \cos\left(\omega t + c_2\right)$$

Recall that $\sin\left(\dfrac{\pi}{2} + \theta\right) = \cos\theta$.

If we write $c_1 = \pi/2 + c_2$, then both solutions can be written in the form

$$x = a \cos\left(\omega t + c_2\right) \tag{10.99}$$

From (10.99), we see that $2\pi/\omega$ is the period of the SHM. The number a is called the *amplitude*, and c_2 is called the *phase* of the SHM.

Exercises 10.8

1. The Slik Company produces x million gidgets and y million gadgets a year. The amounts of x and y are related by

    $$x^2 + y^2 = 169$$

    The company makes a profit of 5¢ on each gidget and 12¢ on each gadget. Find x and y so as to maximize the company's profit.

2. In Problem 1 find x and y if they are related by

    $$x^2 + y^2 + 4x + 6y = 156$$

3. The Bowry Company produces x million gallons of Hangover wine and y million gallons of Delerium wine. The amounts x and y are related by

    $$\frac{(x+1)^2}{4} + \frac{(y-2)^2}{9} = 1$$

    The Company sells the Hangover wine for \$4/gal, and the Delerium wine for \$2/gal. Find x and y so as to maximize the revenue of the company. [*Hint:* Let $x + 1 = 2\cos\theta$, $y - 2 = 3\sin\theta$.]

4. In Problem 3 find x and y if they are related by

    $$9x^2 + 4y^2 + 18x - 24y + 9 = 0$$

5. For the projectile of Example 10.22, find the following:
    (a) the maximum height   (b) time of flight
    (c) horizontal range   (d) θ for maximum horizontal range

6. Show that the equation of the path of the projectile may be written as

$$y = (x \tan \theta)\left(1 - \frac{x}{R}\right)$$

where R is the horizontal range of the projectile.

7. A projectile is fired with velocity 192 ft/sec at an elevation of 30°. Find the following.
    (a) the maximum height   (b) time of flight
    (c) horizontal range.

8. If θ is the angle of projection of a projectile, show that the same horizontal range is obtained when the angle of projection is $(\pi/2) - \theta$.

9. An object projected with a velocity u ft/sec, inclined to the horizontal at an angle of 60°, just clears a wall 12 ft high which is 20 $\sqrt{3}$ ft from the point of projection. Find the velocity u of projection and the distance to the point at which it falls on the ground.

10. The horizontal range of a rifle bullet is 1728 yd when the angle of projection is 45°. If the rifle is fired at the same angle from a car traveling 30 mph (44 ft/sec) toward the target, show that the range is increased by 792 ft.

11. If the displacement x of a moving particle at time t is given by

$$x = a \cos \omega t + b \sin \omega t$$

where a and b are constants, show that the motion is SHM.

12. In Example 10.23 show that the speed of the particle at a distance x is given by $\omega \sqrt{a^2 - x^2}$. Where does the particle have maximum speed?

13. If the displacement, velocity, and acceleration at a particular instant of a particle moving with SHM are 4 ft, 4 ft/sec, and 4 ft/sec², respectively, find the greatest velocity of the particle and the period of motion.

14. A particle moving in SHM of period 8 sec oscillates through a distance of 4 ft on each side of its central position. Find its maximum speed. Also, find the speed when the particle is 2 ft from the center.

Part 1 (*Oral*)   Chapter 10 Review

In Problems 1 through 7, define or discuss each of the following.

1. One degree

2. One radian

3. π radians = _____ degrees

4. The six trigonometric functions

5. Sign of the trigonometric functions in different quadrants

6. Periodic function

7. Inverse trigonometric functions

Part 2 (*Written*)

1. Fill in the blanks.

    (a) $\displaystyle\lim_{x \to 0} \frac{\sin x}{x} =$ _____
    (b) $\displaystyle\lim_{x \to 0} \frac{1 - \cos x}{x} =$ _____

2. Complete Table 10.9.

Table 10.9

y	$\dfrac{dy}{dx} =$ _____
$\sin x$	$\cos x$
$\cos x$	_____
$\tan x$	_____
$\sec x$	_____
$\csc x$	_____
$\cot x$	_____

3. Evaluate the following limits.

    (a) $\displaystyle\lim_{x \to 0} \frac{\sin 3x}{x}$
    (b) $\displaystyle\lim_{x \to 0} \frac{\sin 2x}{\sin 3x}$

    (c) $\displaystyle\lim_{x \to 0^+} \frac{\sin \sqrt{x}}{x}$
    (d) $\displaystyle\lim_{x \to 0} \frac{\sin x}{\tan x}$

In Problems 4 through 24, differentiate with respect to x.

4. $x \cos x$
    5. $e^x \sin x$

6. $\ln x + \cos e^x$
    7. $\sin (x^2 + 3x)$

8. $\sin^2 x \cos x$
    9. $\sin^2 (x^3 + 1) \cos^2 x$

10. $x^2 \ln \sin x$
    11. $e^{\sin x} \ln \cos x$

12. $(\sec x) \ln (\csc x)$
    13. $x \arctan x$

14. $e^x \arcsec x$
    15. $\ln (\arccsc x)$

16. $e^{\sin x - \arctan x}$
    17. $\dfrac{\sin (x - 1)}{\cos (3x + 1)}$

18. $\dfrac{\sec (3x + 2)}{\tan (x - 1)}$
    19. $\dfrac{\arctan x}{1 + \sin x}$

20. $\dfrac{\ln (1 + \sin x)}{\cos x}$
    21. $\sin (\tan^2 x)$

22. $\arccos \sqrt{1 - x^2}$
    23. $(1 + x^2) \arccot x$

24. $\dfrac{\ln \sin x}{\ln \cos x}$

In Problems 25 through 40, integrate.

25. $\displaystyle\int \frac{\cos x}{1 + \sin x}\, dx$

26. $\displaystyle\int \sec^2 x (1 + \tan x)^{1/2}\, dx$

27. $\displaystyle\int x \sin x\, dx$

28. $\displaystyle\int \cos^2 x\, dx$

29. $\displaystyle\int \cos^3 x\, dx$

30. $\displaystyle\int \sin^3 x\, dx$

31. $\displaystyle\int \sec^4 x\, dx$

32. $\displaystyle\int x^2 \cos x\, dx$

33. $\displaystyle\int \frac{1}{x^2 + 4}\, dx$

34. $\displaystyle\int \frac{1}{\sqrt{9 - x^2}}\, dx$

35. $\displaystyle\int \frac{1}{x^2 + 2x + 2}\, dx$

36. $\displaystyle\int \frac{x + 1}{x^2 + 9}\, dx$

37. $\displaystyle\int \frac{x^2 - 3x + 5}{x^2 + 16}\, dx$

38. $\displaystyle\int_0^3 \sqrt{9 - x^2}\, dx$

39. $\displaystyle\int \sqrt{16 - x^2}\, dx$

40. $\displaystyle\int_{-2}^{-1} \frac{\sqrt{x^2 - 1}}{x}\, dx$

Appendices

Mathematical Induction and the Binomial Theorem

The Σ Notation

One of the most important techniques for proving theorems for positive integers is mathematical induction. Let us first introduce the Σ notation and its use in computation.

In this appendix we shall prove some formulas such as

$$1 + 3 + 5 + \cdots + (2n - 1) = n^2 \qquad (\text{A.1})$$

$$1 + 2 + 3 + \cdots + n = \frac{n(n + 1)}{2} \qquad (\text{A.2})$$

$$1^2 + 2^2 + 3^2 + \cdots + n^2 = \frac{n(n + 1)(2n + 1)}{6} \qquad (\text{A.3})$$

Each of the formulas is true for any natural number n. The three dots indicate the missing terms in the sum. We now introduce a short notation for sums.

The symbol Σ is the capital sigma in the Greek alphabet and corresponds to our English S. Thus it reminds us of the word sum. The symbol

$$\sum_{i=1}^{5} x_i$$

(read "the sum of x_i as i assumes values from 1 to 5") means

$$\sum_{i=1}^{5} x_i = x_1 + x_2 + x_3 + x_4 + x_5$$

The sum need not start at 1 and end at 5. Further, any letter can be used instead of i. The following examples indicate various possibilities.

$$\sum_{k=1}^{3} k^2 = 1^2 + 2^2 + 3^2 = 1 + 4 + 9 = 14$$

$$\sum_{j=2}^{4} j(j+3) = 2 \cdot 5 + 3 \cdot 6 + 4 \cdot 7 = 10 + 18 + 28 = 56$$

$$\sum_{i=1}^{6} 1 = 1 + 1 + 1 + 1 + 1 + 1 = 6$$

$$\sum_{i=4}^{7} x_i = x_4 + x_5 + x_6 + x_7$$

$$\sum_{i=1}^{4} 3y_i = 3y_1 + 3y_2 + 3y_3 + 3y_4$$

$$\sum_{n=2}^{6} na_n = 2a_2 + 3a_3 + 4a_4 + 5a_5 + 6a_6$$

Example A.1   Show that

(a) $\displaystyle \sum_{i=1}^{4} x_i + \sum_{i=5}^{7} x_i = \sum_{i=1}^{7} x_i$

(b) $\displaystyle \sum_{i=1}^{4} (x_i + y_i) = \sum_{i=1}^{4} x_i + \sum_{i=1}^{4} y_i$

(c) $\displaystyle \sum_{i=1}^{3} (x_i + y_i)^2 = \sum_{i=1}^{3} x_i^2 + 2 \sum_{i=1}^{3} x_i y_i + \sum_{i=1}^{3} y_i^2$

SOLUTION

(a) $\displaystyle \sum_{i=1}^{4} x_i + \sum_{i=5}^{7} x_i = (x_1 + x_2 + x_3 + x_4) + (x_5 + x_6 + x_7)$

$$= x_1 + x_2 + x_3 + x_4 + x_5 + x_6 + x_7$$

$$= \sum_{i=1}^{7} x_i$$

(b) $\displaystyle\sum_{i=1}^{4}(x_i + y_i) = (x_1 + y_1) + (x_2 + y_2) + (x_3 + y_3) + (x_4 + y_4)$

$$= (x_1 + x_2 + x_3 + x_4) + (y_1 + y_2 + y_3 + y_4)$$

$$= \sum_{i=1}^{4} x_i + \sum_{i=1}^{4} y_i$$

(c) $\displaystyle\sum_{i=1}^{3}(x_i + y_i)^2 = (x_1 + y_1)^2 + (x_2 + y_2)^2 + (x_3 + y_3)^2$

$$= (x_1^2 + 2x_1 y_1 + y_1^2) + (x_2^2 + 2x_2 y_2 + y_2^2)$$

$$+ (x_3^2 + 2x_3 y_3 + y_3^2)$$

$$= (x_1^2 + x_2^2 + x_3^2) + 2(x_1 y_1 + x_2 y_2 + x_3 y_3)$$

$$+ (y_1^2 + y_2^2 + y_3^2)$$

$$= \sum_{i=1}^{3} x_i^2 + 2 \sum_{i=1}^{3} x_i y_i + \sum_{i=1}^{3} y_i^2$$

Example A.2   Use the Σ notation to rewrite formulas (A.1), (A.2), and (A.3).

SOLUTION   With the Σ notation, (A.1), (A.2), and (A.3) become

$$\sum_{k=1}^{n}(2k - 1) = n^2 \qquad\qquad (A.1^*)$$

$$\sum_{i=1}^{n} i = \frac{n(n + 1)}{2} \qquad\qquad (A.2^*)$$

$$\sum_{j=1}^{n} j^2 = \frac{n(n + 1)(2n + 1)}{6} \qquad\qquad (A.3^*)$$

In Problems 1 through 5, write out each of the sums term by term.   Exercises A.1

1. $\displaystyle\sum_{k=1}^{5} k^2$ 2. $\displaystyle\sum_{i=3}^{8}(x_i + y_i)$

3. $\displaystyle\sum_{k=1}^{5} 7x^k$ 4. $\displaystyle\sum_{i=1}^{9} 6$

5. $\displaystyle\sum_{i=2}^{4}(x_i + 1)^2$

In Problems 6 to 10, write each expression in sigma notation.

6. $1 + 2 + 3 + \cdots + 50$

7. $1 \cdot 2 + 2 \cdot 3 + 3 \cdot 4 + \cdots + 16 \cdot 17$

8. $2 + 4 + 8 + 16 + \cdots + 2^n$

9. $x_1^2 + x_2^2 + x_3^2 + \cdots + x_{50}^2$

10. $\dfrac{x_1}{1} + \dfrac{x_2}{2} + \dfrac{x_3}{3} + \cdots + \dfrac{x_n}{n}$

Mathematical Induction

Suppose we wish to find the sum of the first n odd positive integers. Direct computation shows that

$$
\begin{array}{llll}
\text{if} & n = 1, & 1 = 1 & = 1^2 \\
\text{if} & n = 2, & 1 + 3 = 4 & = 2^2 \\
\text{if} & n = 3, & 1 + 3 + 5 = 9 & = 3^2 \\
\text{if} & n = 4, & 1 + 3 + 5 + 7 = 16 & = 4^2 \\
\text{if} & n = 5, & 1 + 3 + 5 + 7 + 9 = 25 & = 5^2 \\
\text{if} & n = 6, & 1 + 3 + 5 + 7 + 9 + 11 = 36 = 6^2
\end{array}
$$

These cases lead us to believe that the sum is always equal to the square of the number of terms in the sum. We observe that 1 is the first term in the sum, and it is easy to see that the nth term of the sum is $2n - 1$. Thus we can express our assertion symbolically as

$$1 + 3 + 5 + \cdots + (2n - 1) = n^2 \tag{A.4}$$

or using the Σ notation, we claim that

$$\sum_{k=1}^{n} (2k - 1) = n^2 \tag{A.5}$$

By direct computation, we have seen that Equation (A.4) [or (A.5)] is true for $n = 1, 2, 3, 4, 5,$ and 6. Does this mean that Equation (A.4) is true for any positive integer n? Can we settle this by continuing our numerical computation? Certainly not! No matter how many cases we check, we can never prove that (A.4) is always true, because there are infinitely many cases and no amount of computation can settle them all. Thus we need some *logical* argument that will prove that (A.4) is always true.

We use the symbol $p(n)$ (read "p of n") to denote some proposition which depends on the positive integer n. For example, $p(n)$ might denote the proposition that (A.4) is true. The proof that $p(n)$ is true for all positive integers n can be given in two steps and these two steps form the

Principle of Mathematical Induction *For a given proposition $p(n)$ let us assume that we can prove the following conditions.*

1. *The proposition is true for $n = 1$.*
2. *If $p(n)$ is true for $n = k$, then it is also true for $n = k + 1$.*

Then the proposition is true for all positive integers n.

Note that the principle of mathematical induction can be used to prove a general proposition for positive integers only. Also the result to be proved should be surmised beforehand.

Example A.3 Prove that Equation (A.4) holds for every positive integer n.

SOLUTION

Step 1 We have already seen that Equation (A.4) is true for $n = 1$.

Step 2 We now assume that $p(n)$ is true for $n = k$:

$$1 + 3 + 5 + \cdots + (2k - 1) = k^2 \tag{A.6}$$

To obtain the sum of the first $k + 1$ odd integers, we merely add the next odd one, $2k + 1$, to both sides of (A.6).

$$(1 + 3 + 5 + \cdots + (2k - 1)) + (2k + 1) = k^2 + (2k + 1)$$
$$= (k + 1)^2 \tag{A.7}$$

Equation (A.7) is precisely Equation (A.4) when $n = k + 1$. Hence we have shown that if $p(n)$ is true for $n = k$, then it is also true for $n = k + 1$.

By the principle of mathematical induction this completes the proof that Equation (A.4) is true for every positive integer n.

Example A.4 Prove that for every positive integer n,

$$1^2 + 2^2 + 3^2 + \cdots + n^2 = \frac{n(n + 1)(2n + 1)}{6} \tag{A.8}$$

SOLUTION

Step 1 For $n = 1$, the assertion of Equation (A.8) is

$$1^2 = \frac{1(1 + 1)(2 + 1)}{6}$$

$$1 = 1$$

which is certainly true.

Step 2 We assume that (A.8) is true for $n = k$:

$$1^2 + 2^2 + 3^2 + \cdots + k^2 = \frac{k(k + 1)(2k + 1)}{6} \tag{A.9}$$

To obtain the left-hand side of (A.8) for $n = k + 1$, we must add $(k + 1)^2$ to the left-hand side of (A.9). This gives

$$1^2 + 2^2 + 3^2 + \cdots + k^2 + (k + 1)^2 = \frac{k(k + 1)(2k + 1)}{6} + (k + 1)^2$$

$$= \tfrac{1}{6}[k(k+1)(2k+1) + 6(k+1)^2] \qquad \text{(A.10)}$$
$$= \tfrac{1}{6}(k+1)[k(2k+1) + 6(k+1)]$$
$$= \tfrac{1}{6}(k+1)(2k^2 + 7k + 6)$$
$$= \tfrac{1}{6}(k+1)(k+2)(2k+3)$$
$$= \tfrac{1}{6}(k+1)[(k+1) + 1][2(k+1) + 1]$$

This is equation (A.8) when $n = k + 1$. Hence by the principle of mathematical induction, formula (A.8) is true for every positive integer n.

Exercises A.2 In Problems 1 through 9, prove that the given assertion is true for all positive integers n.

1. The nth positive even integer is $2n$.

2. $1 + 2 + 3 + \cdots + n = \dfrac{n(n+1)}{2}$

3. $1 \cdot 2 + 2 \cdot 3 + 3 \cdot 4 + \cdots + n(n+1) = \dfrac{n(n+1)(n+2)}{3}$

4. $1 + 2^1 + 2^2 + 2^3 + \cdots + 2^n = 2^{n+1} - 1$

5. $1^3 + 2^3 + 3^3 + \cdots + n^3 = \dfrac{n^2(n+1)^2}{4}$

6. $\dfrac{1}{1 \cdot 2} + \dfrac{1}{2 \cdot 3} + \dfrac{1}{3 \cdot 4} + \cdots + \dfrac{1}{n(n+1)} = \dfrac{n}{n+1}$

7. $1 \cdot 3 + 3 \cdot 5 + 5 \cdot 7 + \cdots + (2n-1)(2n+1) = \tfrac{1}{3}n(4n^2 + 6n - 1)$

8. $\dfrac{1}{1 \cdot 3} + \dfrac{1}{3 \cdot 5} + \dfrac{1}{5 \cdot 7} + \cdots + \dfrac{1}{(2n-1)(2n+1)} = \dfrac{n}{2n+1}$

9. $\dfrac{1}{2 \cdot 5} + \dfrac{1}{5 \cdot 8} + \dfrac{1}{8 \cdot 11} + \cdots + \dfrac{1}{(3n-1)(3n+2)} = \dfrac{n}{6n+4}$

Binomial Theorem

Before we state and prove the binomial theorem, let us define some new notations.

The Factorial Notation We define the symbol $n!$ (read "n factorial"), where n is a nonnegative integer, as

Factorial

$$\begin{cases} 0! = 1 \\ (n+1)! = (n+1) \cdot n! \end{cases} \qquad \text{(A.11)}$$

For example, we have

$$0! = 1$$
$$1! = 1 \cdot 0! = 1 \cdot 1 = 1$$

$$2! = 2 \cdot 1! = 2 \cdot 1 = 2$$
$$3! = 3 \cdot 2! = 3 \cdot 2 = 6$$
$$4! = 4 \cdot 3! = 4 \cdot 6 = 24$$

Thus for any positive integer n, $n!$ can be computed as the product of all positive integers less than or equal to n. For example, $5! = 5 \cdot 4 \cdot 3 \cdot 2 \cdot 1 = 120$.

The Symbol $\binom{n}{r}$ Let n be a nonnegative integer and r a nonnegative integer less than or equal to n. The symbol $\binom{n}{r}$ is defined by

$$\binom{n}{r} = \frac{n!}{(n-r)!r!} \tag{A.12}$$

By (A.12), we observe that

$$\binom{n}{0} = \frac{n!}{(n-0)!0!} = \frac{n!}{n! \cdot 1} = 1 \tag{A.13}$$

$$\binom{0}{0} = \frac{0!}{(0-0)!0!} = 1$$

The following theorem will be used in the proof of the binomial theorem.

Theorem A.1 *If k and r are integers such that $1 \le r \le k$, then* ◄

$$\binom{k}{r} + \binom{k}{r-1} = \binom{k+1}{r} \tag{A.14}$$

PROOF

$$\binom{k}{r} + \binom{k}{r-1} = \frac{k!}{(k-r)!r!} + \frac{k!}{(k-r+1)!(r-1)!}$$

$$= \frac{k!(k-r+1)}{r!(k-r)!(k-r+1)} + \frac{k!r}{(k-r+1)!(r-1)!r}$$

$$= \frac{k!(k-r+1)}{r!(k-r+1)!} + \frac{k!r}{(k-r+1)!r!}$$

$$= \frac{k!(k+1)}{r!(k+1-r)!}$$

$$= \frac{(k+1)!}{(k+1-r)!r!}$$

$$= \binom{k+1}{r}$$

The Binomial Theorem An expression that contains only two terms connected by a positive or a negative sign is called a *binomial* expression. For example, $a + b$, $2x - 7$, $a + (1/a)$, $x + 8y^2$ are all binomial expressions. If n is a positive integer, then $(x + a)^n$ is a power of the binomial expression $x + a$. We are familiar with the expansions of $(x + a)^2$, $(x + a)^3$, and $(x + a)^4$. These expansions can be written using the symbol $\binom{n}{r}$.

$$(x + a)^2 = x^2 + 2xa + a^2 = \binom{2}{0} x^2 + \binom{2}{1} xa + \binom{2}{2} a^2 \quad \text{(A.15)}$$

since $\binom{2}{0} = 1$, $\binom{2}{1} = 2$, and $\binom{2}{2} = 1$.

$$(x + a)^3 = x^3 + 3x^2 a + 3xa^2 + a^3 \quad \text{(A.16)}$$

$$= \binom{3}{0} x^3 + \binom{3}{1} x^2 a + \binom{3}{2} xa^2 + \binom{3}{3} a^3$$

since $\binom{3}{0} = 1$, $\binom{3}{1} = 3$, $\binom{3}{2} = 3$, and $\binom{3}{3} = 1$.

$$(x + a)^4 = x^4 + 4x^3 a + 6x^2 a^2 + 4xa^3 + a^4 \quad \text{(A.17)}$$

$$= \binom{4}{0} x^4 + \binom{4}{1} x^5 a + \binom{4}{2} x^2 a^2 + \binom{4}{3} xa^3 + \binom{4}{4} a^4$$

since $\binom{4}{0} = 1$, $\binom{4}{1} = 4$, $\binom{4}{2} = 6$, $\binom{4}{3} = 4$, and $\binom{4}{4} = 1$.

From (A.15), (A.16), and (A.17) we make the following observations.

(i) In successive terms the exponent of x decreases by 1 and the exponent of a increases by 1.

(ii) The sum of the exponents of x and a remains the same in all the terms of the expansion.

(iii) The coefficients of the successive terms follow a certain rule.

(iv) The number of terms is one more than the exponent of $x + a$.

These facts lead us to state the binomial theorem.

The binomial theorem ▶   *Theorem A.2 For any positive integer n,*

$$(x + a)^n = \binom{n}{0} x^n + \binom{n}{1} x^{n-1} a + \binom{n}{2} x^{n-2} a^2$$

$$+ \cdots + \binom{n}{r} x^{n-r} a^r + \cdots + \binom{n}{n} a^n$$

$$= \sum_{r=0}^{n} \binom{n}{r} x^{n-r} a^r \quad \text{(A.18)}$$

PROOF   We use the principle of mathematical induction.

Step 1   For $n = 1$, we have

$$(x + a)^1 = \binom{1}{0} x^1 + \binom{1}{1} a^1$$

$$= x + a$$

which is certainly true.

Step 2 We now assume that the theorem is true for $n = k$. Thus we assume that

$$(x + a)^k = \sum_{r=0}^{k} \binom{k}{r} x^{k-r} a^r \qquad (A.19)$$

Using Equation (A.19) and the definition of $(x + a)^{k+1}$, we have

$$(x + a)^{k+1} = (x + a)(x + a)^k = (x + a) \sum_{r=0}^{k} \binom{k}{r} x^{k-r} a^r$$

$$= x \sum_{r=0}^{k} \binom{k}{r} x^{k-r} a^r + a \sum_{r=0}^{k} \binom{k}{r} x^{k-r} a^r$$

Thus

$$(x + a)^{k+1} = \sum_{r=0}^{k} \binom{k}{r} x^{k-r+1} a^r + \sum_{r=0}^{k} \binom{k}{r} x^{k-r} a^{r+1} \qquad (A.20)$$

Collecting like terms on the right-hand side of (A.20), we find that the coefficient of $x^{k-r} a^{r+1}$ is

$$\binom{k}{r+1} + \binom{k}{r} \qquad (A.21)$$

By Theorem A.1, this new coefficient is

$$\binom{k}{r+1} + \binom{k}{r} = \binom{k+1}{r+1} \qquad (A.22)$$

But the right-hand side of (A.22) has exactly the form required for the coefficient of $x^{k-r} a^{r+1}$ in the expansion of $(x + a)^{k+1}$. This argument holds if $r = 0, 1, 2, \ldots, k - 1$. The extreme terms, x^{k+1} and a^{k+1}, are not covered by the preceding discussion, but these terms obviously have 1 as coefficients, as required by Theorem A.2.

Example A.5 Expand $(x - 3y^2)^4$.

SOLUTION By Theorem A.2, the coefficients are

$$\binom{4}{0} = \frac{4!}{4!0!} = 1, \qquad \binom{4}{1} = \frac{4!}{3!1!} = 4, \qquad \binom{4}{2} = \frac{4!}{2!2!} = 6$$

$$\binom{4}{3} = \frac{4!}{1!3!} = 4, \qquad \binom{4}{4} = \frac{4!}{0!4!} = 1$$

We can write $(x - 3y^2)^4 = [x + (-3y^2)]^4$. Hence

$$(x - 3y^2)^4 = x^4 + 4x^3(-3y^2) + 6x^2(-3y^2)^2 + 4x(-3y^2)^3 + (-3y^2)^4$$

$$= x^4 - 12x^3 y^2 + 54x^2 y^4 - 108xy^6 + 81y^8$$

Example A.6 Find the sixth term in the expansion of $(x^2 - 2y)^{11}$.

SOLUTION   The $(r + 1)$th term in the expansion of $(x + a)^n$ by Theorem A.2 is $\binom{n}{r} x^{n-r}a^r$. Thus the sixth term in the expansion of $(x^2 - 2y)^{11}$ is

$$\binom{11}{5} (x^2)^{11-5}(-2y)^5 = \frac{11!}{6!5!} (x^2)^6 (-32y^5)$$

$$= \frac{11 \cdot 10 \cdot 9 \cdot 8 \cdot 7 \cdot 6!}{6! \cdot 5 \cdot 4 \cdot 3 \cdot 2 \cdot 1} x^{12}(-32y^5)$$

$$= -32 \cdot 462x^{12}y^5$$

$$= -14784x^{12}y^5$$

Example A.7   Find the term independent of x in the expansion of $(\sqrt{x} - 2/x^2)^{10}$.

SOLUTION   The term independent of x means the term not containing x, and therefore the exponent of x must be zero.

The $(r + 1)$th term in the expansion of $(x^{1/2} - 2/x^2)^{10}$ is

$$\binom{10}{r} (x^{1/2})^{10-r} \left(-\frac{2}{x^2}\right)^r = \binom{10}{r} x^{5-(r/2)} (-2)^r(x^{-2r})$$

$$= \binom{10}{r} \cdot (-2)^r \cdot x^{5-(5r/2)}$$

Since the term independent of x is desired, we must have

$$5 - \frac{5r}{2} = 0$$

or

$$r = 2$$

Therefore the required term is

$$\binom{10}{2} (-2)^2 x^0 = \frac{10!}{8!2!} \cdot 4$$

$$= 45 \cdot 4$$

$$= 180$$

Exercises A.3   In Problems 1 through 8, compute explicitly the given quantity.

1. $7!$

2. $10!$

3. $\binom{8}{3}$

4. $\binom{11}{7}$

5. $\binom{87}{86}\binom{2}{0}$

6. $\dfrac{16}{\binom{6}{4}}$

7. $\left(\begin{array}{c}8\\4\\2\end{array}\right)$ 8. $\left(\left(\begin{array}{c}8\\4\\2\end{array}\right)\right)$

9. Show that $\binom{n}{r}=\binom{n}{n-r}$.

*10. Show that for $0 \le k \le n-2$,

$$\binom{n+2}{k+2}=\binom{n}{k+2}+2\binom{n}{k+1}+\binom{n}{k}$$

[*Hint*: Use (A.14) twice.]

*11. Show that

$$\binom{n}{0}+\binom{n}{1}+\binom{n}{2}+\cdots+\binom{n}{n}=2^n$$

[*Hint*: Let $x=a=1$ in (A.18).]

12. Expand the following.
  (a) $(1+x)^6$ (b) $(2x+1)^5$
  (c) $(x-2)^5$ (d) $\left(x+\dfrac{1}{x}\right)^6$

13. Write the term indicated in the expansion of each of the following.
  (a) Fourth term of $(x-2y)^7$
  (b) Seventh term of $\left(x^2-\dfrac{1}{x}\right)^9$
  (c) Ninth term of $(x-ay^2)^{12}$
  (d) Fifteenth term of $\left(2-\dfrac{1}{x}\right)^{19}$

14. Find the term independent of x in the following expressions.
  (a) $\left(2x^3+\dfrac{1}{4x^8}\right)^{11}$ (b) $\left(2x-\dfrac{1}{x^2}\right)^6$
  (c) $\left(\sqrt{x}-\dfrac{2}{x^2}\right)^{10}$ (d) $\left(x^2-\dfrac{1}{x}\right)^9$

The Polynomial Functions

Appendix 2

In this appendix we deal with a class of functions called polynomials. These are fairly simple functions and are significant for their use in approximating other more complicated functions.

Let n be a nonnegative integer and let $a_0, a_1, \ldots, a_n$ be real numbers with $a_n \neq 0$. Then the function p defined by

$$p(x) = a_n x^n + a_{n-1} x^{n-1} + \cdots + a_1 x + a_0 \qquad (A.23)$$

is called a *polynomial function of degree n* (or simply a *polynomial*). *Polynomial*
The numbers $a_0, a_1, \ldots, a_n$ are called the *coefficients* and a_n the *leading coefficient* of p.

The student has already encountered some special polynomials: the constant function, $p(x) = a_0$; the linear function, $p(x) = a_1 x + a_0$; and the quadratic function, $p(x) = a_2 x^2 + a_1 x + a_0$. These are polynomials of degree zero, one, and two, respectively. No degree is assigned to the zero function, $p(x) \equiv 0$.

Polynomials may be added, subtracted, multiplied, or divided. Thus if p and q are polynomials of degree m and n, respectively, then

(i) $p \pm q$ is a polynomial of degree less than or equal to the maximum of m and n.

(ii) $p \cdot q$ is a polynomial of degree $m + n$.

Example A.8   Let $p(x) = x^3 - 3x + 8$ and $q(x) = x^3 + 5x^2 - x + 1$. Then

(i) $p(x) + q(x) = (x^3 - 3x + 8) + (x^3 + 5x^2 - x + 1)$
$$= 2x^3 + 5x^2 - 4x + 9$$

a polynomial of degree 3

(ii) $p(x) - q(x) = (x^3 - 3x + 8) - (x^3 + 5x^2 - x + 1)$
$$= -5x^2 - 2x + 7$$

a polynomial of degree 2

(iii) $p(x)q(x) = (x^3 - 3x + 8)(x^3 + 5x^2 - x + 1)$
$$= x^3(x^3 + 5x^2 - x + 1) - 3x(x^3 + 5x^2 - x + 1)$$
$$+ 8(x^3 + 5x^2 - x + 1)$$
$$= x^6 + 5x^5 - 4x^4 - 6x^3 + 43x^2 - 11x + 8$$

a polynomial of degree 6

The concept of division involving polynomials is quite similar to that of integers. Thus if p and h are polynomials, then p is *divisible* by h if and only if there is a polynomial q such that

$$p = q \cdot h$$

Example A.9   Let $p(x) = x^3 - 1$ and $h(x) = x - 1$. Find a polynomial q such that $p = q \cdot h$.

SOLUTION   Since

$$x^3 - 1 = (x - 1)(x^2 + x + 1)$$

for

$$q(x) = x^2 + x + 1$$

we have

$$p(x) = q(x)h(x)$$

Analogous to the division algorithm for integers is the division algorithm for polynomials.

Theorem A.3 *If p and h are polynomials and h is of degree greater than zero, then there exist unique polynomials q and r such that*

$$p(x) = q(x) \cdot h(x) + r(x) \tag{A.24}$$

where r is either a polynomial of degree less than the degree of h or the zero function.

The polynomial p is called the *dividend*; h, the *divisor*; q, the *quotient*; and r, the *remainder*.

We illustrate this theorem in

Example A.10 Let $p(x) = x^3 + 5x - 2$ and $h(x) = x + 1$. Find polynomials q and r such that

$$p(x) = q(x) \cdot h(x) + r(x)$$

SOLUTION The procedure is shown in the following sequence of steps (long division).

$$
\begin{array}{r}
x^2 - x + 6 \\
x + 1 \overline{\smash{\big)}\ x^3 + 0 \cdot x^2 + 5x - 2} \\
\underline{x^3 + x^2} \\
-x^2 + 5x \\
\underline{-x^2 - x} \\
6x - 2 \\
\underline{6x + 6} \\
-8
\end{array}
$$

Thus if we let $q(x) = x^2 - x + 6$ and $r(x) = -8$, then

$$x^3 + 5x - 2 = (x + 1)(x^2 - x + 6) - 8$$

The student should observe how the term $0 \cdot x^2$ is used to fill in the missing term.

Theorem A.4 *Let p denote a polynomial of degree $n \geq 1$ and c a real number. Then there exists a polynomial q of degree $n - 1$ such that*

$$p(x) = (x - c)q(x) + p(c) \tag{A.25}$$

for all numbers x.

PROOF We use Theorem A.3 for polynomials $p(x)$ and $x - c$:

$$p(x) = a(x) \cdot (x - c) + r(x)$$

where $r(x)$ has degree less than the degree of $(x - c)$, which means that $r(x)$ has degree zero: $r(x) = r$, a constant. We evaluate p at c:

$$
\begin{aligned}
p(c) &= q(c) \cdot (c - c) + r \\
&= 0 + r \\
&= r
\end{aligned}
$$

◀ The remainder theorem. Comparing the result of this theorem with division involving integers, it should become apparent to the student where the name, remainder theorem, originated.

Thus $p(x) = q(x) \cdot (x - c) + p(c)$, and it is clear that $q(x)$ must have degree $n - 1$. To find q, we let p and q be defined by

$$p(x) = a_n x^n + a_{n-1}x^{n-1} + \cdots + a_1 x + a_0$$

$$q(x) = b_{n-1}x^{n-1} + b_{n-2}x^{n-2} + \cdots + b_1 x + b_0$$

Here a_n, a_{n-1}, ..., a_0 are known constants and we wish to find b_{n-1}, b_{n-2}, ..., b_0 and $p(c)$ such that Equation (A.25) is satisfied. We observe that

$$(x - c)q(x) = b_{n-1}x^n + (b_{n-2} - cb_{n-1})x^{n-1} + \cdots \qquad \text{(A.26)}$$
$$+ (b_0 - cb_1)x + (-cb_0)$$

Hence

$$a_n x^n + a_{n-1}x^{n-1} + \cdots + a_1 x + a_0 \qquad \text{(A.27)}$$
$$= b_{n-1}x^n + (b_{n-2} - cb_{n-1})x^{n-1} + \cdots + (b_0 - cb_1)x + (p(c) - cb_0)$$

We equate coefficients to find those for q.

$$b_{n-1} = a_n$$
$$b_{n-2} = a_{n-1} + cb_{n-1}$$
$$\vdots$$
$$b_0 = a_1 + cb_1$$
$$p(c) = a_0 + cb_0$$

Thus for a given polynomial p, the coefficients of the polynomial q and $p(c)$ can be computed successively, and Equation (A.25) can be obtained.

The student should observe carefully the pattern involved in determining the coefficients of q and $p(c)$. This process provides us with a convenient method for dividing a polynomial p by the special polynomial $x - c$ (called the *monic polynomial*). The method is called *synthetic division*. For convenience, we arrange the work as follows.

c	a_n	a_{n-1}	a_{n-2}	$\cdots$	a_1	a_0
		$+$	$+$		$+$	$+$
		cb_{n-1}	cb_{n-2}	$\cdots$	$c \cdot b_1$	cb_0
	b_{n-1}	b_{n-2}	b_{n-3}		b_0	$p(c)$

Here we are provided with an alternative method to long division involving polynomials when the divisor is a monic polynomial. We illustrate this procedure in the following example.

Example A.11   Use synthetic division to divide

$$p(x) = 3x^4 - 8x^2 - 3x + 7 \text{ by } x - 2.$$

SOLUTION

$$
\begin{array}{r|rrrrr}
2 & 3 & 0 & -8 & -3 & 7 \\
  &   & 2\cdot 3 & 2\cdot 6 & 2\cdot 4 & 2\cdot 5 \\
\hline
  & 3 & 6 & 4 & 5 & 17
\end{array}
$$

Hence $q(x) = 3x^3 + 6x^2 + 4x + 5$ and $p(2) = 17$, that is,

$$3x^4 - 8x^2 - 3x + 7 = (x - 2)(3x^3 + 6x^2 + 4x + 5) + 17$$

We observe that

$$p(2) = 3(2)^4 - 8(2)^2 - 3(2) + 7 = 17$$

The student will do well to work through this problem using long division.

If $p(c) = 0$, then from Theorem A.4 we have

Theorem A.5   Let p denote a polynomial and c a real number. Then $x - c$ is a factor of p (or p is divisible by $x - c$) if and only if $p(c) = 0$.   ◀ *The factor theorem*

Example A.12   Show that $x + 1$ is a factor of

$$p(x) = x^3 - 6x^2 - 6x + 1$$

SOLUTION   By inspection we find that $p(-1) = 0$. Hence $x + 1$ is a factor of p. Using synthetic division, we find

$$
\begin{array}{r|rrrr}
-1 & 1 & -6 & -6 & 1 \\
   &   & -1 & 7 & -1 \\
\hline
   & 1 & -7 & 1 & 0
\end{array}
$$

Therefore

$$p(x) = (x + 1)(x^2 - 7x + 1)$$

In Problems 1 through 10, use long division to divide p by h and determine a quotient q such that the remainder r is a real number.   Exercises A.4

1. $p(x) = 4x^3 + 2x^2 + x + 3, \quad h(x) = 2x + 3$
2. $p(x) = 4x^3 - 5x^2 - 4x + 1, \quad h(x) = 3x - 2$
3. $p(x) = 6x^3 + x^2 + x + 4, \quad h(x) = 3x + 1$
4. $p(x) = 9x^3 - x^2 - 13x + 6, \quad h(x) = x + 5$
5. $p(x) = x^3 - 75x^2 - 5x + 1, \quad h(x) = x - 1$
6. $p(x) = 3x^4 - 6x^2 + 3x - 7, \quad h(x) = x - 2$

7. $p(x) = -x^5 + 4x^3 + 2, \quad h(x) = x + 2$

8. $p(x) = x^5 - 1, \quad h(x) = x - 1$

9. $p(x) = x^6 + 2x^4 + x^3 + 5, \quad h(x) = x + 1$

10. $p(x) = x^6 + 3x^2 + 10, \quad h(x) = 2x + 3$

In Problems 11 through 15, use synthetic division to divide p by h and find $p(c)$.

11. $p(x) = x^3 + 3x^2 - x + 12, \quad h(x) = x + 2$

12. $p(x) = 2x^3 + 4x^2 + x - 15, \quad h(x) = x - 3$

13. $p(x) = x^4 + 2x^2 - x - 7, \quad h(x) = x + 1$

14. $p(x) = 2x^4 - 2x^3 + 9, \quad h(x) = x - 1$

15. $p(x) = 3x^4 + 2x^3 + x + 1, \quad h(x) = x - 2$

16. If $p(x) = 2x^2 - x + 4$, find a polynomial q such that for all x
$$p(x) = (x - 1)q(x) + p(1)$$

17. If $p(x) = x^3 + x^2 + x + 1$, find a polynomial q such that for all x
$$p(x) = (x + 2)q(x) + p(-2)$$

18. If $p(x) = x^2 - 3cx + c$, find c such that $x - 1$ is a factor of p.

19. If $p(x) = 2x^3 - cx^2 + x - c$, find c such that $x + 1$ is a factor of p.

20. If $p(x) = 2x^3 - x^2 - 4cx - 5c$, find c such that $x + 3$ is a factor of p.

Zeros of Polynomial Functions

We have seen from the factor theorem that if p is a polynomial of degree $n \geq 1$ and c is a number, then $p(c) = 0$ implies that $x - c$ is a factor of p. The number c is called a zero (or *root*) of p. Geometrically, c represents the point where the graph of p intersects the x-axis.

Clearly, since a polynomial p of degree n can have no more than n factors which are monic polynomials, then p has at *most* n zeros. For the rational zeros of a polynomial we state

▶   *Theorem A.7   If a/b, a rational number in lowest terms, is a zero of the polynomial*

$$p(x) = a_n x^n + a_{n-1} x^{n-1} + \cdots + a_1 x + a_0$$

where the a_i's ($i = 0, 1, \ldots, n$) are integers and $a_n \neq 0$, then a is an integral factor of a_0 and b is an integral factor of a_n.

Note that this theorem does not guarantee the existence of rational zeros of a polynomial. It merely enables us to identify the possible rational zeros, which we can then check using synthetic division or otherwise.

Example A.13   Find all rational zeros of the polynomial

$$p(x) = 2x^3 + 5x^2 - 4x - 3$$

SOLUTION   If a/b is a rational zero of p, then by Theorem A.7, a must be an integral factor of 3 and b must be an integral factor of 2. That is,

$$a \in \{-1, -3, 1, 3\}, \qquad b \in \{-1, -2, 1, 2\}$$

and the set of possible rational zeros of p is

$$\{-3, -\tfrac{3}{2}, -1, -\tfrac{1}{2}, \tfrac{1}{2}, 1, \tfrac{3}{2}, 3\}$$

Using synthetic division we check each of the possible candidates:

-3	2	5	-4	-3
		-6	3	3
	2	-1	-1	0

From the remainder theorem we see that -3 is a zero of p. For the remaining possible rational zeros, we set up the following array.

	2	5	-4	-3
$-\tfrac{3}{2}$	2	2	-7	$\tfrac{15}{2}$
-1	2	3	-7	4
$-\tfrac{1}{2}$	2	4	-6	0
$\tfrac{1}{2}$	2	6	-1	$-\tfrac{7}{2}$
1	2	7	3	0

Thus far we have found three zeros of p: -3, $-\tfrac{1}{2}$, and 1. Since p is of degree 3, we discontinue our investigations of the remaining numbers. The set of all rational zeros of p is $\{-3, -\tfrac{1}{2}, 1\}$.

We now consider an alternative procedure. Having found that -3 is a zero of p, we use the factor theorem and write

$$2x^3 + 5x^2 - 4x - 3 = (x + 3)(2x^2 - x - 1)$$

The quotient is a quadratic function whose zeros can be found by using the quadratic formula or simply by factoring. Thus

$$2x^3 + 5x^2 - 4x = (x + 3)(2x + 1)(x - 1)$$

and the zeros of p are easily obtained.

Example A.14   Find the rational zeros of

$$p(x) = x^3 + x^2 + 2x + 2$$

SOLUTION The set of possible rational zeros of p is

$$\{-2, -1, 1, 2\}$$

Using synthetic division, we find

$$
\begin{array}{r|rrrr}
-2 & 1 & 1 & 2 & 2 \\
   &   & -2 & 2 & -8 \\
\hline
   & 1 & -1 & 4 & -6
\end{array}
$$

Thus -2 is not a zero of p. We try -1.

$$
\begin{array}{r|rrrr}
-1 & 1 & 1 & 2 & 2 \\
   &   & -1 & 0 & -2 \\
\hline
   & 1 & 0 & 2 & 0
\end{array}
$$

Hence -1 is a zero of p, and by the factor theorem we have

$$x^3 + x^2 + 2x + 2 = (x + 1)(x^2 + 2)$$

Since $x^2 + 2$ has no real zeros (use the quadratic formula), we find that -1 is the only rational zero of p.

It should be clear to the student that by the zeros of a polynomial p we mean the solutions of the equation $p(x) = 0$. In 1799, Gauss presented his doctoral dissertation in which he proved an important algebraic result which today is known as the fundamental theorem of algebra.

The fundamental theorem of algebra ▶

Theorem A.8 *Let p denote the polynomial*

$$p(z) = a_n z^n + a_{n-1} z^{n-1} + \cdots + a_1 z + a_0$$

of degree $n \geq 1$, where the a_i's ($i = 0, 1, \ldots, n$) and z are complex numbers. Then p has at least one zero (real or complex).

In fact, as a consequence of this theorem and the factor theorem, we can write p as

$$p(z) = a_n(z - c_1)(z - c_2) \cdots (z - c_n)$$

where $c_1, c_2, \ldots, c_n$ are complex numbers not necessarily all distinct. Thus the polynomial p of degree n has n zeros exactly. Since some of the factors may occur more than once, we refer to the corresponding zero as a *multiple zero*.

Example A.15 Find all the zeros of

$$p(z) = 2x^4 - 4z^3 + 4z^2 - 4z + 2$$

SOLUTION Using synthetic division, we can write

$$p(z) = 2(z - 1)(z - 1)(z^2 + 1)$$

Hence the set of all zeros is $\{1, 1, i, -i\}$. Here we find that 1 is a multiple zero of p. We say that 1 is a zero of p of *multiplicity* 2.

In Problems 1 through 6, find all the rational zeros of p.

1. $p(x) = x^3 - 2x^2 + x + 4$

2. $p(x) = 3x^3 - x^2 - 4x + 2$

3. $p(x) = 2x^3 - 3x^2 - 7x - 6$

4. $p(x) = 2x^4 + x^3 + x^2 + x - 1$

5. $p(x) = 5x^3 + 10x^2 - 2x - 8$

6. $p(x) = x^4 + 4x^3 + 2x^2 - 4x - 3$

For each of the polynomials in Problems 7 through 12, find the zeros and their multiplicity.

7. $p(x) = x^4 - 8x^2 + 16$

8. $p(x) = x^4 - 2x^2 - 15$

9. $p(x) = 64x^4 - 1$

10. $p(x) = x^4 + 2x^3 - 3x^2 - 4x + 4$

11. $p(x) = x^4 - 2x^3 + 2x - 1$

12. $p(x) = x^6 - 2x^4 + x^2$

13. Show that the polynomial of degree 2 has at most two zeros. [*Hint*: Assume that there are three distinct roots and show that this leads to a contradiction.]

14. If r_1 and r_2 are two zeros of $p(x) = ax^2 + bx + c$, show that

    (a) $r_1 + r_2 = -\dfrac{b}{a}$ (b) $r_1 r_2 = \dfrac{c}{a}$

15. Find the sum and product of the zeros of each of the following quadratic polynomials.
    (a) $p(x) = x^2 + 3x + 7$ (b) $p(x) = 2x^2 - 3x - 7$
    (c) $p(x) = 3x^2 + 4x - 2$ (d) $p(x) = x^2 - 10x + 19$

16. If r_1 and r_2 are two zeros of a quadratic polynomial p, find p given that
    (a) $r_1 + r_2 = 5$, $r_1 r_2 = 9$ (b) $r_1 + r_2 = -2$, $r_1 r_2 = 3$
    (c) $r_1 + r_2 = \frac{3}{2}$, $r_1 r_2 = \frac{4}{3}$ (d) $r_1 + r_2 = \frac{5}{3}$, $r_1 r_2 = 3$

17. Find the values of c such that the graph of the given quadratic polynomial touches the x-axis at only one point.
    (a) $p(x) = x^2 + 3x + c$ (b) $p(x) = x^2 - 5x - c$
    (c) $p(x) = x^2 - 3cx + 4$

*18. If $u + iv$ is a zero of the quadratic polynomial

$$p(z) = az^2 + bz + c$$

show that $u - iv$ is also a zero.

*19. Find all the zeros of p if one of the zeros is the given number.
  (a) $p(z) = 3z^3 - 7z^2 + 27z - 63; -3i$
  (b) $p(z) = z^4 - 2z^3 + 6z^2 - 2z + 5; 1 - 2i$

*20. If $p(z) = a_3z^3 + a_2z^2 + a_1z + a_0$, where $a_0, a_1, a_2,$ and a_3 are real numbers, show that p has at least one real zero.

Carl Friedrich Gauss

Carl Friedrich Gauss (1777–1855), one of the greatest of all mathematicians, was born in Brunswick, Germany. Gauss was born in an extremely poor family. His father was a gardener and bricklayer who had neither the inclination nor means to help educate young Carl. Gauss' outstanding talent and genius manifested itself at an early age. It is said that at the age of three Carl pointed out an error in his father's payroll accounts.

Gauss' outstanding achievements in school attracted the attention of the Duke of Brunswick who took an interest in young Carl and sent him first to the Collegium Carolinum and later to the University of Göttingen (1795–1798). Gauss' first big mathematical discovery involved the ruler and compass construction of regular polygons—a problem that had not been resolved since the ancient Greeks. This discovery helped Gauss decide to study mathematics rather than philology. While at Göttingen Gauss started a diary recording his mathematical and scientific ideas. Many of his discoveries originated in this diary which also contained facts about elliptic functions and non-Euclidean geometry.

Gauss was a prolific writer. His early reputation was made by his work in the theory of numbers. Gauss tackled all of mathematics. He developed the theory of surfaces, conceived the normal law of distribution in probability theory, and invented the method of least squares. The latter helped Gauss rediscover the asteroid Ceres after all the experts had failed. In 1807, the Duke of Brunswick appointed Gauss as professor of astronomy and director of the new observatory at Göttingen. He spent most of his life there. Gauss disliked administrative work and regarded teaching as a waste of time. However he performed his duties thoroughly and taught superbly. Among his many celebrated students were Dedekind and Riemann.

Table 1   Squares and Square Roots   **A21**

Table 1 Squares and Square Roots

x	x^2	$\sqrt{x}$	x	x^2	$\sqrt{x}$
1	1	1.000	51	2,601	7.141
2	4	1.414	52	2,704	7.211
3	9	1.732	53	2,809	7.280
4	16	2.000	54	2,916	7.348
5	25	2.236	55	3,025	7.416
6	36	2.449	56	3,136	7.483
7	49	2.646	57	3,249	7.550
8	64	2.828	58	3,364	7.616
9	81	3.000	59	3,481	7.681
10	100	3.162	60	3,600	7.746
11	121	3.317	61	3,721	7.810
12	144	3.464	62	3,844	7.874
13	169	3.606	63	3,969	7.937
14	196	3.742	64	4,096	8.000
15	225	3.873	65	4,225	8.062
16	256	4.000	66	4,356	8.124
17	289	4.123	67	4,489	8.185
18	324	4.243	68	4,624	8.246
19	361	4.359	69	4,761	8.307
20	400	4.472	70	4,900	8.367
21	441	4.583	71	5,041	8.426
22	484	4.690	72	5,184	8.485
23	529	4.796	73	5,329	8.544
24	576	4.899	74	5,476	8.602
25	625	5.000	75	5,625	8.660
26	676	5.099	76	5,776	8.718
27	729	5.196	77	5,929	8.775
28	784	5.292	78	6,084	8.832
29	841	5.385	79	6,241	8.888
30	900	5.477	80	6,400	8.944
31	961	5.568	81	6,561	9.000
32	1,024	5.657	82	6,724	9.055
33	1,089	5.745	83	6,889	9.110
34	1,156	5.831	84	7,056	9.165
35	1,225	5.916	85	7,225	9.220
36	1,296	6.000	86	7,396	9.274
37	1,369	6.083	87	7,569	9.327
38	1,444	6.164	88	7,744	9.381
39	1,521	6.245	89	7,921	9.434
40	1,600	6.325	90	8,100	9.487
41	1,681	6.403	91	8,281	9.539
42	1,764	6.481	92	8,464	9.592
43	1,849	6.557	93	8,649	9.644
44	1,936	6.633	94	8,836	9.695
45	2,025	6.708	95	9,025	9.747
46	2,116	6.782	96	9,216	9.798
47	2,209	6.856	97	9,409	9.849
48	2,304	6.928	98	9,604	9.899
49	2,401	7.000	99	9,801	9.950
50	2,500	7.071	100	10,000	10.000

Table 2
Exponential Functions

x	e^x	e^{-x}	x	e^x	e^{-x}
0.00	1.0000	1.0000	1.5	4.4817	0.2231
0.01	1.0101	0.9901	1.6	4.9530	0.2019
0.02	1.0202	0.9802	1.7	5.4739	0.1827
0.03	1.0305	0.9705	1.8	6.0496	0.1653
0.04	1.0408	0.9608	1.9	6.6859	0.1496
0.05	1.0513	0.9512	2.0	7.3891	0.1353
0.06	1.0618	0.9418	2.1	8.1662	0.1225
0.07	1.0725	0.9324	2.2	9.0250	0.1108
0.08	1.0833	0.9331	2.3	9.9742	0.1003
0.09	1.0942	0.9139	2.4	11.0230	0.0907
0.10	1.1052	0.9048	2.5	12.182	0.0821
0.11	1.1163	0.8958	2.6	13.464	0.0743
0.12	1.1275	0.8869	2.7	14.880	0.0672
0.13	1.1388	0.8781	2.8	16.445	0.0608
0.14	1.1503	0.8694	2.9	18.174	0.0550
0.15	1.1618	0.8607	3.0	20.086	0.0498
0.16	1.1735	0.8521	3.1	22.198	0.0450
0.17	1.1853	0.8437	3.2	24.533	0.0408
0.18	1.1972	0.8353	3.3	27.113	0.0369
0.19	1.2092	0.8270	3.4	29.964	0.0334
0.20	1.2214	0.8187	3.5	33.115	0.0302
0.21	1.2337	0.8106	3.6	36.598	0.0273
0.22	1.2461	0.8025	3.7	40.447	0.0247
0.23	1.2586	0.7945	3.8	44.701	0.0224
0.24	1.2712	0.7866	3.9	49.402	0.0202
0.25	1.2840	0.7788	4.0	54.598	0.0183
0.30	1.3499	0.7408	4.1	60.340	0.0166
0.35	1.4191	0.7047	4.2	66.686	0.0150
0.40	1.4918	0.6703	4.3	73.700	0.0136
0.45	1.5683	0.6376	4.4	81.451	0.0123
0.50	1.6487	0.6065	4.5	90.017	0.0111
0.55	1.7333	0.5769	4.6	99.484	0.0101
0.60	1.8221	0.5488	4.7	109.950	0.0091
0.65	1.9155	0.5220	4.8	121.510	0.0082
0.70	2.0138	0.4966	4.9	134.290	0.0074
0.75	2.1170	0.4724	5.0	148.41	0.0067
0.80	2.2255	0.4493	5.5	244.69	0.0041
0.85	2.3396	0.4274	6.0	403.43	0.0025
0.90	2.4596	0.4066	6.5	665.14	0.0015
0.95	2.5857	0.3867	7.0	1096.60	0.0009
1.0	2.7183	0.3679	7.5	1808.0	0.0006
1.1	3.0042	0.3329	8.0	2981.0	0.0003
1.2	3.3201	0.3012	8.5	4914.8	0.0002
1.3	3.6693	0.2725	9.0	8103.1	0.0001
1.4	4.0552	0.2466	10.0	22026.0	0.00005

Table 3   Common Logarithms   **A23**

Table 3
Common
Logarithms

x	0	1	2	3	4	5	6	7	8	9
1.0	.0000	.0043	.0086	.0128	.0170	.0212	.0253	.0294	.0334	.0374
1.1	.0414	.0453	.0492	.0531	.0569	.0607	.0645	.0682	.0719	.0755
1.2	.0792	.0828	.0864	.0899	.0934	.0969	.1004	.1038	.1072	.1106
1.3	.1139	.1173	.1206	.1239	.1271	.1303	.1335	.1367	.1399	.1430
1.4	.1461	.1492	.1523	.1553	.1584	.1614	.1644	.1673	.1703	.1732
1.5	.1761	.1790	.1818	.1847	.1875	.1903	.1931	.1959	.1987	.2014
1.6	.2041	.2068	.2095	.2122	.2148	.2175	.2201	.2227	.2253	.2279
1.7	.2304	.2330	.2355	.2380	.2405	.2430	.2455	.2480	.2504	.2529
1.8	.2553	.2577	.2601	.2625	.2648	.2672	.2695	.2718	.2742	.2765
1.9	.2788	.2810	.2833	.2856	.2878	.2900	.2923	.2945	.2967	.2989
2.0	.3010	.3032	.3054	.3075	.3096	.3118	.3139	.3160	.3181	.3201
2.1	.3222	.3243	.3263	.3284	.3304	.3324	.3345	.3365	.3385	.3404
2.2	.3424	.3444	.3464	.3483	.3502	.3522	.3541	.3560	.3579	.3598
2.3	.3617	.3636	.3655	.3674	.3692	.3711	.3729	.3747	.3766	.3784
2.4	.3802	.3820	.3838	.3856	.3874	.3892	.3909	.3927	.3945	.3962
2.5	.3979	.3997	.4014	.4031	.4048	.4065	.4082	.4099	.4116	.4133
2.6	.4150	.4166	.4183	.4200	.4216	.4232	.4249	.4265	.4281	.4298
2.7	.4314	.4330	.4346	.4362	.4378	.4393	.4409	.4425	.4440	.4456
2.8	.4472	.4487	.4502	.4518	.4533	.4548	.4564	.4579	.4594	.4609
2.9	.4624	.4639	.4654	.4669	.4683	.4698	.4713	.4728	.4742	.4757
3.0	.4771	.4786	.4800	.4814	.4829	.4843	.4857	.4871	.4886	.4900
3.1	.4914	.4928	.4942	.4955	.4969	.4983	.4997	.5011	.5024	.5038
3.2	.5051	.5065	.5079	.5092	.5105	.5119	.5132	.5145	.5159	.5172
3.3	.5185	.5198	.5211	.5224	.5237	.5250	.5263	.5276	.5289	.5302
3.4	.5315	.5328	.5340	.5353	.5366	.5378	.5391	.5403	.5416	.5428
3.5	.5441	.5453	.5465	.5478	.5490	.5502	.5514	.5527	.5539	.5551
3.6	.5563	.5575	.5587	.5599	.5611	.5623	.5635	.5647	.5658	.5670
3.7	.5682	.5694	.5705	.5717	.5729	.5740	.5752	.5763	.5775	.5786
3.8	.5798	.5809	.5821	.5832	.5843	.5855	.5866	.5877	.5888	.5899
3.9	.5911	.5922	.5933	.5944	.5955	.5966	.5977	.5988	.5999	.6010
4.0	.6021	.6031	.6042	.6053	.6064	.6075	.6085	.6096	.6107	.6117
4.1	.6128	.6138	.6149	.6160	.6170	.6180	.6191	.6201	.6212	.6222
4.2	.6232	.6243	.6253	.6263	.6274	.6284	.6294	.6304	.6314	.6325
4.3	.6335	.6345	.6355	.6365	.6375	.6385	.6395	.6405	.6415	.6425
4.4	.6435	.6444	.6454	.6464	.6474	.6484	.6493	.6503	.6513	.6522
4.5	.6532	.6542	.6551	.6561	.6571	.6580	.6590	.6599	.6609	.6618
4.6	.6628	.6637	.6646	.6656	.6665	.6675	.6684	.6693	.6702	.6712
4.7	.6721	.6730	.6739	.6749	.6758	.6767	.6776	.6785	.6794	.6803
4.8	.6812	.6821	.6830	.6839	.6848	.6857	.6866	.6875	.6884	.6893
4.9	.6902	.6911	.6920	.6928	.6937	.6946	.6955	.6964	.6972	.6981
5.0	.6990	.6998	.7007	.7016	.7024	.7033	.7042	.7050	.7059	.7067
5.1	.7076	.7084	.7093	.7101	.7110	.7118	.7126	.7135	.7143	.7152
5.2	.7160	.7168	.7177	.7185	.7193	.7202	.7210	.7218	.7226	.7235
5.3	.7243	.7251	.7259	.7267	.7275	.7284	.7292	.7300	.7308	.7316
5.4	.7324	.7332	.7340	.7348	.7356	.7364	.7372	.7380	.7388	.7396
x	0	1	2	3	4	5	6	7	8	9

Table 3

(continued)

x	0	1	2	3	4	5	6	7	8	9
5.5	.7404	.7412	.7419	.7427	.7435	.7443	.7451	.7459	.7466	.7474
5.6	.7482	.7490	.7497	.7505	.7513	.7520	.7528	.7536	.7543	.7551
5.7	.7559	.7566	.7574	.7582	.7589	.7597	.7604	.7612	.7619	.7627
5.8	.7634	.7642	.7649	.7657	.7664	.7672	.7679	.7686	.7694	.7701
5.9	.7709	.7716	.7723	.7731	.7738	.7745	.7752	.7760	.7767	.7774
6.0	.7782	.7789	.7796	.7803	.7810	.7818	.7825	.7832	.7839	.7846
6.1	.7853	.7860	.7868	.7875	.7882	.7889	.7896	.7903	.7910	.7917
6.2	.7924	.7931	.7938	.7945	.7952	.7959	.7966	.7973	.7980	.7987
6.3	.7993	.8000	.8007	.8014	.8021	.8028	.8035	.8041	.8048	.8055
6.4	.8062	.8069	.8075	.8082	.8089	.8096	.8102	.8019	.8116	.8122
6.5	.8129	.8136	.8142	.8149	.8156	.8162	.8169	.8176	.8182	.8189
6.6	.8195	.8202	.8209	.8215	.8222	.8228	.8235	.8241	.8248	.8254
6.7	.8261	.8267	.8274	.8280	.8287	.8293	.8299	.8306	.8312	.8319
6.8	.8325	.8331	.8338	.8344	.8351	.8357	.8363	.8370	.8376	.8382
6.9	.8388	.8395	.8401	.8407	.8414	.8420	.8426	.8432	.8439	.8445
7.0	.8451	.8457	.8463	.8470	.8476	.8482	.8488	.8494	.8500	.8506
7.1	.8513	.8519	.8525	.8531	.8537	.8543	.8549	.8555	.8561	.8567
7.2	.8573	.8579	.8585	.8591	.8597	.8603	.8609	.8615	.8621	.8627
7.3	.8633	.8639	.8645	.8651	.8657	.8663	.8669	.8675	.8681	.8686
7.4	.8692	.8698	.8704	.8710	.8716	.8722	.8727	.8733	.8739	.8745
7.5	.8751	.8756	.8762	.8768	.8774	.8779	.8785	.8791	.8797	.8802
7.6	.8808	.8814	.8820	.8825	.8831	.8837	.8842	.8848	.8854	.8859
7.7	.8865	.8871	.8876	.8882	.8887	.8893	.8899	.8904	.8910	.8915
7.8	.8921	.8927	.8932	.8938	.8943	.8949	.8954	.8960	.8965	.8971
7.9	.8976	.8982	.8987	.8993	.8998	.9004	.9009	.9015	.9020	.9025
8.0	.9031	.9036	.9042	.9047	.9053	.9058	.9063	.9069	.9074	.9079
8.1	.9085	.9090	.9096	.9101	.9106	.9112	.9117	.9122	.9128	.9133
8.2	.9138	.9143	.9149	.9154	.9159	.9165	.9170	.9175	.9180	.9186
8.3	.9191	.9196	.9201	.9206	.9212	.9217	.9222	.9227	.9232	.9238
8.4	.9243	.9248	.9253	.9258	.9263	.9269	.9274	.9279	.9284	.9289
8.5	.9294	.9299	.9304	.9309	.9315	.9320	.9325	.9330	.9335	.9340
8.6	.9345	.9350	.9355	.9360	.9365	.9370	.9375	.9380	.9385	.9390
8.7	.9395	.9400	.9405	.9410	.9415	.9420	.9425	.9430	.9435	.9440
8.8	.9445	.9450	.9455	.9460	.9465	.9469	.9474	.9479	.9484	.9489
8.9	.9494	.9499	.9504	.9509	.9513	.9518	.9523	.9528	.9533	.9538
9.0	.9542	.9547	.9552	.9557	.9562	.9566	.9571	.9576	.9581	.9586
9.1	.9590	.9595	.9600	.9605	.9609	.9614	.9619	.9624	.9628	.9633
9.2	.9638	.9643	.9647	.9652	.9657	.9661	.9666	.9671	.9675	.9680
9.3	.9685	.9689	.9694	.9699	.9703	.9708	.9713	.9717	.9722	.9727
9.4	.9731	.9736	.9741	.9745	.9750	.9754	.9759	.9763	.9768	.9773
9.5	.9777	.9782	.9786	.9791	.9795	.9800	.9805	.9809	.9814	.9818
9.6	.9823	.9827	.9832	.9836	.9841	.9845	.9850	.9854	.9859	.9863
9.7	.9868	.9872	.9877	.9881	.9886	.9890	.9894	.9899	.9903	.9908
9.8	.9912	.9917	.9921	.9926	.9930	.9934	.9939	.9943	.9948	.9952
9.9	.9956	.9961	.9965	.9969	.9974	.9978	.9983	.9987	.9991	.9996
x	0	1	2	3	4	5	6	7	8	9

Table 4   Natural Logarithms of Numbers   **A25**

Table 4
Natural
Logarithms of
Numbers

x	ln x	x	ln x	x	ln x
		4.5	1.5041	9.0	2.1972
0.1	7.6974*	4.6	1.5261	9.1	2.2083
0.2	8.3906	4.7	1.5476	9.2	2.2192
0.3	8.7960	4.8	1.5686	9.3	2.2300
0.4	9.0837	4.9	1.5892	9.4	2.2407
0.5	9.3069	5.0	1.6094	9.5	2.2513
0.6	9.4892	5.1	1.6292	9.6	2.2618
0.7	9.6433	5.2	1.6487	9.7	2.2721
0.8	9.7769	5.3	1.6677	9.8	2.2824
0.9	9.8946	5.4	1.6864	9.9	2.2925
1.0	0.0000	5.5	1.7047	10	2.3026
1.1	0.0953	5.6	1.7228	11	2.3979
1.2	0.1823	5.7	1.7405	12	2.4849
1.3	0.2624	5.8	1.7579	13	2.5649
1.4	0.3365	5.9	1.7750	14	2.6391
1.5	0.4055	6.0	1.7918	15	2.7081
1.6	0.4700	6.1	1.8083	16	2.7726
1.7	0.5306	6.2	1.8245	17	2.8332
1.8	0.5878	6.3	1.8405	18	2.8904
1.9	0.6419	6.4	1.8563	19	2.9444
2.0	0.6931	6.5	1.8718	20	2.9957
2.1	0.7419	6.6	1.8871	25	3.2189
2.2	0.7885	6.7	1.9021	30	3.4012
2.3	0.8329	6.8	1.9169	35	3.5553
2.4	0.8755	6.9	1.9315	40	3.6889
2.5	0.9163	7.0	1.9459	45	3.8067
2.6	0.9555	7.1	1.9601	50	3.9120
2.7	0.9933	7.2	1.9741	55	4.0073
2.8	1.0296	7.3	1.9879	60	4.0943
2.9	1.0647	7.4	2.0015	65	4.1744
3.0	1.0986	7.5	2.0149	70	4.2485
3.1	1.1314	7.6	2.0281	75	4.3175
3.2	1.1632	7.7	2.0412	80	4.3820
3.3	1.1939	7.8	2.0541	85	4.4427
3.4	1.2238	7.9	2.0669	90	4.4998
3.5	1.2528	8.0	2.0794	100	4.6052
3.6	1.2809	8.1	2.0919	110	4.7005
3.7	1.3083	8.2	2.1041	120	4.7875
3.8	1.3350	8.3	2.1163	130	4.8676
3.9	1.3610	8.4	2.1282	140	4.9416
4.0	1.3863	8.5	2.1401	150	5.0106
4.1	1.4110	8.6	2.1518	160	5.0752
4.2	1.4351	8.7	2.1633	170	5.1358
4.3	1.4586	8.8	2.1748	180	5.1930
4.4	1.4816	8.9	2.1861	190	5.2470

*Subtract 10 for $x < 1$. Thus ln $0.3 = 8.7960 - 10 = -1.2040$.

Table 5 Values of Trigonometric Functions

| Angle θ | | | | | | | | | Angle θ | |
Degrees	Radians	sin θ	csc θ	tan θ	cot θ	sec θ	cos θ			
0° 00′	.0000	.0000	No value	.0000	No value	1.000	1.0000	1.5708	90° 00′	
10	029	029	343.8	029	343.8	000	000	679	50	
20	058	058	171.9	058	171.9	000	000	650	40	
30	087	087	114.6	087	114.6	000	1.0000	621	30	
40	116	116	85.95	116	85.94	000	.9999	592	20	
50	145	145	68.76	145	68.75	000	999	563	10	
1° 00′	.0175	.0175	57.30	.0175	57.29	1.000	.9998	1.5533	89° 00′	
10	204	204	49.11	204	49.10	000	998	504	50	
20	233	233	42.98	233	42.96	000	997	475	40	
30	262	262	38.20	262	38.19	000	997	446	30	
40	291	291	34.38	291	34.37	000	996	417	20	
50	320	320	31.26	320	31.24	001	995	388	10	
2° 00′	.0349	.0349	28.65	.0349	28.64	1.001	.9994	1.5359	88° 00′	
10	378	378	26.45	378	26.43	001	993	330	50	
20	407	407	24.56	407	24.54	001	992	301	40	
30	436	436	22.93	437	22.90	001	990	272	30	
40	465	465	21.49	466	21.47	001	989	243	20	
50	495	494	20.23	495	20.21	001	988	213	10	
3° 00′	.0524	.0523	19.11	.0524	19.08	1.001	.9986	1.5184	87° 00′	
10	553	552	18.10	553	18.07	002	985	155	50	
20	582	581	17.20	582	17.17	002	983	126	40	
30	611	610	16.38	612	16.35	002	981	097	30	
40	640	640	15.64	641	15.60	002	980	068	20	
50	669	669	14.96	670	14.92	002	978	039	10	
4° 00′	.0698	.0698	14.34	.0699	14.30	1.002	.9976	1.5010	86° 00′	
10	727	727	13.76	729	13.73	003	974	981	50	
20	756	765	13.23	758	13.20	003	971	952	40	
30	785	785	12.75	787	12.71	003	969	923	30	
40	814	814	12.29	816	12.25	003	967	893	20	
50	844	843	11.87	846	11.83	004	964	864	10	
5° 00′	.0873	.0872	11.47	.0875	11.43	1.004	.9962	1.4835	85° 00′	
10	902	901	11.10	904	11.06	004	959	806	50	
20	931	929	10.76	934	10.71	004	957	777	40	
30	960	958	10.43	963	10.39	005	954	748	30	
40	.0989	.0987	10.13	.0992	10.08	005	951	719	20	
50	.1018	.1016	9.839	.0122	9.788	005	948	690	10	
6° 00′	.1047	.1045	9.567	.1051	9.514	1.006	.9945	1.4661	84° 00′	
10	076	074	9.309	080	9.255	006	942	632	50	
20	105	103	9.065	110	9.010	006	939	603	40	
30	134	132	8.834	139	8.777	006	936	573	30	
40	164	161	8.614	169	8.556	007	932	544	20	
50	193	190	8.405	198	8.345	007	929	515	10	
7° 00′	.1222	.1219	8.206	.1228	8.144	1.008	.9925	1.4486	83° 00′	
		cos θ	sec θ	cot θ	tan θ	csc θ	sin θ	Radians	Degrees	
								Angle θ		

Table 5 Values of Trigonometric Functions **A27**

Table 5   (continued)

Angle θ									
Degrees	Radians	sin θ	csc θ	tan θ	cot θ	sec θ	cos θ		
7° 00′	.1222	.1219	8.206	.1228	8.144	1.008	.9925	1.4486	83° 00′
10	251	248	8.016	257	7.953	008	922	457	50
20	280	276	7.834	287	7.770	008	918	428	40
30	309	305	7.661	317	7.596	009	914	399	30
40	338	334	7.496	346	7.429	009	911	370	20
50	367	363	7.337	376	7.269	009	907	341	10
8° 00′	.1396	.1392	7.185	.1405	7.115	1.010	.9930	1.4312	82° 00′
10	425	421	7.040	435	6.968	010	899	283	50
20	454	449	6.900	465	827	011	894	254	40
30	484	478	765	495	691	011	890	224	30
40	513	507	636	524	561	012	886	195	20
50	542	536	512	554	435	012	881	166	10
9° 00′	.1571	.1564	6.392	.1584	6.314	1.012	.9877	1.4137	81° 00′
10	600	593	277	614	197	013	872	108	50
20	629	622	166	644	6.084	013	868	079	40
30	658	650	6.059	673	5.976	014	863	050	30
40	687	679	5.955	703	871	014	858	1.4021	20
50	716	708	855	733	769	015	853	1.3992	10
10° 00′	.1745	.1736	5.759	.1763	5.671	1.015	.9848	1.3963	80° 00′
10	774	765	665	793	576	016	843	934	50
20	804	794	575	823	485	016	838	904	40
30	833	822	487	853	396	017	833	875	30
40	862	851	403	883	309	018	827	846	20
50	891	880	320	914	226	018	822	817	10
11° 00′	.1920	.1908	5.241	.1944	5.145	1.019	.9816	1.3788	79° 00′
10	949	937	164	.1974	5.066	019	811	759	50
20	.1978	965	089	.2004	4.989	020	805	730	40
30	.2007	.1994	5.016	035	915	020	799	701	30
40	036	.2022	4.945	065	843	021	793	672	20
50	065	051	876	095	773	022	787	643	10
12° 00′	.2094	.2079	4.810	.2126	4.705	1.022	.9781	1.3614	78° 00′
10	123	108	745	156	638	023	775	584	50
20	153	136	682	186	574	024	769	555	40
30	182	164	620	217	511	024	763	526	30
40	211	193	560	247	449	025	757	497	20
50	240	221	502	278	390	026	750	468	10
13° 00′	.2269	.2250	4.445	.2309	4.331	1.026	.9744	1.3439	77° 00′
10	298	278	390	339	275	027	737	410	50
20	327	306	336	370	219	028	730	381	40
30	356	334	284	401	165	028	724	352	30
40	385	363	232	432	113	029	717	323	20
50	414	391	182	462	061	030	710	294	10
14° 00′	.2443	.2419	4.134	.2493	4.011	1.031	.9703	1.3265	76° 00′
		cos θ	sec θ	cot θ	tan θ	csc θ	sin θ	Radians	Degrees
								Angle θ	

Table 5 (continued)

Angle θ Degrees	Radians	sin θ	csc θ	tan θ	cot θ	sec θ	cos θ		
14° 00′	.2443	.2419	4.134	.2493	4.011	1.031	.9703	1.3265	76° 00′
10	473	447	086	524	3.962	031	696	235	50
20	502	476	4.039	555	914	032	689	206	40
30	531	504	3.994	586	867	033	681	177	30
40	560	532	950	617	821	034	674	148	20
50	589	560	906	648	776	034	667	119	10
15° 00′	.2618	.2588	3.864	.2679	3.732	1.035	.9659	1.3090	75° 00′
10	647	616	822	711	689	036	652	061	50
20	676	644	782	742	647	037	644	032	40
30	705	672	742	773	606	038	636	1.3003	30
40	734	700	703	805	566	039	628	1.2974	20
50	763	728	665	836	526	039	621	945	10
16° 00′	.2793	.2756	3.628	.2867	3.487	1.040	.9613	1.2915	74° 00′
10	822	784	592	899	450	041	605	886	50
20	851	812	556	931	412	042	596	857	40
30	880	840	521	962	376	043	588	828	30
40	909	868	487	.2944	340	044	580	799	20
50	938	896	453	.3026	305	045	572	770	10
17° 00′	.2967	.2924	3.420	.3057	3.271	1.046	.9563	1.2741	73° 00′
10	.2996	952	388	089	237	047	555	712	50
20	.3025	.2979	357	121	204	048	546	683	40
30	054	.3007	326	153	172	048	537	654	30
40	083	035	295	185	140	049	528	625	20
50	113	062	265	217	108	050	520	595	10
18° 00′	.3142	.3090	3.236	.3249	3.078	1.051	.9511	1.2566	72° 00′
10	171	118	207	281	047	052	502	537	50
20	200	145	179	314	3.018	053	492	508	40
30	229	173	152	346	2.989	054	483	479	30
40	258	201	124	378	960	056	474	450	20
50	287	228	098	411	932	057	465	421	10
19° 00′	.3316	.3256	3.072	.3443	2.904	1.058	.9455	1.2392	71° 00′
10	345	283	046	476	877	059	446	363	50
20	374	311	3.021	508	850	060	436	334	40
30	403	338	2.996	541	824	061	426	305	30
40	432	365	971	574	798	062	417	275	20
50	462	393	947	607	773	063	407	246	10
20° 00′	.3491	.3420	2.924	.3640	2.747	1.064	.9397	1.2217	70° 00′
10	520	448	901	673	723	065	387	188	50
20	549	475	878	706	699	066	377	159	40
30	578	502	855	739	675	068	367	130	30
40	607	529	833	772	651	069	356	101	20
50	636	557	812	805	628	070	346	072	10
21° 00′	.3665	.3584	2.790	.3839	2.605	1.071	.9336	1.2043	69° 00′
		cos θ	sec θ	cot θ	tan θ	csc θ	sin θ	Radians	Degrees
								Angle θ	

Table 5 Values of Trigonometric Functions   **A29**

Table 5   (continued)

Angle θ Degrees	Radians	sin θ	csc θ	tan θ	cot θ	sec θ	cos θ	Radians	Degrees
21° 00′	.3665	.3584	2.790	.3839	2.605	1.071	.9336	1.2043	69° 00′
10	694	611	769	872	583	072	325	1.2014	50
20	723	638	749	906	560	074	315	1.1985	40
30	752	665	729	939	539	075	304	956	30
40	782	692	709	.3973	517	076	293	926	20
50	811	719	689	.4006	496	077	283	897	10
22° 00′	.3840	.3746	2.669	.4040	2.475	1.079	.9272	1.1868	68° 00′
10	869	773	650	074	455	080	261	839	50
20	898	800	632	108	434	081	250	810	40
30	927	827	613	142	414	082	239	781	30
40	956	854	595	176	394	084	228	752	20
50	985	881	577	210	375	085	216	723	10
23° 00′	.4014	.3907	2.559	.4245	2.356	1.086	.9205	1.1694	67° 00′
10	043	934	542	279	337	088	194	665	50
20	072	961	525	314	318	089	182	636	40
30	102	.3987	508	348	300	090	171	606	30
40	131	.4014	491	383	282	092	159	577	20
50	160	041	475	417	264	093	147	548	10
24° 00′	.4189	.4067	2.459	.4452	2.246	1.095	.9135	1.1519	66° 00′
10	218	094	443	487	229	096	124	490	50
20	247	120	427	522	211	097	112	461	40
30	276	147	411	557	194	099	100	432	30
40	305	173	396	592	177	100	088	403	20
50	334	200	381	628	161	102	075	374	10
25° 00′	.4363	.4226	2.366	.4663	2.145	1.103	.9063	1.1345	65° 00′
10	392	253	352	699	128	105	051	316	50
20	422	279	337	734	112	106	038	286	40
30	451	305	323	770	097	108	026	257	30
40	480	331	309	806	081	109	013	228	20
50	509	358	295	841	066	111	.9001	199	10
26° 00′	.4538	.4384	2.281	.4877	2.050	1.113	.8988	1.1170	64° 00′
10	567	410	268	913	035	114	975	141	50
20	596	436	254	950	020	116	962	112	40
30	625	462	241	.4986	2.006	117	949	083	30
40	654	488	228	.5022	1.991	119	936	054	20
50	683	514	215	059	977	121	923	1.1025	10
27° 00′	.4712	.4540	2.203	.5095	1.963	1.122	.8910	1.0996	63° 00′
10	741	566	190	132	949	124	897	966	50
20	771	592	178	169	935	126	884	937	40
30	800	617	166	206	921	127	870	908	30
40	829	643	154	243	907	129	857	879	20
50	858	669	142	280	894	131	843	850	10
28° 00′	.4887	.4695	2.130	.5317	1.881	1.133	.8829	1.0821	62° 00′
		cos θ	sec θ	cot θ	tan θ	csc θ	sin θ	Radians	Degrees
									Angle θ

Table 5 (continued)

Angle θ Degrees	Angle θ Radians	sin θ	csc θ	tan θ	cot θ	sec θ	cos θ		
28° 00′	.4887	.4695	2.130	.5317	1.881	1.133	.8829	1.0821	62° 00′
10	916	720	118	354	868	134	816	792	50
20	945	746	107	392	855	136	802	763	40
30	.4974	772	096	430	842	138	788	734	30
40	.5003	797	085	467	829	140	774	705	20
50	032	823	074	505	816	142	760	676	10
29° 00′	.5061	.4848	2.063	.5543	1.804	1.143	.8746	1.0647	61° 00′
10	091	874	052	581	792	145	732	617	50
20	120	899	041	619	780	147	718	588	40
30	149	924	031	658	767	149	704	559	30
40	178	950	020	696	756	151	689	530	20
50	207	.4975	010	735	744	153	675	501	10
30° 00′	.5236	.5000	2.000	.5774	1.732	1.155	.8660	1.0472	60° 00′
10	265	025	1.990	812	720	157	646	443	50
20	294	050	980	851	709	159	631	414	40
30	323	075	970	890	698	161	616	385	30
40	352	100	961	930	686	163	601	356	20
50	381	125	951	.5969	675	165	587	327	10
31° 00′	.5411	.5150	1.942	.6009	1.664	1.167	.8572	1.0297	59° 00′
10	440	175	932	048	653	169	557	268	50
20	469	200	923	088	643	171	542	239	40
30	498	225	914	128	632	173	526	210	30
40	527	250	905	168	621	175	511	181	20
50	556	275	896	208	611	177	496	152	10
32° 00′	.5585	.5299	1.887	.6249	1.600	1.179	.8480	1.0123	58° 00′
10	614	324	878	289	590	181	465	094	50
20	643	348	870	330	580	184	450	065	40
30	672	373	861	371	570	186	434	036	30
40	701	398	853	412	560	188	418	1.0007	20
50	730	422	844	452	550	190	403	.9977	10
33° 00′	.5760	.5446	1.836	.6494	1.540	1.192	.8387	.9948	57° 00′
10	789	471	828	536	530	195	371	919	50
20	818	495	820	577	520	197	355	890	40
30	847	519	812	619	511	199	339	861	30
40	876	544	804	661	501	202	323	832	20
50	905	568	796	703	492	204	307	803	10
34° 00′	.5934	.5592	1.788	.6745	1.483	1.206	.8290	.9774	56° 00′
10	963	616	781	787	473	209	274	745	50
20	.5992	640	773	830	464	211	258	716	40
30	.6021	664	766	873	455	213	241	687	30
40	050	688	758	916	446	216	225	657	20
50	080	712	751	.6959	437	218	208	628	10
35° 00′	.6109	.5736	1.743	.7002	1.428	1.221	.8192	.9599	55° 00′
		cos θ	sec θ	cot θ	tan θ	csc θ	sin θ	Radians	Degrees
								Angle θ	

Table 5   Values of Trigonometric Functions   **A31**

Table 5   (continued)

Degrees	Radians	sin θ	csc θ	tan θ	cot θ	sec θ	cos θ	Radians	Degrees
Angle θ									**Angle θ**
35° 00′	.6109	.5736	1.743	.7002	1.428	1.221	.8192	.9599	55° 00′
10	138	760	736	046	419	223	175	570	50
20	167	783	729	089	411	226	158	541	40
30	196	807	722	133	402	228	141	512	30
40	225	831	715	177	393	231	124	483	20
50	254	854	708	221	385	233	107	454	10
36° 00′	.6283	.5878	1.701	.7265	1.376	1.236	.8090	.9425	54° 00′
10	312	901	695	310	368	239	073	396	50
20	341	925	688	355	360	241	056	367	40
30	370	948	681	400	351	244	039	338	30
40	400	972	675	445	343	247	021	308	20
50	429	.5995	668	490	335	249	.8004	279	10
37° 00′	.6458	.6018	1.662	.7536	1.327	1.252	.7986	.9250	53° 00′
10	487	041	655	581	319	255	696	221	50
20	516	065	649	627	311	258	951	192	40
30	545	088	643	673	303	260	934	163	30
40	574	111	636	720	295	263	916	134	20
50	603	134	630	766	288	266	898	105	10
38° 00′	.6632	.6157	1.624	.7813	1.280	1.269	.7880	.9076	52° 00′
10	661	180	618	860	272	272	862	047	50
20	690	202	612	907	265	275	844	.9018	40
30	720	225	606	.7954	257	278	826	.8988	30
40	749	248	601	.8002	250	281	808	959	20
50	778	271	595	050	242	284	790	930	10
39° 00′	.6807	.6293	1.589	.8098	1.235	1.287	.7771	.8901	51° 00′
10	836	316	583	146	228	290	753	872	50
20	865	338	578	195	220	293	735	843	40
30	894	361	572	243	213	296	716	814	30
40	923	383	567	292	206	299	698	785	20
50	952	406	561	342	199	302	679	756	10
40° 00′	.6981	.6428	1.556	.8391	1.192	1.305	.7660	.8727	50° 00′
10	.7010	450	550	441	185	309	642	698	50
20	039	472	545	491	178	312	623	668	40
30	069	494	540	541	171	315	604	639	30
40	098	517	535	591	164	318	585	610	20
50	127	539	529	642	157	322	566	581	10
41° 00′	.7156	.6561	1.524	.8693	1.150	1.325	.7547	.8552	49° 00′
10	185	583	519	744	144	328	528	523	50
20	214	604	514	796	137	332	509	494	40
30	243	626	509	847	130	335	490	465	30
40	272	648	504	899	124	339	470	436	20
50	301	670	499	.8952	117	342	451	407	10
42° 00′	.7330	.6691	1.494	.9004	1.111	1.346	.7431	.8378	48° 00′
		cos θ	sec θ	cot θ	tan θ	csc θ	sin θ	Radians	Degrees

Table 5   (continued)

Angle θ Degrees	Angle θ Radians	sin θ	csc θ	tan θ	cot θ	sec θ	cos θ		
42° 00′	.7330	.6691	1.494	.9004	1.111	1.346	.7431	.8378	48° 00′
10	359	713	490	057	104	349	412	348	50
20	389	734	485	110	098	353	392	319	40
30	418	756	480	163	091	356	373	290	30
40	447	777	476	217	085	360	353	261	20
50	476	799	471	271	079	364	333	232	10
43° 00′	.7505	.6820	1.466	.9325	1.072	1.367	.7314	.8203	47° 00′
10	534	841	462	380	066	371	294	174	50
20	563	862	457	435	060	375	274	145	40
30	592	884	453	490	054	379	254	116	30
40	621	905	448	545	048	382	234	087	20
50	650	926	444	601	042	386	214	058	10
44° 00′	.7679	.6947	1.440	.9657	1.036	1.390	.7193	.8029	46° 00′
10	709	967	435	713	030	394	173	.7999	50
20	738	.6988	431	770	024	398	153	970	40
30	767	.7009	427	827	018	402	133	941	30
40	796	030	423	884	012	406	112	912	20
50	825	050	418	.9942	006	410	092	883	10
45° 00′	.7854	.7071	1.414	1.000	1.000	1.414	.7071	.7854	45° 00′
		cos θ	sec θ	cot θ	tan θ	csc θ	sin θ	Radians	Degrees
								Angle θ	

Answers to Selected Exercises

1. (a) P is a proper subset of Q (b) y is an element of A
   (c) S is not a proper subset of T (d) x is not an
   element of Q (e) Empty set   (f) Set containing zero
2. (a) T (b) F (c) T (d) T (e) T (f) F
3. (a) $\{x \mid x$ is an even integer and $2 \le x \le 8\}$ (b) $\{x^2 \mid x$ is
   an integer and $1 \le x \le 6\}$ (c) $\{x \mid x$ is an integer and
   $0 \le x \le 8\}$ (d) $\{x \mid x$ is a prime number and $2 \le x \le 19\}$
   (e) $\{x \mid x$ is an odd integer and $1 \le x \le 91\}$ (f) $\{x \mid x$ is
   a positive odd integer$\}$ (g) $\{x^2 \mid x$ is an integer and
   $5 \le x \le 9\}$ (h) $\{x \mid x^2 = 1\}$
5. (a) $x = 2$ and $y = 8$ or $x = 4$ and $y = 4$ (b) $x = 5$
   (c) $x = 10$ (d) $x = 2$
6. Yes
8. (a) $\emptyset$, $\{1\}$, $\{3\}$, $\{5\}$, $\{1,3\}$, $\{1,5\}$, $\{3,5\}$, $\{1,3,5\}$ (b) $\emptyset$, $\{0\}$
   (c) $\emptyset$, $\{1\}$, $\{2\}$, $\{1,2\}$ (d) $\emptyset$, $\{0\}$, $\{1\}$, $\{2\}$, $\{3\}$, $\{4\}$,
   $\{0,1\}$, $\{0,2\}$, $\{0,3\}$, $\{0,4\}$, $\{1,2\}$, $\{1,3\}$, $\{1,4\}$, $\{2,3\}$, $\{2,4\}$,
   $\{3,4\}$, $\{0,1,2\}$, $\{0,1,3\}$, $\{0,1,4\}$, $\{0,2,3\}$, $\{0,2,4\}$, $\{0,3,4\}$,
   $\{1,2,3\}$, $\{1,2,4\}$, $\{1,3,4\}$, $\{2,3,4\}$, $\{0,1,2,3\}$, $\{0,1,2,4\}$,
   $\{0,1,3,4\}$, $\{0,2,3,4\}$, $\{1,2,3,4\}$, $\{0,1,2,3,4\}$.
9. 2^n 10. B 11. $\{4,6\}$ 12. $\emptyset$ 13. A 14. B
   15. $\{4,6\}$ 16. $\{1,2,3,4,5,6\}$ 18. $\{1,2,...,8\}$
19. $\{1,2,3,4,5,6\}$ 20. A 21. $\{1,2,3,4,5,6,7,8\}$
22. $\{1,2,3,4,5,6\}$ 23. $\emptyset$ 24. $C = \{1,2,3\}$

25. {1,3,4,6,8,9}   26. {2,4,6,8,10}   27. {1,3,5,7,9,10}
29. {4,6,8}   31. {1,2,3,4,6,8,9,10}   32. {1,3,5,7,9,10}
33. *B*   34. ∅   35. *A*   36. (a) 4   (b) 4
38. (a) 22   (b) 34   (c) 13

Exercises 1.2
2. Closure property for multiplication   4. Additive identity
7. Commutative law for multiplication   8. Additive inverse
9. Multiplicative inverse, reciprocal   11. Multiplicative
identity   16. 12   17. −1   18. 1   19. 2
21. $\frac{1}{6}$ 22. 0

23.

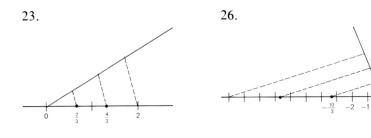

26.

29. Yes   30. No

Exercises 1.3
1. $5 < 8$ 2. $-3 \le 2$ 4. $8 > 0$ 7. $x \le 0$
8. $9 > 2x - 5 \ge 5$ 12. $\{x \mid 2 < x < 5\}$
15. $\{x \mid -3 \le x < 6\}$ 17. $\{x \mid x \le -3\}$ 19. $[-3,5)$
21. $(3, \infty)$ 22. $(-\infty, 4]$

29.

31.

34.

35.

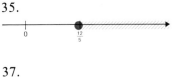

36.

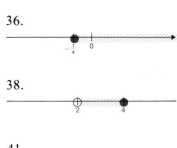

37.

38.

40.

41.

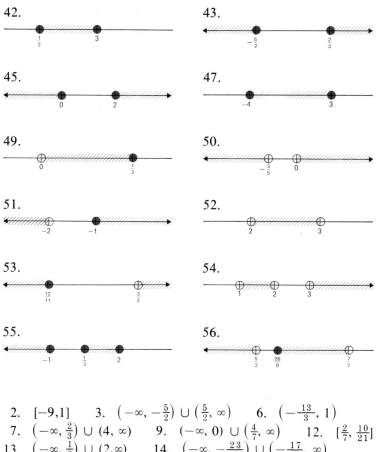

42.

43.

45.

47.

49.

50.

51.

52.

53.

54.

55.

56.

2. $[-9,1]$ 3. $\left(-\infty, -\frac{5}{2}\right) \cup \left(\frac{5}{2}, \infty\right)$ 6. $\left(-\frac{13}{3}, 1\right)$

7. $\left(-\infty, \frac{2}{3}\right) \cup (4, \infty)$ 9. $(-\infty, 0) \cup \left(\frac{4}{7}, \infty\right)$ 12. $[\frac{2}{7}, \frac{10}{21}]$

13. $\left(-\infty, \frac{1}{2}\right) \cup (2, \infty)$ 14. $\left(-\infty, -\frac{23}{4}\right) \cup \left(-\frac{17}{6}, \infty\right)$

15. $\left(-\infty, \frac{7}{5}\right] \cup [\frac{5}{2}, \infty)$ 17. $\{x | x < 0 \text{ and } x \neq -1\}$

18. $\{x | \frac{11}{5} \leq x \leq 13 \text{ and } x \neq 4\}$

19. $(-\infty, -2] \cup [\frac{1}{5}, \infty)$ 21. $[-\frac{7}{2}, \frac{7}{8}]$ 22. $(-\infty, -7] \cup [\frac{1}{3}, \infty)$

23. $\{x | -1 < x \leq \frac{1}{2} \text{ or } x < -1\}$

24. $\{x | 2 \leq x < \frac{5}{2} \text{ or } \frac{5}{2} < x \leq \frac{14}{5}\}$

1. b^{10} 2. $-\frac{1}{8}$ 3. $\frac{4}{25}$ 4. x^{3+2n} 5. x^{5n} 6. b^n

7. $\frac{a^8}{y^2}$ 8. 1   10. $\frac{8}{81}$ 12. 0   13. 1   15. $\frac{y^2}{x^4}$

16. $\frac{-z^6}{4x^{10}y}$ 18. $\frac{2^4}{5^2 y^{20} z^{24}}$ 19. $1 + x$ 22. $\frac{2xy + x^2 + y^2}{xy}$

23. $\frac{1}{x}$ 24. 8   25. 9   27. $\frac{1}{4}$ 28. $\frac{9}{2}$ 29. x^3

30. $b^{13/6}$ 32. $x^{12/7}$ 33. $\frac{a^2}{b}$ 34. $\frac{4}{xy^2}$ 35. $x^{3n/2}$

36. a^4b^n **37.** $\frac{2}{3}a^4b^5$ **38.** $\dfrac{5}{-2x}$ **39.** $3\sqrt[4]{3}$

40. $6\sqrt{2}$ **41.** $2^{3/4}$ **43.** $b^2 - b^{9/2}$ **44.** $\dfrac{1}{2x+1}$

Exercises 1.6   **1.** $\dfrac{2}{x}$ **3.** $\dfrac{3}{x+1}$ **4.** $\dfrac{2(x+2)}{x+3}$ **5.** $\dfrac{2}{a}$ **6.** 1

7. $\dfrac{1}{a+3}$ **8.** $\dfrac{y}{y-1}$ **10.** 1   **11.** $\dfrac{x-1}{x}$ **13.** $\dfrac{x-1}{x^2}$

15. $\dfrac{x+2}{1-x^2}$ **17.** $\dfrac{x^2-3}{x+2}$ **18.** $\dfrac{-2}{2x-y}$ **20.** $\dfrac{3x^2-x+1}{2x^2-x-1}$

22. $\dfrac{x-1}{3x+1}$ **23.** $2x+h$ **25.** $\dfrac{-1}{x^2+xh}$ **26.** $\dfrac{-2x-h}{x^2(x+h)^2}$

27. $\dfrac{x+1}{2x^2+1}$ **28.** $\dfrac{2y-3}{2}$

Exercises 1.7   **1.** Yes   **2.** No   **3.** No   **5.** $x=\frac{6}{5}$ **7.** $x=-3$
10. $x=\frac{9}{2}$ **12.** $x=4$ **13.** $x=\frac{42}{5}$ **14.** $x=-\frac{1}{6}$
16. $x=-\frac{6}{7}$ **17.** $x=4$ **18.** $x=-\frac{13}{4}$ **19.** $x=-\frac{3}{5}$
21. No solution   **22.** $x=\frac{9}{4}$ **24.** No solution
25. $x=-\frac{31}{4}$ **26.** $x=-\frac{55}{12}$ **27.** 17, 18, 19   **28.** 4 and 9
29. Mario is 31, Anna is 13   **30.** In 24 years   **31.** $4,500
at 5% and $10,500 at 6%   **32.** 10 quarters, 12 dimes,
5 nickels   **33.** 13 dimes and 8 quarters   **34.** 12.5 gal
of 40% alcohol solution must be mixed with 37.5 gal of
20% alcohol solution   **35.** $\frac{9}{2}$ ounces   **36.** He had
$\frac{100}{3}$ pounds of sugar for 165¢/lb and he had $\frac{50}{3}$ pounds of
sugar for 180¢/lb.   **37.** 32 mph, 64 mph.   **38.** 300 mi
39. $\frac{24}{7}$ hours

Chapter 1 Review   **Part 2**
2. $\{a,b,c,d\}$ **3.** $\{2\}$ **4.** (a) 7   (b) 3   **5.** (a) R
(b) $\emptyset$ **6.** 7   **8.** 1   **9.** $\frac{1}{6}$ **10.** $\frac{8}{15}$ **11.** $\frac{10}{3}$
12. $x>-7$ **13.** $x>\frac{8}{3}$ **16.** $x\le-5$ or $x\ge\frac{1}{2}$
17. $-1<x<\frac{3}{2}$ **18.** $x<0$ or $x>\frac{5}{7}$
21. $3\ge x\ge-\frac{3}{2}$ **22.** $x<-3$ or $x>5$
23. $\{x\mid x\in(-\infty,0)\cup(0,1]\}$ **24.** 2   **26.** $x^{47/12}$
27. $\dfrac{-2y}{3}$ **28.** $\dfrac{2}{x-1}$ **30.** $\dfrac{-x^2+6x-2}{x(x-1)(x+2)}$

31. $x=\dfrac{100}{3}$ lb   **32.** Width is 9 and length is 18   **33.** 24

34. $21,000 at 6% bonds; $37,000 at 8% mortgages; $42,000
at 10% stocks

1.   2. $\{(1,-1), (1,-2), (2,-1), (2,-2)\}$ Exercises 2.1

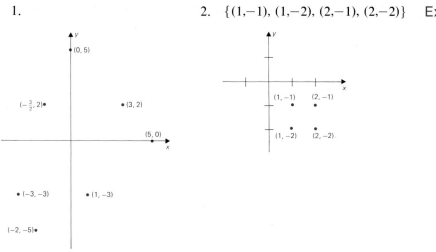

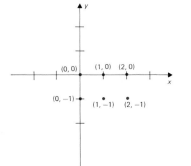

3. $\{(0,-1), (0,0), (1,-1), (1,0), (2,-1), (2,0)\}$

6.   (a)   12   (b)   12   7. $\sqrt{34}, \left(\frac{5}{2}, \frac{11}{2}\right)$
10. $5\sqrt{2}, \left(-\frac{3}{2}, \frac{3}{2}\right)$ 12. $\sqrt{58}/2, \left(\frac{5}{4}, -\frac{5}{4}\right)$ 14.   8, (1,2)
15. $146 = d(B,C)^2, 73 = d(A,C)^2 = d(A,B)^2$
17. $d(A,B)^2 = 13, d(B,C)^2 = 65, d(A,C)^2 = 52$
20. $d(A,B)^2 = 5 = d(B,C)^2, d(A,C)^2 = 10$
21. $d(A,B) = \sqrt{17} = d(A,C), d(B,C) = 3\sqrt{2}$, isosceles triangle
23.   Since $d(A,B) = d(B,C) = d(A,C) = 2\sqrt{2}$, the triangle is
      equilateral.   24. $d(A,B) = \sqrt{17}, d(B,C) = 3\sqrt{2}$,
      $d(A,C) = \sqrt{17}$; the triangle is isosceles   25.   collinear,
      not collinear, collinear   26. $(-4,\pm3)$
27. $d(C,D) = d(A,B) = d(B,C) = d(A,D)$ implies that $ABCD$ is a
      square, $d(A,C) = d(B,D), = 17\sqrt{2}$ 28. $M_1(A,B) = $
      $\left(-\frac{3}{2}, \frac{7}{2}\right), M_2(B,C) = \left(-\frac{1}{2}, 6\right), d(A,M_2) = \frac{1}{2}\sqrt{73}$,
      $M_3(A,C) = \left(2, \frac{9}{2}\right), d(B,M_3) = \frac{1}{2}\sqrt{145}$ 29. $d(D,M) = $

1000 miles, $d(M,P) = 1500$ miles, total distance 2500 miles, $d(D,P) = 500 \sqrt{13}$ miles

Exercises 2.2

1. 3, 4   2. 7, 2   4. 7, 8   5. 7, 2
6. $2x^4 - 2x^2 + 3, 2x^2 + 4$
7. $2a^2 + 4ah + 2h^2 - 2a - 2h + 3, 2(a + h) + 4$
8. $4a + 2h - 2$ 9. 2   10. $8x^2 + 28x + 27, 4x^2 - 4x + 10$
11. Function
12. Not a function $D = \{1,2\}$ 13. Function

14. Function   15. Not a function   16. Function
17. Not a function   18. Not a function

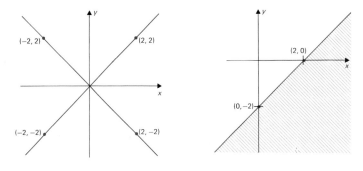

19. $\{x | x \in R\}$ 20. $\{x | x \in R \text{ and } x \neq 0\}$
21. $\{x | x \in R\}$ 22. $\{t | t \in R \text{ and } t \neq 0, 5\}$
24. $\{s | s \in R \text{ and } s \geq 0\}$ 25. $\{r | r \in R$
  and $r \neq 1, -2\}$ 26. $\{x | x \leq 2\}$ 27. $\{x | x \in R\}$
28. $\{z | -2 \leq z \leq 2\}$ 29. $\{x | \, |x| \geq 4\}$ 30. $\{t | t \neq 1\}$
31. Domain $g = \{t | t \leq -2\} \cup \{t | t \geq -1\}$ 32. Domain $h = $
  $\{x | x < -2\} \cup \{x | x > 3\}$

33.

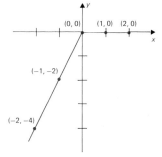

35.

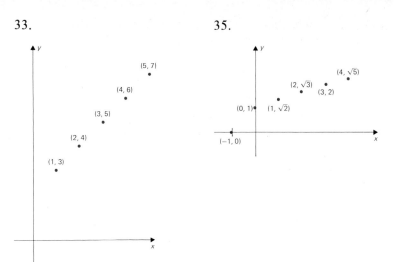

38.   (a), (d)

39. $-4, -2, 0, 0, 0, D = \{x|x \in R\}, \text{range} = \{y|y \leq 0\}$

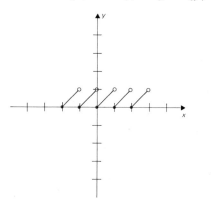

41. $D = \{x|x \in R\}, \text{range} = \{y|y \geq 0\}$

43. $D = \{x|x \in R\}, \text{range} = \{y|0 \leq y < 1\}$

44. $D = \{x|x \in R\}$, range $= \{y|y \in I\}$

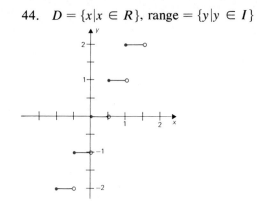

45. $D = \{x|x \in R\}$,
    range $= \{y|k < y \leq k + 1,$ where k is odd and $k \geq 1\}$

46. $D = \{x|x \in R\}$, range $= \{y|y \geq 1\}$ 47. (c) $x = \frac{1}{2}(c - 1)$
48. (a) ± 2 (b) $x = \pm\sqrt{7}$ (c) No real solution
50. (a) Even   (b) Odd   (c) Even   (d) Neither
51. Yes

Exercises 2.3

1. Domain of $f = \{x|x \in R\} =$ domain of g
   $(f + g)(x) = 2x + 5 + x^2$, $D(f + g) = \{x|x \in R\}$
   $(f - g)(x) = 2x + 5 - x^2$, $D(f - g) = \{x|x \in R\}$
   $(f \cdot g)(x) = x^2(2x + 5)$, $D(f \cdot g) = \{x|x \in R\}$
   $\left(\dfrac{f}{g}\right)(x) = \dfrac{2x + 5}{x^2}$, $D\left(\dfrac{f}{g}\right) = \{x|x \in R, x \neq 0\}$

3. Domain $f = \{x||x| \geq 2\}$, domain $g = \{x|x \in R\}$,
   $(f + g)(x) = \sqrt{x^2 - 4} + x^2 + 2x - 3$, $(f - g)(x) = \sqrt{x^2 - 4} - x^2 - 2x + 3$, $(fg)(x) = (x^2 + 2x - 3)\sqrt{x^2 - 4}$, domain $f + g$,
   $f - g$, $fg = \{x||x| \geq 2\}$ domain $(f/g) = \{x/|x| \geq 2, x \neq -3\}$.

4. $(f + g)(x) = \dfrac{x^2 + 3x - 1}{(x + 1)(2x - 1)}$, domain $f + g$, $f - g$, $fg =$
   $\{x|x \neq -1, \frac{1}{2}\}$, domain $f/g = \{x|x \neq -1, 0, \frac{1}{2}\}$

5. Domain $f = \{x|x > 0\}$, $g = \{x|x \neq 0\}$, $(f + g)(x) =$
   $\dfrac{1}{\sqrt{x}} + \dfrac{1}{x^2}$, $(f - g)(x) = \dfrac{1}{\sqrt{x}} - \dfrac{1}{x^2}$,

$(f \cdot g)(x) = \dfrac{1}{x^2\sqrt{x}}$, domain $f + g$, $f - g$

$f \cdot g = \{x | x > 0\}$, $\left(\dfrac{f}{g}\right)(x) = x^{3/2}$, domain $\left(\dfrac{f}{g}\right) =$

$\{x | x > 0\}$ 7. One-to-one, $f^{-1} = \{(4,1), (2,3), (6,5), (7,0)\}$

9. Not one-to-one because $f(2) = f(3) = -1$ 10. One-to-one, $f^{-1} = \{(2,0), (4,1), (6,2), (1,4)\}$

11. $y = x + 1$ 12. $y = \dfrac{x - 6}{4}$ 13. $y = \dfrac{5 - x}{2}$

14. $y = \dfrac{4 - 2x}{3}$ 15. $y = \sqrt[3]{x}$ 16. $y = \sqrt[3]{\dfrac{x + 1}{2}}$

17. $\{x | x \geq 0\}$ or $\{x | x \leq 0\}$ 18. $(f \circ g)(x) = 2 + 3x$, $(g \circ f)(x) = 6 + 3x$ 20. $(f \circ g)(x) = x + 4$, $(g \circ f)(x) = \sqrt{x^2 + 4}$ 21. $(f \circ g)(x) = \sqrt[3]{2x^3 + 2}$, $(g \circ f)(x) = 2x + 3$

23. $(f \circ g)(x) = \dfrac{x}{1 + 2x^2}$, $(g \circ f)(x) = \dfrac{x^2 + 2}{x}$

24. $(f \circ g)(x) = \dfrac{\sqrt{3x^4 + 8}}{x^2}$, $(g \circ f)(x) = \dfrac{2}{2x^2 + 3}$

25. $(f \circ g)(x) = \dfrac{1}{\sqrt{x - 2}}$, domain $f \circ g = \{x | x > 2\}$,

$(g \circ f)(x) = \dfrac{\sqrt{x} - 2x}{x}$, domain $g \circ f = \{x | x > 0\}$

26. (a) 16   (b) -52 (c) -2 (d) 7   (e) 56
(f) 242   27. (a) Undefined   (b) $2(\sqrt{5} - \sqrt{10})$
(c) Undefined   (d) Undefined   (e) Undefined
(f) Undefined

28. (c) $(f \circ f)(x) = \sqrt{4 + (\sqrt{4 + x})}$, domain $f \circ f =$ domain f

(d) $(f \circ f)(x) = \dfrac{2 + x}{5 + 2x}$, domain $f \circ f = \{x | x \neq -2, -\tfrac{5}{2}\}$

(e) $(f \circ f)(x) = \dfrac{8x^4 + 9x^2 + 2}{2x^3 + x}$, domain $f \circ f =$ domain f

32. $2x^2 + 9$ 34. $4x^2 + 20x + 25$

1. $2x + 5y - 16 = 0$ 2. $3x + y - 4 = 0$ **Exercises 2.4**
4. $8x + y + 11 = 0$ 5. $2x - 3y = 0$ 7. $4x + 7y + 29 = 0$
8. $2x + y - 3 = 0$ 9. $3x - 12y + 2 = 0$
10. $3x - 6y + 17 = 0$ 12. $x - y + 1 = 0$
13. $x + y + 6 = 0$ 14. $2x + y - 3 = 0$ 17. $y = -3$
18. $-\tfrac{3}{4}, \tfrac{3}{2}, 2$ 19. $\tfrac{1}{2}, 2, -4$ 24. $c = -21$
25. (a) \$2600/3   (b) \$200   (c) \$2000/3   (d) \$100/3
(e) \$1300/3, \$700/3   26. (a) \$350   (b) \$0.50
(c) \$600   (d) \$250   27. (a) $p = 5t - 20$, for
$t \geq 8$ (c) 20, 40, 70,   (d) 5
28. (a) $25t + 6p - 420 = 0$, where t is measured from 8:00 P.M.
(b) $53\tfrac{1}{3}$, $p(6) = 45$, $p(8) = 36\tfrac{2}{3}$ (c) $-\dfrac{25}{6}$

Exercises 2.5   1. $\left(-\frac{5}{3}, -\frac{7}{3}\right)$ 2. $\left(-\frac{11}{3}, \frac{37}{3}\right)$ 3. $\left(-\frac{5}{7}, -\frac{1}{7}\right)$
4. $\left(-\frac{19}{6}, -\frac{17}{6}\right)$ 5. $\left(-\frac{3}{16}, -\frac{7}{16}\right)$ 6. $\left(\frac{50}{23}, \frac{112}{23}\right)$
8. $\left(\frac{5}{2}, \frac{7}{6}\right)$ 11. Parallel   12. Perpendicular
14. Parallel   16. Parallel   18. Parallel   19. (a) 0
(b) -3 (c) -8 (d) $-\frac{5}{2}$ 21. (a) Right
triangle   (b) Rectangle   (c) Right triangle
(d) Rectangle   24. (a) $y = \frac{22}{35}x + 22{,}000$

(b)

(c) $\approx \$59{,}230$ (d) $\approx \$7{,}714$ 25. (a) $\$24{,}000$
(b) $\approx \$16{,}667$ (c) $\$36{,}000$ (d) $\approx \$2{,}667$
26. (a) $C = \frac{5}{9}F - \frac{160}{9}$ (b) $F = \frac{9}{5}C + 32$ (c) -40

Exercises 2.6   2. $x = \frac{3}{2}$ is a line of symmetry; lowest pt is $\left(\frac{3}{2}, -\frac{1}{4}\right)$; $x = 1, 2$
5. $x = \frac{3}{2}$ is a line of symmetry; highest pt is $\left(\frac{3}{2}, \frac{1}{4}\right)$; $x = 1, 2$
7. $x = -\frac{3}{4}$ is a line of symmetry; highest pt is $\left(-\frac{3}{4}, \frac{25}{8}\right)$;
$x = -2, \frac{1}{2}$ 11. 5, 5   12. $x = 5$, $y = 10$
13. (a) 100,000   (b) $\$250{,}000$ 14. $x = \dfrac{36}{4 + \pi}$

$y = \dfrac{18}{4 + \pi}$ 15. (b) $x = 90$ (c) $I = \$44{,}100$
16. (b) 27 or 28   (c) $\$11{,}024$ (d) $\$218$ or $\$222$

Chapter 2 Review   **Part 2**

1. $A \times B = \{(1,c), (1,d), (3,c), (3,d), (5,c), (5,d)\}$, $B \times A =$
$\{(c, 1), (c, 3), (c,5), (d,1), (d,3), (d,5)\}$ 2. $d(A,B) =$
$4\sqrt{2}$, $M_{AB} = (5,7)$ 5. $d(A,B) = 6$, $M_{AB} = \left(-\frac{3}{2}, -\frac{3}{2}\right)$
8. $(x + 1)^2 + (y - 2)^2 = 16$ 9. $(x - 3)^2 + (y + 4)^2 = 97$
10. $(x - 5)^2 + (y - 3)^2 = 8$ 12. $(x - 2)^2 + (y - 2)^2 = 9$
13. $4x - 3y + 1 = 0$ 15. $x - 2y - 5 = 0$ 16. $y = -5$
17. $x = 3$ 18. $2x + 3y - 12 = 0$ 19. $3x - 2y + 6 = 0$
20. y-intercept is 3, x-intercept is -4 21. (a) $-\frac{2}{3}$
(b) $\frac{3}{2}$ 23. $5x - 12y + 26 = 0$
25. (a) $4x + 3y = 5$ (b) $3x + 4y = 2$
26. (a) $x - 2y + 1 = 0$ (b) $x - y - 4 = 0$
(c) $3x - 2y - 17 = 0$ (d) (9,5)

27.

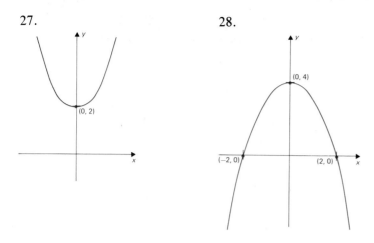

28.

30.

 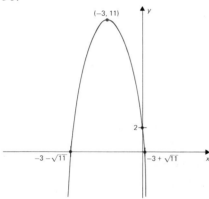

32. (b) $22,400   33. (a) $450   (b) $400   (c) $40
    (d) $0.50   34. 25 and 25   35. 256   36. 400
37. $35, $2450

1. (a) 1   (b) 2   (c) 3   (d) 1   (e) 4   (f) 4   Exercises 3.1
   (g) 2   (h) Does not exist   (i) 1   (j) 5   2. −8
4. 5   5. 12   7. 4   8. −1   9. −2   10. 4
11. −3   12. 0   13. Does not exist   15. −19   17. 2
19. 8   21. −12.7   22. 1.29   23. 12   26. 4

2. −21   3. −7   4. 35   5. 24   6. −3   7. 0   Exercises 3.2
8. −98   9. $\frac{16}{7}$ 10. $\frac{13}{10}$ 11. $-\frac{2}{7}$ 12. $\frac{25}{67}$

13. −4   14. $\frac{7}{2}$ 18. 0^3 19. $\frac{3}{2}$ 20. $\sqrt[3]{\frac{1}{3}}$

21. Does not exist   22. $\dfrac{1}{2\sqrt{x}}$ 25. 1   26. 1

27. -3 29. $\dfrac{1}{2\sqrt{x}}$ 32. $\dfrac{-2}{\sqrt{3-4x}}$ 34. $\dfrac{3}{2\sqrt{x}}$

36. $3x^2 + 4x - 3$ 38. 0

Exercises 3.3   1. -5 2. $\frac{3}{4}$ 3. $\frac{1}{3}$ 4. 0 6. -1 7. $-\infty$
8. $\frac{2}{5}$ 10. $+\infty$ 11. Does not exist   14. $-\infty$
16. $+\infty$ 17. $+\infty$ 18. $-\infty$ 20. Does not exist
21. $-\infty$ 22. Does not exist

Exercises 3.4   3. Discontinuous   4. Discontinuous   6. Discontinuous
8. Discontinuous   10. Continuous   11. Continuous
12. Continuous   14. f is continuous at $c = -1$
but discontinuous at $c = 1$ 15. Continuous
16. Continuous   17. Discontinuous   21. Continuous
23. 3   25. -2

Chapter 3 Review   **Part 2**

1. 10   2. 15   4. 98   5. 0   7. 2   9. 2
12. 0   13. 1   14. $\frac{5}{3}$ 15. $-\frac{7}{4}$ 16. $\frac{2}{3}$ 21. 2
22. -1 24. Always continuous   25. 3
27. Discontinuous at $x = 0, \pm 2$ 28. Discontinuous at all the
integers

Exercises 4.1   1. $1, R$ 2. $2x, R$ 3. $10x, R$ 4. $0, R$ 7. $\dfrac{-2}{x^2}$,

$x \neq 0$ 9. $1 - \dfrac{1}{x^2}, x \neq 0$ 11. $\dfrac{-1}{2x^{3/2}}, x > 0$

12. $1 - \dfrac{1}{2\sqrt{x}}, x > 0$ 13. $\dfrac{x}{\sqrt{x^2+1}}, x \in R.$

14. $3x^2 - \dfrac{1}{2x\sqrt{x}}, x > 0$ 15. 0   16. 1   19. 0

20. None

Exercises 4.2   1. (a) 100 ft/min   (b) 100 ft/min   (c) 2:02.4 P.M.
2. (a) 50 ft/min,   110 ft/min, 170 ft/min   (b) 8:40 A.M.
(c) 35,000 ft   3. (b) 7 ft/sec   4. (a) $4.00
(b) $2.00   (c) $2.00, $2.00   5. (a) $160.00
(b) $4x + 2h$ (c) $200, $320   6. (a) $\sqrt{10}/4$
(b) 2500 gal, 625 gal   7. (a) $375,000   (b) $120,
$140, $200   (c) 1250   8. (a) 6×10^7/min
(b) 6×10^7/min, 4×10^7/min   (c) $t^* = 40$ min   (d) 0

1. 7   2. $12x^3$ 3. $-10x$ 4. $4x - 3$ 5. $x^2 + 3x$
6. $3x^2 + 14x - 9$ 7. $-16x^3 - 9x^2$

8. $20x^3 - 9x^2 - 20x + 8$ 10. $\dfrac{-3}{4\sqrt{x}}$ 13. $-8x^{-5}$

14. $6x^{-3}$ 15. $\dfrac{3}{x^2}$ 16. $-\dfrac{1}{\sqrt{x}} + \dfrac{3\sqrt{x}}{2}$

17. $505x^{100} + 17$ 19. $-\dfrac{1}{x^2}$ 20. $-\dfrac{12}{x^4}$

21. $-\dfrac{1}{x^2} + \dfrac{1}{2\sqrt{x}}$ 24. $-\dfrac{1}{x^2}$ 25. $\frac{3}{2}t^{-1/2} - 27t^2$

28. $2u - \dfrac{2}{u^3}$ 29. $2x + x^{1/2} - x^{-2}$ 32. $10x + 3 + \dfrac{2}{x^2}$

33. $-\frac{9}{2}x^{1/2}$ 34. $\frac{1}{6}x^{-5/6} + \frac{1}{12}x^{-13/12}$

35. $\frac{5}{4}x^{3/2} + \frac{3}{2}x^{-1/2} - \frac{27}{4}x^{-5/2}$ 38. (a) $1, \frac{3}{4}, \frac{1}{2}$ b. $\frac{9}{4}$ miles
     c. $81\pi/16$ sq mi   39. 216   41. $x = 1,000$
42. 21,000 gal/min, 18,000 gal/min, 12,000 gal/min, 6,000 gal/min
     (b) 40 min

2. $16x + 2$ 3. $-6x - 1$ 4. $-30x + 8$
6. $9x^2 - 8x - 3$ 8. $30x^4 - 8x^3 - 12x + 2$ 10. $4x^3 - 4x$

12. $3\sqrt{x} - \dfrac{1}{\sqrt{x}}$ 13. $7x^{5/2} - 10x^{3/2}$ 16. $\dfrac{-1}{(x + 1)^2}$

17. $\dfrac{12}{(x + 4)^2}$ 20. $\dfrac{4}{(3 - 2x)^2}$ 21. $\dfrac{-(36x + 25)}{x^6}$

23. $\dfrac{-2x}{(x^2 - 5)^2}$ 24. $\dfrac{x^2 + 14x + 18}{(x + 7)^2}$ 25. $\dfrac{2x^2 + 24x - 3}{(2x^2 + 3)^2}$

26. $\dfrac{-24x(x + 1)}{(2x^3 + 3x^2 + 5)^2}$ 28. $\dfrac{-12x^{-1}}{(x^3 - x^{-3})^2}$ 29. $\dfrac{7x^2 - 10x - 35}{x^2(x + 7)^2}$

32. $\dfrac{\sqrt{x}(18x^2 + 37x - 12)}{2(3x + 4)^2}$ 34. $-8x^3 + 27x^2 - 12x + 9$

35. $-25x^9 - 21x^8 + 5x^7 + 215x^6 - 6x^5 - 6x^4 - 240x^3 + 36x^2$
36. $12x^{1/2} - 54x^{7/2} - 144x^2 + x^{-7/2} + 4x^{-3/2} - 6x^{3/2} + 16x^{-3} + \frac{15}{2}x^{-1/2} + 16x^{-5} + 12x^{-2} - 24$

1. $12x - y - 16 = 0$ 2. $x - 4y + 4 = 0$
4. $7x - 4y - 2 = 0$ 5. $y = -4, 4x - y - 4 = 0$
6. $\left(-\frac{3}{2}, -\frac{21}{4}\right)$ 7. 24, 12   8. (2, 4) and (−2, −4)
9. (5, 5)   10. (2, 60)   11. (2, 0)   12. $0 < c < 2$
13. $c = 1$ 14. $\frac{4}{3}$ 15. $x = \frac{3}{2}$ 16. $c = -2 + \sqrt{10}$

2. $2\left(x - \dfrac{1}{x}\right)\left(1 + \dfrac{1}{x^2}\right)$ 5. $4(x - 5)^3$ 6. $-12(2 - 3x)^3$

8. $\dfrac{2x}{\sqrt{2x^2 + 1}}$ 10. $12x(4x^2 + 5)^{1/2}$

12. $3(1 - 4x^2)^2(28x^2 - 1)$ 15. $\dfrac{4x^2 + x + 10}{\sqrt{x^2 + 5}}$

17. $\dfrac{8x^2 - x}{(2x^2 - x)^{5/2}}$ 18. $\dfrac{-(4x + 1)}{2(2x^2 + x - 1)^{3/2}}$

19. $2x - x(x^2 + 3)^{-3/2}$

21. $-\frac{1}{3}(5x^2 + 7x + 3)^{-4/3}(x^2 + 2x + 4)^{-2/3}(3x^2 + 34x + 22)$

24. $\dfrac{-3(12x^2 - 4x^{-1} + x^{-4})}{2(3x^2 - x^{-1})^{5/2}(4x - x^{-2})^{-1/2}}$

25. $80\,\pi$ ft²/sec, 200π ft²/sec, 800π ft²/sec   26. 25 in.²/min

27. (a) $\dfrac{1}{125\pi}$ in/hr   (b) $\frac{4}{5}$ in.²/hr   28. 15%/min,

increasing   29. $\dfrac{8}{25\pi}$ ft/min

Exercises 4.7   2. $\dfrac{9x^2}{2y}$ 5. $\dfrac{9x^2 + 10x}{2y}$ 8. $\dfrac{-xb^2}{ya^2}$ 10. $-\left(\dfrac{y}{x}\right)^{1/3}$

12. $\dfrac{5(x + 4)^4}{6y^2}$ 13. $\dfrac{6x^2 - 4xy - 1}{1 + 2x^2}$ 16. $\sqrt{3}x - y + 4 = 0$

18. $x + 8y - 5 = 0$ 20. 8 in./sec   21. $\dfrac{85}{\sqrt{13}}$ knots

Exercises 4.8   1. $f'(x) = 3x^2, f''(x) = 6x$ 2. $f'(x) = 10x^4 - 6x, f''(x) = 40x^3 - 6$ 4. $f'(x) = \frac{1}{2}x^{-1/2} - x^{-2}, f''(x) = -\frac{1}{4}x^{-3/2} + 2x^{-3}$

7. $f'(x) = -2(x - 1)^{-2}, f''(x) = 4(x - 1)^{-3}$ 8. $f'(x) = \frac{5}{2}x^{3/2} - 3,$ $f''(x) = \frac{15}{4}x^{1/2}$

10. $f'(x) = -3x(3x^2 + 1)^{-3/2}, f''(x) = 27x^2(3x^2 + 1)^{-5/2} - 3(3x^2 + 1)^{-3/2}$

12. $f'(x) = \dfrac{-1}{\sqrt{x}(1 + \sqrt{x})^2}, f''(x) = \frac{1}{2}(x)^{-3/2}(1 + \sqrt{x})^{-3}(1 + 3\sqrt{x})$

13. $\dfrac{-12}{(12y - 1)^3}$ 15. $\dfrac{1}{3}\left(\dfrac{1}{x^4 y}\right)^{1/3}$ 17. $f'''(x) = 24x + 18x^{-4}$

19. $f'''(x) = -6(x - 1)^{-4}$ 20. $\dfrac{6(x + 3)}{(1 - x)^8}$ 23. $v(t) = s'(t) = -20 - 32t, a(t) = v'(t) = -32$ 26. $v(t) = s'(t) = 3t^2 + 12t - 21, a(t) = v'(t) = 6t + 12$

27. (a) $-20, 56, 24, -21$ (c) $-32, 32, 0, \pm 6\sqrt{11}$

28. (b) $\frac{3}{16}, \frac{2}{9}, \frac{1}{4}, \frac{2}{9}, \frac{3}{16}$

Chapter 4 Review   **Part 2**

1. 8   2. 0   5. $4x^3 - \frac{2}{3}x^{-2/3}$ 6. $x^{-1/2} - \frac{1}{2}x^{-2}$

9. $31(3x^3 - 7x + 1)^{30}(9x^2 - 7)$

11. $(2x + 1)^{-1/2}(5x^2 + 2x + 1)$ 12. $\dfrac{2 - 7x^2}{(x^2 + 2)^5}$

14. $3\left(\dfrac{x + 1}{(1 - x)^5}\right)^{1/2}$ 16. $C'(x) = -2x^{-2} + 3$ (b) $\dfrac{2399}{800}$

(c)   270.05   18.   (a)   32 ft/sec   (b)   480 ft from the ground
(c) $t = 8\frac{1}{2}$ 20.   (2, 3)   21. $x = 1$
25. $y = -x, y = 2x$ 26. $-24(6y - 1)^{-3}$ 27.   0
28. $\dfrac{18}{\sqrt{13}}$ km/min   29. 20π m²/sec   30. $\dfrac{3.45}{\sqrt{98.5}}$ m/sec
31. 24π cm²/min, 144π cm³/min   32. $\dfrac{5}{2\pi}$ ft/min

1. f is increasing in $(-\infty, \infty)$ 2. f is decreasing in $(-\infty, \infty)$ **Exercises 5.1**
3. f is decreasing in $(-\infty, \infty)$ 5. f is increasing in $(-\infty, 0)$;
     f is decreasing in $(0, \infty)$ 6. f is increasing in $(1, \infty)$;
     f is decreasing in $(-\infty, 1)$ 10. f is increasing in $(-\infty, 4)$;
     f is decreasing in $(4, \infty)$ 11. f is increasing for $x < -1$ or
     $x > 3$; f is decreasing for $-1 < x < 3$ 12. f is increasing for $x > 2$
     or $x < -2$, f is decreasing for $-2 < x < 2$
13. f is increasing for $x < -1$ or $x > 1$;
     f is decreasing for $-1 < x < 1$
15. f is increasing for $-\frac{1}{2} < x < 0$ or $x > \frac{1}{2}$ and f is decreasing
     for $x < -\frac{1}{2}$ or $0 < x < \frac{1}{2}$.
17. f is increasing for $1 < x < 2$ or $x > 3$, and f is decreasing
     for $x < 1$ or $2 < x < 3$.   18. f is decreasing on
     $(-\infty, 0) \cup (0, \infty)$ 20. f is increasing on $(-\infty, -1)$
     $\cup (-1, \infty)$ 22. f is decreasing on $(-\infty, 0) \cup (0, \infty)$
24. R is increasing on $[0, 2)$, and decreasing on $(2, 4]$
25.   Decreasing

1.   Relative maximum at x_2, x_4, x_5, x_7; relative minimum at   **Exercises 5.2**
     x_1, x_3, x_5, x_6, x_8; absolute maximum at x_7; absolute
     minimum at x_3 2.   No critical values, $f(0) = -5$ absolute
     minimum, $f(2) = 7$ absolute maximum   4. $x = 0, 2$;
     at $x = 0$ is relative maximum, $x = 2$ is relative minimum
6. $x = \pm 1$; at $x = 1$ is relative minimum, $x = -1$ is relative
     maximum   7. $x = \pm 1$; at $x = 1$ is relative minimum, $x = -1$ is
     relative maximum   10. $x = 0, \pm\frac{1}{2}$; at $x = \pm 1$ is absolute
     maximum, $x = \pm\frac{1}{2}$ is absolute minimum, $x = 0$ is relative
     maximum   11. $x = 0$; absolute minimum is at $x = -1$,
     absolute maximum is at $x = 1$
15.   No critical values;
     no extrema   19. $x = \pm 2$; at $x = 2$ is relative maximum,
     $x = -2$ is relative minimum   20.   No critical values;
     $x = 0$ gives absolute minimum
21.   No critical points; no extrema   23. $x = 0$; at $x = 0$ is
     relative minimum, $x = -2$ is relative maximum
24. $x = \frac{1}{5}, 1$; at $x = \frac{1}{5}$ is relative maximum, $x = 1$ is relative

minimum   27. (a) $a = -3$, $b = -24$ (b) Relative maximum at $x = -2$ and a relative minimum at $x = 4$

30. (c) $x = 20$ gives maximum profit

Exercises 5.3

1. Square of side $P/4$ 2. Cube with side 4   3. $10''$
4. $2\sqrt{3}$ ft $\times$ $3\sqrt{3}$ ft   5. $8'' \times 8'' \times 8''$
7. $r\sqrt{2} \times r/\sqrt{2}$ 8. $a\sqrt{2} \times b\sqrt{2}$ 9. 3×2
10. $\dfrac{20\pi}{4 + \pi}, \dfrac{80}{\pi + 4}$ 12. $12' \times 12' \times 2'$ 13. $300\sqrt{2}$ ft
14. $10' \times 10' \times 10'$

Exercises 5.4

1. $x = 100$; 120,000   2. $t = 2.5$; 332,500   3. $10^{-5/2}$
4. $\frac{2}{3}r_0$ 5. (a) $\frac{5}{3}$ (b) 250   7. (b) 49,900
(c) \$37,349,641   8. 7000, \$11 per item   9. The marginal revenue equals the marginal cost.   12. (b) 112
(c) \$451

Exercises 5.5

1. Increasing on $\left(\frac{3}{2}, \infty\right)$; decreasing on $\left(-\infty, \frac{3}{2}\right)$; no inflection point; concave up for every x; relative minimum is $\left(\frac{3}{2}, \frac{7}{4}\right)$ 2. Increasing on $\left(-\infty, \frac{3}{4}\right)$; decreasing on $\left(\frac{3}{4}, \infty\right)$; no inflection point; concave down for every x; relative maximum is $\left(\frac{3}{4}, -\frac{7}{8}\right)$
5. Increasing on $(0, \infty)$; decreasing on $(-\infty, 0)$; concave down in $(-\infty, 0) \cup (0, \infty)$; absolute minimum at $(0, 0)$
6. Decreasing on $(-\infty, \infty)$; concave down in $(0, \infty)$; concave up in $(-\infty, 0)$; $x = 0$ is point of inflection   7. Increasing on $(-\infty, -6) \cup (0, \infty)$; decreasing on $(-6, 0)$; $x = -3$ is point of inflection; concave up in $(-3, \infty)$; concave down in $(-\infty, -3)$; at $x = 0$ is relative minimum, at $x = -6$ is relative maximum   9. Increasing on $(-\infty, -1) \cup \left(-\frac{1}{3}, \infty\right)$, decreasing on $\left(-1, -\frac{1}{3}\right)$; concave up in $\left(-\frac{2}{3}, \infty\right)$; concave down in $\left(-\infty, -\frac{2}{3}\right)$; $x = -\frac{2}{3}$ is point of inflection; at $x = -\frac{1}{3}$ is relative minimum; $x = -1$ is relative maximum
12. Decreasing on $(-\infty, \infty)$; $x = 1$ is point of inflection; concave up in $(-\infty, 1)$; concave down in $(1, \infty)$ 14. Decreasing on $(-\infty, 0) \cup (0, \infty)$; concave up in $(0, \infty)$; concave down in $(-\infty, 0)$ 17. Concave up in $(0, \infty)$; concave down in $(-\infty, 0)$; decreasing for every $x \neq 0$ 20. Increasing on $(-3, 3)$; decreasing on $(-\infty, -3) \cup (3, \infty)$; $x = 6$ point of inflection; concave up in $(6, \infty)$; concave down in

$(-\infty, -3) \cup (-3, 6)$; $x = 3$ is relative maximum

21. $3x + y - 1 = 0$, $18x + 27y - 11 = 0$, $9x + 27y - 17 = 0$,
     $60x - 432y - 3 = 0$ and $3x - 27y = 0$ 22. -8

25. Concave down at $x = 100$; has an inflection point at $x = 200$;
     and is concave up for $x = 300, 400$

2. 44.5, 50.75, 6.25   3. 6.5, 6.75, 0.25;   6. -0.03125,
    -0.0286, 0.00265   8. 11.1364   11. 15.9687

15. 0.5031   16. $\frac{8}{5}\pi$ sq in.   17. 2.4π sq in., 1.7%

18. ≈ 0.007 ft   20. 100π cu cm   21. (b) 7

(c) 0.01   (d) 0.875%   22. (b) $\left(\frac{4}{5}x - 35 - \frac{x^2}{500}\right) dx$

(c) 0

Part 2

1. $f'(x) = 6x - 7$; f is increasing on $\left(\frac{7}{6}, \infty\right)$; decreasing on
    $\left(-\infty, \frac{7}{6}\right)$; $f''(x) = 6$; f is concave up in $(-\infty, \infty)$; at $x = \frac{7}{6}$ is
    relative minimum   2. $f'(x) = 4x + 1$; f is increasing on
    $\left(-\frac{1}{4}, \infty\right)$; decreasing on $\left(-\infty, -\frac{1}{4}\right)$; $f''(x) = 4$; f is concave
    up in $(-\infty, \infty)$; at $x = -\frac{1}{4}$ is relative minimum
    4. $f'(x) = 3x^2 - 27$; f is increasing on $(-\infty, -3) \cup (3, \infty)$;
    decreasing on $(-3, 3)$; $f''(x) = 6x$; $x = 0$ is point of
    inflection; f is concave up in $(0, \infty)$; concave down in
    $(-\infty, 0)$; at $x = -3$ is relative maximum; at $x = 3$ is
    relative minimum   6. $f'(x) = 3x^2 - 6x + 3$; f is increasing
    on $(-\infty, \infty)$; $f''(x) = 6x - 6$; f is concave up in $(1, \infty)$;
    concave down in $(-\infty, 1)$; $x = 1$ is point of inflection

9. $f'(x) = x^{-2/3}$; f is increasing on $(-\infty, \infty)$; $f''(x) = -\frac{2}{3} x^{-5/3}$;
    f is concave up in $(-\infty, 0)$; concave down in $(0, \infty)$; $x = 0$ is
    point of inflection

11. $f'(x) = 2 - \frac{1}{x^2}$, f is increasing on $\left(-\infty, -\frac{\sqrt{2}}{2}\right) \cup \left(\frac{\sqrt{2}}{2}, \infty\right)$;

     decreasing on $\left(-\frac{\sqrt{2}}{2}, \frac{\sqrt{2}}{2}\right)$; at $x = \frac{\sqrt{2}}{2}$ is relative

     maximum; at $x = \frac{-\sqrt{2}}{2}$ is relative minimum; $f''(x) = \frac{2}{x^3}$;

     f is concave up in $(0, \infty)$; concave down in $(-\infty, 0)$

13. $f'(x) = \frac{2}{(x + 3)^2}$, f is increasing on $(-\infty, -3) \cup (-3, \infty)$;

     $f''(x) = \frac{-4}{(x + 3)^3}$; f is concave up in $(-\infty, -3)$; concave

     down in $(-3, \infty)$ 15. $f'(x) = \frac{-8}{x^3}$, f is increasing on

$(-\infty, 0)$; decreasing on $(0, \infty)$; $f''(x) = \dfrac{24}{x^4}$; f is concave

up in $(-\infty, 0) \cup (0, \infty)$ 16. $f'(x) = x^4$; f is increasing on
$(-\infty, \infty)$; $f''(x) = 4x^3$; f is concave up in $(0, \infty)$; concave
down in $(-\infty, 0)$; $x = 0$ is point of inflection

18. f is increasing on $(-3, 1) \cup (2, \infty)$; decreasing on
$(-\infty, -3) \cup (1, 2)$; at $x = -3, 2$ is relative minimum;
at $x = 1$ is relative maximum; f is concave

up in $\left(-\infty, -\sqrt{\tfrac{7}{3}}\right) \cup \left(\sqrt{\tfrac{7}{3}}, \infty\right)$; concave down in
$\left(-\sqrt{\tfrac{7}{3}}, \sqrt{\tfrac{7}{3}}\right)$; $x = \pm\sqrt{\tfrac{7}{3}}$ are points of inflection   21.   No
relative nor absolute extrema   (b)   Absolute and relative
minimum at $x = 1$ (c)   No relative nor absolute extrema

22.   (a)   No relative nor absolute extrema   (b)   Relative and
absolute maximum at $x = 1$.   (c)   No relative nor absolute
extrema   23.   25, 25   24.   40, 40   25.   (a)   Square
20×20 (b)   Square, 20×20

26. $\dfrac{40}{\pi}\left(\dfrac{\pi}{20}\right)^{2/3}, r = \left(\dfrac{20}{\pi}\right)^{1/3}$ 27. $30 \times 60 \times 30$ 28. $2 \times$
$(2 + \pi)$, $\dfrac{\sqrt{5(\pi + 4)}}{2}$ 30.   4 miles   31. $d = \dfrac{c}{2}$

32. $x = 3$ 35.   0.01 in.

Exercises 6.1   1.      2.

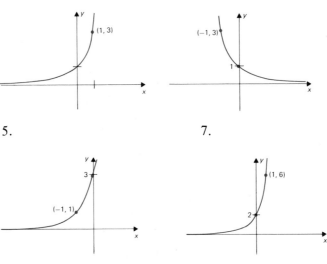

5.   7.

8.

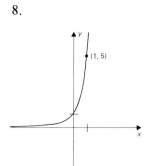

9.

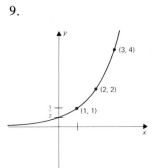

10.

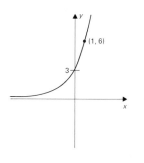

11.

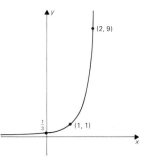

12.

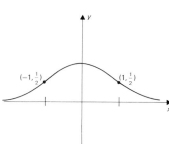

15.

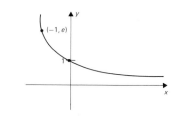

19. 3   21. $\frac{1}{2}$ 23. $\frac{1}{5}$ 24. 4

1. $\log_3 81 = 4$ 2. $\log_2 32 = 5$ 3. $\log_5 \frac{1}{125} = -3$

4. $\log_{1/3} 9 = -2$ 6. $\log_{27} 9 = \frac{2}{3}$ 7. $2^5 = 32$

10. $16^{-3/4} = \frac{1}{8}$ 11. 3   12. 4   13. $\frac{5}{6}$ 16. 3

17. 2   19. 2   21. -3 22. 16   23. 5

25. 3   26. $\dfrac{-1 + \sqrt{73}}{4}$ 31. 1.5441

32. 1.1761   35. 0.4226   38. -0.1461

Exercises 6.2

Exercises 6.3   1. $4e^{4x}$ 2. $32xe^{4x^2+5}$ 4. $(6x-4)e^{3x^2-4x+1}$

5. $x^2e^{2x}(3+2x)$ 7. $-4xe^{-2x^2}$ 10. $\dfrac{e^x}{e^x+1}$

13. $\dfrac{4xe^{2x^2+4}}{1+e^{2x^2+4}}$ 16. $-5^{-x}\ln 5$ 19. $2e^x + 2^e e(x)^{e-1}$

21. $\dfrac{1}{x}\log_{10} e$ 23. $\dfrac{(2x+e^x)\log_3 e}{x^2+e^x}$ 25. $\dfrac{16}{x+1} +$

$\dfrac{8(4x+1)}{2x^2+x} - \dfrac{x}{x^2+4}$ 27. $\dfrac{14xe^{2x}}{e^{2x}+6} + \dfrac{1}{2(x+4)} - \dfrac{5(e^x - e^{-x})}{e^{-x}+e^x}$

28. $\dfrac{(3x^2+1)^5(x^2+1)^{3/2}}{(x^2+4)^{10}}\left[\dfrac{30x}{3x^2+1} + \dfrac{3x}{x^2+1} - \dfrac{20x}{x^2+4}\right]$

Chapter 6 Review   **Part 2**

1. (a)

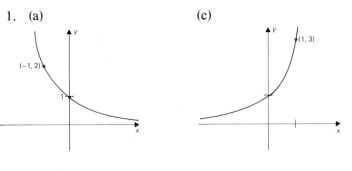

(e)

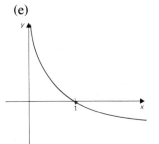

2. (a) 2^{x+1} (b) -2^{1-x}
    (e) $2^{2x^2} - 2^{-2x^2}$ 3. (a) 3   (d) 7   6. $3e^{3x}$
7. $-5e^{-5x}$ 8. $2(2x-1)e^{x^2-x}$ 9. $e^{-x} - xe^{-x}$

10. $e^{1/x} - \dfrac{1}{x}e^{1/x}$ 11. $1 + \ln x$ 17. $e^x + exe^{-1}$

18. $\dfrac{1}{x}\log_5 e$ 19. $\dfrac{1}{x^3 - x^2}(\log_2 e)(3x^2 - 2x)$

20. $\log e\left[\dfrac{5}{x-1}+\dfrac{7}{x+2}-\dfrac{4}{x-2}\right]$

24. $\left[\dfrac{80(x-1)}{x^2-2x+3}+\dfrac{1}{2(x-1)}-\dfrac{3}{4(x+5)}\right]\dfrac{(x^2-2x+3)^{40}(x-1)^{1/2}}{(x+5)^{3/4}}$

26. $x^x(\ln x+1)$ 27. $(\ln 3)3^x+x^x(\ln x+1)+3x^2$

Exercises 7.1

1. $\frac{1}{2}x+C$ 2. $\frac{1}{4}x^2+C$ 3. $\frac{3}{4}x^4+C$. 4. $\dfrac{x^2}{2}+\dfrac{x^4}{4}+C$

7. $-\dfrac{1}{x}+C$ 10. $-\frac{1}{2}x^{-2}-\frac{4}{3}x^{-3}+C$ 11. $\frac{4}{3}x^{3/2}+C$

13. $\frac{3}{2}x^{2/3}+C$ 14. $\frac{9}{4}x^{4/3}+3x^{2/3}+C$ 16. $\frac{3}{2}x^{4/3}+$

$\frac{3}{5}x^{5/3}+C$ 17. $\frac{2}{7}x^{7/2}-4x^{1/2}+C$ 18. $\dfrac{x^3}{3}+\dfrac{3}{2}x^2+$

$9x+C$ 20. $\frac{3}{5}x^{5/3}-\frac{3}{4}x^{4/3}+x+C$ 24. $\frac{2}{3}x^3+$

$\frac{5}{2}x^2-3x+5000$ 26. $x^3-\frac{5}{2}x^2+8x-\frac{25}{2}$

Exercises 7.2

2. $\frac{15}{8}$, $\frac{17}{8}$ 3. $\frac{35}{2}$, $\frac{51}{2}$ 6. $\dfrac{3}{8}\sum_{k=1}^{8}\sqrt{\dfrac{45+3k}{8}}$,

$\dfrac{3}{8}\sum_{k=1}^{8}\sqrt{\dfrac{48+3k}{8}}$ 11. 2   12. 3.5   13. $\frac{1}{3}$

16. 12   18. $\frac{26}{3}$ 20. $\frac{1}{4}$

Exercises 7.3

1. $\frac{8}{3}$ 2. 12   6. $\frac{4}{3}$ 7. 30   9. $-\frac{32}{3}$

12. $\frac{256}{5}$ 14. 10   15. 36   18. $\frac{9}{2}$ 19. $\frac{1}{3}$ 20. $\frac{8}{3}$

21. $\frac{4}{3}$ 22. 36   25. $(x-1)^2$

26. $2\sqrt{x}$ 28. $8x^2$ 30. 0   33. 2   34. $-\frac{3}{2}$, 2

Exercises 7.4

1. $t-\dfrac{3t^2}{2}+4$ 2. $x^4+\dfrac{x^3}{3}-\dfrac{10}{3}$ 4. $\frac{2}{3}t^{3/2}+$

$4t-\frac{55}{3}$ 6. $2x-\frac{2}{5}x^{5/2}+\frac{406}{5}$

9. $-\frac{2}{5}x^{5/2}+2x^{1/2}+\frac{83}{10}$ 10. $\frac{9}{4}x^{4/3}-\frac{9}{5}x^{5/3}+\frac{103}{5}$

11. (b) $-2x^2+360x-500$ (c) 90 units   12. $C(x)=$

$\frac{20}{3}x^{3/2}+350$ 13. $p(x)=\dfrac{R(x)}{x}=\dfrac{1}{100}x+20$

14. (a) 9 min (b) 48 in. (c) 18 min   15. (b) ≈ 674

min   16. Yes, by 6 ft   17. (a) $v(t)=\dfrac{t^2}{2}+t+2$,

$a(t)=\dfrac{t^3}{6}+\dfrac{t^2}{2}+2t+1$ (b) $v(t)=\dfrac{t^3}{3}-5t-1,\ a(t)=\dfrac{t^4}{12}-$

$\frac{5}{2}t^2-t+4$ 18. (a) $v(t)=-32t+288,\ s(t)=-16t^2+$

$288t$ (b) 1296 ft   19. $3\sqrt{2}$ sec   20. $t=4$ sec

Exercises 7.5

1. $\dfrac{(x+4)^3}{3}+C$ 2. $\dfrac{(x-5)^7}{7}+C$ 3. $\frac{2}{9}(3x+4)^{3/2}+C$

6. $\frac{1}{12}(x-3)^3+C$ 7. $\dfrac{-1}{4(4x-1)}+C$

8. $\dfrac{-1}{2(x^2+3x-1)^2}+C$ 9. $\dfrac{(4x^5+6)^{101}}{2020}+C$

12. $-\frac{1}{3}(4-x^2)^{3/2}+C$ 14. $\frac{2}{5}(2+x^{5/3})^{3/2}+C$ 15. $\frac{1}{2}$

17. $2(\sqrt{6}-\sqrt{2})$ 19. $\frac{1}{4}$

Exercises 7.6

2. $-\frac{1}{5}e^{-5x}+C$ 3. $2e^{1/2x}+C$ 5. $\frac{1}{2}e^{x^2}+C$

9. $-2e^{1/x}+C$ 10. $e^x-e^{-x}+C$ 11. $-e^{-x}+\ln|x|+C$

12. $\ln|x+1|+C$ 15. $\ln|x-2|+C$ 16. $\frac{1}{2}\ln(x^2+4)+C$

18. $\frac{5}{2}-\ln 2$ 19. $\frac{3}{2}-\ln 3$ 20. $\frac{1}{3}x^3+\frac{3}{2}x^2+4x-$
$16\ln|x+2|+C$

Exercises 7.7

1. $\frac{2}{3}x^3-\frac{1}{2}x^2+C$ 3. $\frac{1}{5}x^2(1+x^2)^{5/2}-\frac{2}{35}(1+x^2)^{7/2}+C$

5. xe^x-e^x+C 7. $\dfrac{x^2}{2}\ln x-\dfrac{x^2}{4}+C$ 8. $\dfrac{x^3}{3}\ln x-\dfrac{x^3}{9}+C$

10. $\frac{2}{3}x^{3/2}(\ln x-\frac{2}{3})+C$ 11. $e^x(2-x)+C$ 12. $e^x(x^2-$
$2x+4)+C$ 13. $\dfrac{e^{2x}}{4}(2x-1)+C$ 18. $-(1-x^2)e^{-x}+$
$2xe^{-x}+2e^{-x}+C$ 19. $\ln x[\ln(\ln x)]-\ln x+C$
20. $-xe^x+e^x+C$

Exercises 7.8

1. 4   4. Does not exist   5. $2\sqrt{2}$ 6. $2\sqrt{6}$
7. Does not exist   11. Does not exist   12. $\sqrt{3}$
13. 1   14. Does not exist   16. $\frac{1}{2}$ 19. $\frac{1}{2}$ 20. $-\frac{1}{2}$
21. Does not exist   23. 2   25. Does not exist
26. 0   30. (a) \$50,000

Chapter 7 Review   Part 2

1. $\dfrac{x^6}{6}+C$ 2. $\dfrac{2x^{3/2}}{3}+C$ 3. $\dfrac{x^{-2}}{-2}+C$ 4. x^6+C

5. $\frac{3}{5}x^{5/3}+C$ 6. $\dfrac{x^{1-n}}{1-n}+C$ 7. $\dfrac{x^6}{6}+\dfrac{3}{5}x^5+x+C$

8. $\dfrac{3x^2}{2}+2x+5\ln|x|+C$ 9. $\frac{2}{7}x^{7/2}+\frac{6}{5}x^{5/2}+8x^{1/2}+C$

10. $\dfrac{x^4}{4}+\dfrac{3x^2}{2}+C$ 13. $\dfrac{(x+2)^4}{4}+C$ 17. $\dfrac{-1}{2(x-1)^2}+C$

19. $x^2+\ln|x|+C$ 22. $x+\ln|x|+C$ 24. $-x-$
$\ln|1-x|+C$ 26. $\ln\left|\dfrac{x+1}{1-x}\right|+C$ 28. $\dfrac{(1+2x^2)^8}{32}+C$

30. $\frac{7}{22}(4 + x^2)^{11/7} + C$ 31. $e^{x^3+1} + C$ 32. $-\frac{1}{2}e^{-(x^2+4)} + C$

33. $\frac{-1}{a}e^{-ax} + C$ 35. $-\frac{1}{7}\ln|3 - 7x| + C$

36. $\frac{-1}{20(5x^2 + 2)^4} + C$ 38. $\frac{x^2}{2}[2\ln|x| - 1] + C$

42. $(\ln x)\left(x + \frac{x^3}{3}\right) - x - \frac{x^3}{9} + C$ 43. $f(x) = x^2 -$

  $3x + 4$ 45. $t = 8$, 1024 ft   47. (a) $\frac{81}{4}$ (b) $8\frac{2}{3}$
  (c) $\frac{74}{3}$ (e) 36   (f) 36

1. Second order, linear   2. Second order, nonlinear   **Exercises 8.1**
3. First order, nonlinear   4. Second order, nonlinear
5. Fourth order, linear   6. Third order, linear   7. First
  order, nonlinear   8. Third order, nonlinear
9. Second order, linear   10. Second order, nonlinear
11. Second order, nonlinear   12. Second order, linear
29. (b) $y = -3e^{-2t}$ 30. (b) $c = 2$ (c) $k = \ln c$
  (d) $y(2) = ce^{-2k}$ 31. (b) $y = 9e^{2x} - 6e^{3x}$
32. (b) $(e^3 - 1)^{-1}[(e^3 - 2e)e^{-x} + (2e - 1)e^{2x}]$

34. (b) $y = \dfrac{1}{\sqrt{2x - \frac{7}{4}}}$

1. $y = ce^x$ 2. $y = ce^{-2x}$ 3. $y = ce^{-\sqrt{3}x}$ 4. $y =$ **Exercises 8.2**
  $ce^{-3x/2}$ 5. $y = cx^4$ 6. $y = ce^{3x}$

7. $y = \dfrac{-1}{\ln|x + 1| + C}$ 8. $\dfrac{y^2}{2} - 4y = \dfrac{x^2}{2} + 3x + C$

10. $y = -\frac{1}{2}e^{-2x} + c$ 11. $y = -\ln|-e^x + c|$

16. $y = ce^{-2t^2}$ 17. $y = \dfrac{-1}{xe^{-x} + e^{-x} + c}$

19. $y = \frac{3}{2} + ce^{-2x}$ 20. $y = -10 + ce^x$

21. $y = \dfrac{x}{3} - \dfrac{1}{9} + ce^{-3x}$ 22. $y = \dfrac{x^3}{2} + \dfrac{c}{x^3}$

24. $y = \frac{1}{2}t^2 - \frac{1}{2} + ce^{-t^2}$ 26. $y = 2x^2 + cx^{-2}$ 28. $y =$

  $te^{-t} + 2e^{-t}$ 29. $y^2 = \dfrac{t^4}{2} + \dfrac{199}{4}$ 35. (a) $m = 1, 3$

36. $m = -1, -2$ 37. $m = \pm 2$ 38. $m = \pm\sqrt{2}$
40. $m = \dfrac{1 \pm \sqrt{7}}{2}$

1. $20{,}000\left(1 - \left(\frac{4}{5}\right)^t\right)$, ≈ 3 2. $\approx 38.1°$ 3. $t \approx 1.35$ hr   **Exercises 8.3**
4. (a) $21.8°$ (b) ≈ 9.4 min   5. (a) ≈ -0.0875

(b) ≈ 16 min   6. (a) 2.96×10^8 (b) 4.28×10^8
(c) Will double in $t \approx 17.7$ yr   (d) Will triple in $t \approx$
28.1 yr   7. ≈ 1990 8. In the year 2022
9. (a) 1.2528   (b) ≈ 2229 10. (a) About 65%
(b) 74.91%   12. 63.496 ft   13. (7, 4 ln 2)

Chapter 8 Review   Part 2

1. $y = -\dfrac{3x^5}{5} + 4x + C$ 2. $y = \dfrac{3x^2}{2} + \dfrac{4x^3}{3} + c_1 x + c_2$

3. $y = \frac{3}{2}x - \frac{3}{4} + \frac{11}{4}e^{-2x}$ 4. $y = \dfrac{x^3}{3} + \dfrac{x^2}{2} - \dfrac{5}{6}x + 2$

6. (a) $k = 50$ 7. (b) $y = 100e^{(-11/120t)}$ 8. 85, 110, 85
9. 302, 500   11. (a) $P(x) = \frac{4}{3}x^3 - 15x + \frac{508}{3}$ (b) $\frac{284}{3}$

13. $y^2 = \dfrac{2x^3}{3} + 2x^2 + c$ 14. $y = ce^{-x^2}$ 15. $y = ce^{x^3/3}$
17. $y = \frac{1}{2} + ce^{-x^2}$ 19. $y = xe^{-x} + ce^{-x}$
21. $y = c_1 e^{5x} + c_2 e^{-5x}$ 24. $y = c_1 + c_2 e^x$

Exercises 9.1

1. 1, 5, 13   2. $-2, 1, e - 1$ 5. $\{(x,y \mid x,y \in R\}$
7. $\{(x,y) \mid 1 \le x \le 2 \text{ and } 3 \le y \le 4\}$ 9. $\{x,y \mid (x,y) \ne (0,0)\}$
11. (a) yz-plane   (b) xz-plane   (c) At $z = 3$ plane
parallel to xy-plane   (d) z-axis   (e) Straight line
parallel to z-axis going through the point (1,0,0)
(f) Plane parallel to xy-plane 2 units below (at $z = -2$)
(g) Straight line through (0,0,0) and (1,1,1)   (h) Point
$(1,-1,2)$ 12. (a) $\sqrt{a^2 + b^2 + c^2}$ (c) 5   (e) 13

Exercises 9.2   4.

6.

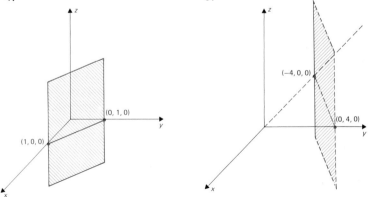

11.

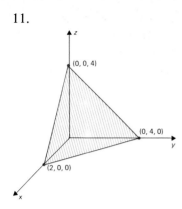

12.

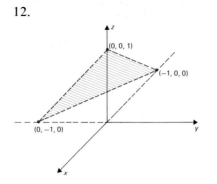

14.

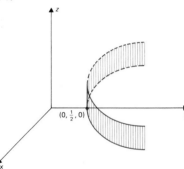

16.

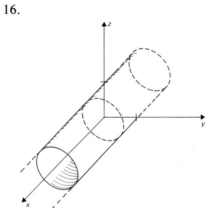

19.

22. Sphere with origin at (0,0,0) and radius $r = 4$
23. Sphere with origin at (1,0,0) and radius 3
26. Sphere with origin at $(-3,1,-2)$ and radius 4

29.

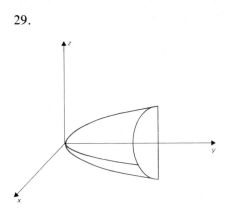

Exercises 9.3

1. (a) 3, f is continuous at (2, 1)   (b) $\frac{3}{2}$, f is continuous at (1, 1)   (c) 3, f is discontinuous at (2, 2)

2. (b) $3x^2y^3$, $3x^3y^2$ (d) $\dfrac{-2y}{(x-y)^2}$, $2x(x-y)^{-2}$

3. (b) $2xe^{x^2+y^2}$, $2ye^{x^2+y^2}$ (c) $e^{x+y}[\ln(x^2+y^2) + 2x(x^2+y^2)^{-1}]$, $e^x e^y [\ln(x^2+y^2) + 2y(x^2+y^2)^{-1}]$

4. (b) $f_{xx}(1, 2) = 48$, $f_{xy}(1, 2) = 36 = f_{yx}(1, 2)$, $f_{yy}(1, 2) = 12$
    (c) $f_{xx}(1, 2) = 0$, $f_{xy}(1, 2) = f_{yx}(1, 2) = -\frac{1}{4}$, $f_{yy}(1, 2) = \frac{1}{4}$
    (d) $f_{xx}(1, 2) = -8$, $f_{yx}(1, 2) = f_{xy}(1, 2) = 6$, $f_{yy}(1, 2) = -4$

5. (a) $f_x = 2xe^{u+v}$, $f_y = 3y^2e^{u+v}$, $f_u = f_v = (x^2 + y^3)e^{u+v}$
    (c) $f_{xyu} = f_{uyx} = f_{uxy} = 0$

Exercises 9.4

1. $\dfrac{\partial P}{\partial X}\bigg| = 12c + 25$, $\dfrac{\partial P}{\partial y}\bigg|_{(2,3)} = 10$ 3. (a) $\dfrac{\partial P}{\partial X} =$
    $\dfrac{[y(x^2 + y^2 + y) - 2x]e^{xy}}{(x^2+y^2+y)^2}$, $\dfrac{\partial P}{\partial y} = \dfrac{[x(x^2+y^2+y) - (2y+1)]e^{xy}}{(x^2+y^2+y)^2}$ (3,5)

5. (a) $f_x = 50y$, $f_y = 50x$ (b) $f_x|_{y=5} = 250$
    (c) $f_y|_{x=2} = 100$ 6. (a) $f_x = 2x + 20y$, $f_y = 2y + 20x$
    (b) $f_x|_{y=5} = 2x + 100$ (c) $f_y|_{x=2} = 2y + 40$

7. (a) $(y^2 + e^y)\, dx + x(2y + e^y)\, dy$ (b) $(1 + y/x)\, dx + \ln x\, dy$ 9. 3%   10. 4%   11. 1%   12. 27

13. (a) $\dfrac{\partial C}{\partial x} = 8x + y$, $\dfrac{\partial C}{\partial y} = x - 26y$ (b) $\dfrac{\partial C}{\partial x} = \dfrac{y}{x} + 6xy$,

    $\dfrac{\partial C}{\partial y} = \ln x + 3x^2$ (c) $\dfrac{\partial C}{\partial x} = e^y$, $\dfrac{\partial C}{\partial y} = xe^y + 1$

    (d) $\dfrac{\partial C}{\partial x} = \ln(16x + e^y) + x\dfrac{16}{16x + e^y}$, $\dfrac{\partial C}{\partial y} = x\dfrac{e^y}{16x + e^y}$

Exercises 9.5

1. At (0, 0) is a saddle point
2. At (0, 0) is the relative minimum
5. At (0, 0) is the relative minimum   7. None   8. None

9. (0, 0, 1)   10. $4' \times 4' \times 2'$ 11. $6'' \times 6'' \times 3''$
14. $l \approx 3.495$, $w \approx 3.495$, $h \approx 5.241$ 15. $l \approx 3.319$,
     $w \approx 3.319$, $h \approx 5.810$ 17. $y = 10$, $x = 6$

Exercises 9.6

3. f attains maximum value $\sqrt{2}$ at $(1/\sqrt{2}, 1/\sqrt{2})$; f attains
    minimum value $-\sqrt{2}$ at $(-1/\sqrt{2}, -1/\sqrt{2})$ 4. f attains
    maximum value 1 at $(\pm 1, 0)$; f attains minimum value -1 at
    $(0, \pm 1)$ 5. $\frac{1}{2}$ is minimum value at $\left(\frac{1}{2}, \frac{1}{2}\right)$ 11. $\frac{134}{75}$

Exercises 9.7

1. 3   2. 52   5. $\frac{11}{6}$ 7. $-\frac{5}{6}$ 10. 9
12. $\frac{1}{2}(e^3 - e^2 - e + 1)$ 13. $\frac{5}{3}$ 16. $\frac{23}{15}$ 18. $\frac{1}{2}(e^4 - 1)$
21. $2 - \frac{4}{3}\sqrt{2}$ 22. $\frac{21}{2}$ 24. $3 \ln 3 - 2$ 25. $\frac{3}{2}$
26. $\frac{1}{6}$

Part 2

Chapter 9 Review

2. (a) $2\sqrt{3}$ (b) $\sqrt{30}$ 3. (a) $z = 0$ (c) $y = 0$
    (d) $2x + y + 2z = 2$ 6. $f_x = 2$, $f_y = -10y$, $f_{xx} = 0$,
    $f_{yy} = -10$, $f_{yx} = 0$ 7. $f_x = y^2$, $f_y = 2xy$, $f_{xx} = 0$, $f_{yy} = 2x$,
    $f_{yx} = 2y$ 8. $f_x = 8 - 10yx + 7y^3$, $f_y = -5x^2 + 21xy^2$,
    $f_{xx} = -10y$, $f_{yy} = 42xy$, $f_{yx} = -10x + 21y^2$ 11. $f_{xx} =$
    $e^x e^y - e^{y^2 + 2y}$, $f_y = e^x e^y - x(2y + 2) e^{y^2 + 2y}$, $f_{xx} = e^x e^y$,
    $f_{yy} = e^x e^y - 2xe^{y^2 + 2y}(2y^2 + 4y + 3)$, $f_{yx} = e^x e^y - (2y + 2)e^{y^2 + 2y}$
12. $f_x = \dfrac{2}{x}$, $f_y = \dfrac{3}{y}$, $f_{xx} = \dfrac{-2}{x^2}$, $f_{yy} = \dfrac{-3}{y^2}$, $f_{yx} = 0$ 16. $f_x(1, 2) = 3$,
    $f_{xy}(2, 3) = 0$, $f_y(-1, 3) = -24$ 17. $f_x(1, 2) = 8$,
    $f_y(-1, 3) = -4$, $f_{xy}(2, 3) = 4$ 19. $f_x(1, 2) = e^2$,
    $f_y(-1, 3) = -e^3$, $f_{xy}(2, 3) = e^3$ 21.   1%   23. $\left(\frac{2}{3}, \frac{4}{3}\right)$ is
    relative maximum   25. $\left(\frac{2}{3}, \frac{4}{3}\right)$ is relative maximum
30. $10\sqrt{3}$ 31. $z = 1$ is the minimum distance; $z = 2$ is the
    maximum distance   32. $\left(\frac{61}{62}, \frac{3}{31}, \frac{5}{62}\right)$ is the relative
    minimum

Exercises 10.1

1. (a) IV   (b) II   (c) I   (d) III   2.   (b) I, II
    (c) I, IV   3.   (c) $\sin \theta = \frac{4}{5}$, $\cos \theta = -\frac{3}{5}$, $\tan \theta = -\frac{4}{3}$
4. (b) $\sin \theta = -\frac{3}{4}$, $\cot \theta = \mp\dfrac{\sqrt{7}}{3}$, $\sec \theta = \pm\dfrac{4}{\sqrt{7}}$
5. (a) $\sin 0 = 0$, $\cos 0 = 1$, $\csc 0$ undefined, $\cot 0$ undefined,
    $\tan 0 = 0$, $\sec 0 = 1$ (b) $\sin \pi = 0$, $\cos \pi = -1$,
    $\sec \pi = -1$, $\tan \pi = 0$, $\cot \pi$ undefined, $\csc \pi$ undefined

(c) $\sin \dfrac{3\pi}{2} = -1$, $\csc \dfrac{3\pi}{2} = -1$, $\cos \dfrac{3\pi}{2} = 0$, $\sec \dfrac{3\pi}{2}$

and $\tan \dfrac{3\pi}{2}$ undefined, $\cot \dfrac{3\pi}{2} = 0$

Exercises 10.2   1. (a) $\dfrac{\sqrt{2}}{2}, -\dfrac{\sqrt{2}}{2}$ (b) $-\dfrac{\sqrt{3}}{2}, -\dfrac{1}{2}$

2. (a) $\cos \dfrac{\pi}{26}$ (d) $-\cot \dfrac{\pi}{4}$ 5. $\dfrac{23}{4}$

9.

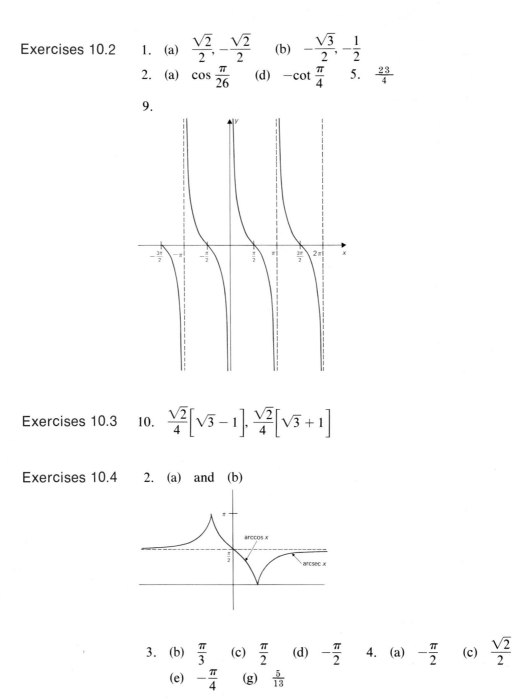

Exercises 10.3   10. $\dfrac{\sqrt{2}}{4}\left[\sqrt{3} - 1\right], \dfrac{\sqrt{2}}{4}\left[\sqrt{3} + 1\right]$

Exercises 10.4   2. (a) and (b)

3. (b) $\dfrac{\pi}{3}$ (c) $\dfrac{\pi}{2}$ (d) $-\dfrac{\pi}{2}$ 4. (a) $-\dfrac{\pi}{2}$ (c) $\dfrac{\sqrt{2}}{2}$

(e) $-\dfrac{\pi}{4}$ (g) $\dfrac{5}{13}$

1. 2   4. 1   5. 1

1. $\sin x + x \cos x$

4. $\dfrac{3 \cos(2x + 3) \cos(3x + 1) + 2 \sin(3x + 1) \sin(2x + 3)}{\cos^2(2x + 3)}$

5. $4(14x + 3) \tan^3(7x^2 + 3x + 9) \sec^2(7x^2 + 3x + 9)$

7. $e^{\sin x - \tan x}(\cos x - \sec^2 x)$

9. $e^x \sin(\cos x) [2 \cos(\cos x)(-\sin x) + \sin(\cos x)]$

10. $\dfrac{2}{\sqrt{1 - 4x^2}}$ 13. $e^{\sin x + \arctan x}\left(\cos x + \dfrac{1}{1 + x^2}\right)$

14. $\dfrac{x\sqrt{x^2 - 1}(2x + \sec^2 x) - 1}{x\sqrt{x^2 - 1}(x^2 + \tan x + \text{arccsc } x)}$ 15. $3 + \dfrac{\pi}{2}$

17. $\dfrac{2}{\pi}$ ft/min   19. 100 ft

1. $\ln|1 + \tan x| + C$ 3. $e^{\sin x} + C$ 5. $\frac{1}{4}(x^2 + 1) +$

$\frac{1}{8} \sin 2(x^2 + 1) + C$ 6. $e^{(\sin 2x)/2} + C$ 9. $\frac{1}{8}x -$

$\frac{1}{32} \sin 4x + C$ 10. $x \sin x + \cos x + C$ 11. $e^{\tan x} + C$

12. $-x^2 \cos x + 2(x \sin x + \cos x) + C$ 15. $\frac{1}{3}$

16. $-\frac{1}{4} + \ln 2$

17. $\dfrac{\sin^2 x^2}{4} + C$ 18. $\dfrac{a^2}{2} \sin^{-1}\left(\dfrac{x}{a}\right) + \dfrac{1}{2}x\sqrt{a^2 - x^2} + C$

20. $-\frac{1}{3}(9 - x^2)^{3/2} + C$ 21. $\frac{1}{2} \ln|\sec(x^2 + 2x)| + C$

22. $\frac{1}{2} \ln|\tan x^2| + C$ 24. $\dfrac{1}{a} \tan^{-1} \dfrac{x}{a} + C$

1. 5, 12   2. 3, 9   4. $\frac{3}{5}, \frac{24}{5}$ 5. (a) $\dfrac{u^2}{2g} \sin^2 \theta$

(b) $\dfrac{2u \sin \theta}{g}$ (c) $\dfrac{u^2}{g} \sin 2\theta$ (d) $\pi/4$

7. (a) 144 ft   (b) 6 sec   (c) 576 $\sqrt{3}$ ft

9. 40 ft/sec, 25 $\sqrt{3}$ ft   13. $4\sqrt{2}$ ft/sec, 2π sec

14. π ft/sec, $\dfrac{\sqrt{3}\,\pi}{2}$ ft/sec

Part 2

1. (a) 1   (b) 0   3. (a) 3   (d) 1   4. $\cos x -$

$x \sin x$ 5. $e^x \sin x + e^x \cos x$ 6. $\dfrac{1}{x} - e^x \sin e^x$

7. $(2x + 3) \cos(x^2 + 3x)$ 8. $2 \sin x \cos^2 x - \sin^3 x$

10. $x(2 \ln \sin x + x \cot x)$ 11. $e^{\sin x}(\cos x \ln \cos x - \tan x)$

14. $e^x \left(\text{arcsec } x + \dfrac{1}{x\sqrt{x^2 - 1}} \right)$ 16. $e^{\sin x - \arctan x} \left(\cos x - \dfrac{1}{1 + x^2} \right)$

19. $\dfrac{(1 + \sin x)\dfrac{1}{1 + x^2} - \arctan x (\cos x)}{(1 + \sin x)^2}$ 20. $\dfrac{1}{1 + \sin x} +$

$\dfrac{\sin x}{\cos^2 x} \ln(1 + \sin x)$ 22. $\dfrac{1}{\sqrt{1 - x^2}}$ 23. $2x \text{ arccot } x - 1$

25. $\ln|1 + \sin x| + C$ 26. $\frac{2}{3}(1 + \tan x)^{3/2} + C$

27. $\sin x - x \cos x + C$ 29. $\frac{1}{3} \sin x \cos^2 x + \frac{2}{3} \sin x + C$

31. $\frac{1}{3} \sec^2 x \tan x + \frac{2}{3} \tan x + C$ 32. $x^2 \sin x - 2 \sin x +$

$2x \cos x + C$ 33. $\dfrac{1}{2} \tan^{-1} \dfrac{x}{2} + C$ 34. $\sin^{-1} \dfrac{x}{3} + C$

37. $x - \dfrac{3}{2} \ln (x^2 + 16) - \dfrac{11}{4} \tan^{-1} \dfrac{x}{4} + C$ 39. $\dfrac{x}{4} \sqrt{16 - x^2} +$

$4 \arcsin \dfrac{x}{4} + C$ 40. $-\dfrac{\pi}{3} - \sqrt{3}$

Exercises A.1

1. $1^2 + 2^2 + 3^2 + 4^2 + 5^2 = 1 + 4 + 9 + 16 + 25$
2. $(x_3 + y_3) + (x_4 + y_4) + (x_5 + y_5) + (x_6 + y_6) + (x_7 + y_7) + (x_8 + y_8)$
3. $7x^1 + 7x^2 + 7x^3 + 7x^4 + 7x^5$
4. $6 + 6 + 6 + 6 + 6 + 6 + 6 + 6 + 6$
5. $(x_2 + 1)^2 + (x_3 + 1)^2 + (x_4 + 1)^2$

6. $\displaystyle\sum_{i=1}^{50} i$ 7. $\displaystyle\sum_{n=1}^{16} n(n + 1)$ 8. $\displaystyle\sum_{i=1}^{n} 2^i$

9. $\displaystyle\sum_{i=1}^{50} x^{2i}$ 10. $\displaystyle\sum_{i=1}^{n} \dfrac{x_i}{i}$

Exercises A.3

1. 5,040 2. 3,628,800 3. 56 4. 330
5. 87 6. $\frac{16}{15}$ 7. 28 8. 2415
12. (a) $x^6 + 6x^5 + 15x^4 + 20x^3 + 15x^2 + 6x + 1$
    (b) $1 + 10x + 40x^2 + 80x^3 + 80x^4 + 32x^5$
    (c) $-32 + 80x - 80x^2 + 40x^3 - 10x^4 + x^5$
    (d) $\dfrac{1}{x^6} + \dfrac{6}{x^4} + \dfrac{15}{x^2} + 20 + 15x^2 + 6x^4 + x^6$

13. (a) $-280x^4y^3$ (b) 84 (c) $495a^8x^4y^{16}$ (d) $372096x^{-14}$
14. (a) 4th term $= 660$ (b) 3rd term $= 240$
    (c) 3rd term $= 180$ (d) 7th term $= 84$

Exercises A.4

1. $2x^2 - 2x + \frac{7}{2}, -\frac{15}{2}$ 2. $\frac{4}{3}x^2 - \frac{7}{9}x - \frac{50}{27}, -\frac{73}{27}$
3. $2x^2 - \frac{1}{3}x + \frac{4}{9}, \frac{32}{9}$ 4. $9x^2 - 46x + 217, -1079$
5. $x^2 - 74x - 79, -78$ 6. $3x^3 + 6x^2 + 6x + 15, 23$
7. $-x^4 + 2x^3, 2$ 8. $x^4 + x^3 + x^2 + x + 1, 0$
9. $x^5 - x^4 + 3x^3 - 2x^2 + 2x - 2, 7$
10. $\frac{1}{2}x^5 - \frac{3}{4}x^4 + \frac{9}{8}x^3 - \frac{27}{16}x^2 + \frac{129}{32}x - \frac{387}{62}, \frac{1325}{64}$
11. $p(-2) = 18$ 12. $p(3) = 78$ 13. $p(-1) = -3$
14. $p(1) = 9$ 15. $p(2) = 67$ 16. $2x + 1$
17. $x^2 - x + 3$ 18. $\frac{1}{2}$ 19. $-\frac{3}{4}$ 20. 9

Exercises A.5

1. -1 2. 1 3. 3 4. $\frac{1}{2}$
5. None   6. $1, -1, -3$ 7. $2, 2, -2, -2$ 8. $\pm\sqrt{5}, \pm i\sqrt{3}$
9. $\pm\frac{\sqrt{2}}{4}, \pm i\frac{\sqrt{2}}{4}$ 10. $1, 1, -2, -2$
11. $1, 1, -1, -1$ 12. $0, 0, 1, 1, -1, -1$
15. (a) $-3, 7$ (b) $\frac{3}{2}, -\frac{7}{2}$ (c) $-\frac{4}{3}, -\frac{2}{3}$ (d) $10, 19$
16. (a) $x^2 - 5x + 9$ (b) $x^2 + 2x + 3$
    (c) $6x^2 - 9x + 8$ (d) $3x^2 - 5x + 9$
17. (a) $\frac{9}{4}$ (b) $-\frac{25}{4}$ (c) $\frac{4}{3}$ or $-\frac{4}{3}$
19. (a) $3i, \frac{7}{3}$ (b) $1 + 2i, \pm i$

Index